163
Advances in Polymer Science

Editorial Board:
A. Abe · A.-C. Albertsson · H.-J. Cantow
K. Dušek · S. Edwards · H. Höcker
J. F. Joanny · H.-H. Kausch · S. Kobayashi
K.-S. Lee · I. Manners · O. Nuyken
S. I. Stupp · U. W. Suter · -G. Wegner

Springer
Berlin
Heidelberg
New York
Hong Kong
London
Milan
Paris
Tokyo

Liquid Chromatography
FTIR Microspectroscopy
Microwave Assisted Synthesis

With contributions by
R. Bhargava, D. Bogdal, T. Chang, D. Hunkeler,
J. L. Koenig, T. Macko, P. Penczek, J. Pielichowski,
A. Prociak, S.-Q. Wang

Springer

This series presents critical reviews of the present and future trends in polymer and biopolymer science including chemistry, physical chemistry, physics and materials science. It is addressed to all scientists at universities and in industry who wish to keep abreast of advances in the topics covered.

As a rule, contributions are specially commissioned. The editors and publishers will, however, always be pleased to receive suggestions and supplementary information. Papers are accepted for „Advances in Polymer Science" in English.

In references Advances in Polymer Science is abbreviated Adv Polym Sci and is cited as a journal.

Springer APS home page: http://www.springerlink.com/series/aps/
Springer-Verlag home page: http://www.springer.de

ISSN 0065-3195
ISBN 3-540-00525-0
DOI 10.1007/b10932
Springer-Verlag Berlin Heidelberg New York

Library of Congress Catalog Card Number 61642

This work is subject to copyright. All rights are reserved, whether the whole or part of the material is concerned, specifically the rights of translation, reprinting, re-use of illustrations, recitation, broadcasting, reproduction on microfilms or in other ways, and storage in data banks. Duplication of this publication or parts thereof is only permitted under the provisions of the German Copyright Law of September 9, 1965, in its current version, and permission for use must always be obtained from Springer-Verlag. Violations are liable for prosecution under the German Copyright Law.

Springer-Verlag Berlin Heidelberg New York
a member of BertelsmannSpringer Science+Business Media GmbH
http://www.springer.de

© Springer-Verlag Berlin Heidelberg 2003
Printed in Germany

The use of registered names, trademarks, etc. in this publication does not imply, even in the absence of a specific statement, that such names are exempt from the relevant protective laws and regulations and therefore free for general use.

Typesetting: medio Technologies AG, Berlin
Cover: Design & Production, Heidelberg
Printed on acid-free paper   02/3020kk - 5 4 3 2 1 0

Editorial Board

Prof. Akihiro Abe
Department of Industrial Chemistry
Tokyo Institute of Polytechnics
1583 Iiyama, Atsugi-shi 243-02, Japan
E-mail: aabe@chem.t-kougei.ac.jp

Prof. Ann-Christine Albertsson
Department of Polymer Technology
The Royal Institute of Technology
S-10044 Stockholm, Sweden
E-mail: aila@polymer.kth.se

Prof. Hans-Joachim Cantow
Freiburger Materialforschungszentrum
Stefan Meier-Str. 21
79104 Freiburg i. Br., Germany
E-mail: cantow@fmf.uni-freiburg.de

Prof. Karel Dušek
Institute of Macromolecular Chemistry, Czech
Academy of Sciences of the Czech Republic
Heyrovský Sq. 2
16206 Prague 6, Czech Republic
E-mail: dusek@imc.cas.cz

Prof. Sam Edwards
Department of Physics
Cavendish Laboratory
University of Cambridge
Madingley Road
Cambridge CB3 OHE, UK
E-mail: sfe11@phy.cam.ac.uk

Prof. Hartwig Höcker
Lehrstuhl für Textilchemie
und Makromolekulare Chemie
RWTH Aachen
Veltmanplatz 8
52062 Aachen, Germany
E-mail: hoecker@dwi.rwth-aachen.de

Prof. Jean-François Joanny
Institute Charles Sadron
6, rue Boussingault
F-67083 Strasbourg Cedex, France
E-mail: joanny@europe.u-strasbg.fr

Prof. Hans-Henning Kausch
c/o IGC I, Lab. of Polyelectrolytes
and Biomacromolecules
EPFL-Ecublens
CH-1015 Lausanne, Switzerland
E-mail: kausch.cully@bluewin.ch

Prof. S. Kobayashi
Department of Materials Chemistry
Graduate School of Engineering
Kyoto University
Kyoto 606-8501, Japan
E-mail: kobayasi@mat.polym.kyoto-u.ac.jp

Prof. Kwang-Sup Lee
Department of Polymer Science & Engineering
Hannam University
133 Ojung-Dong
Teajon 300-791, Korea
E-mail: kslee@mail.hannam.ac.kr

Prof. Ian Manners
Department of Chemistry
University of Toronto
80 St. George St.
M5S 3II6
Ontario, Canada
E-mail: imanners@chem.utoronto.ca

Prof. Oskar Nuyken
Lehrstuhl für Makromolekulare Stoffe
TU München
Lichtenbergstr. 4
85747 Garching
E-mail: oskar.nuyken@ch.tum.de

Prof. Samuel I. Stupp
Department of Measurement Materials Science
and Engineering
Northwestern University
2225 North Campus Drive
Evanston, IL 60208-3113, USA
E-mail: s-stupp@nwu.edu

Prof. Ulrich W. Suter
Vice President for Research
ETH Zentrum, HG F 57
CH-8092 Zürich, Switzerland
E-mail: ulrich.suter@sl.ethz.ch

Prof. Gerhard Wegner
Max-Planck-Institut für Polymerforschung
Ackermannweg 10
Postfach 3148
55128 Mainz, Germany
E-mail: wegner@mpip-mainz.mpg.de

Preface

By polymers for biological use we understand biopolymers and living matter. Biomaterials are man-made or -modified materials which repair, reinforce or replace damaged functional parts of the (human) body. Hip joints, cardiovascular tubes or skin adhesives are just a few examples. Such materials are principally chosen for their mechanical performance (stiffness, strength, fatigue resistance). All mechanical and biological interactions between an implant and the body occur across the interface, which has to correspond as nearly as possible to its particular function. A natural surface is a complex (three-dimensional) structure, which has to fulfil many roles: recognition, adhesion (or rejection), transport or growth. We have to admit that at present biomaterials are far removed from such performance although new strategies in surface engineering have been adopted in which man tries to learn from nature.

Much of the progress in adapting polymer materials for use in a biological environment has been obtained through irradiation techniques. For this reason the most recent developments in 4 key areas are reviewed in this special volume. All surface engineering necessarily begins with an analysis of the topology and the elemental composition of a functional surface and of the degree of assimilation obtained by a particular modification. X-ray photoelectron spectroscopy (XPS) and time-of-flight secondary ion mass spectroscopy (ToF-SIMS) play a prominent role in such studies and these are detailed by H.J. Mathieu and his group from the Ecole Polytechnique Fédérale de Lausanne (EPFL). Generally, the first step towards procuring desired physico-chemical properties in a biomaterial substrate is a chemical modification of the surface. As pointed out by B. Gupta and N. Anjum from the Indian Institute of Technology (IIT), plasma- and radiation-induced grafting treatments are widely used since they have the particular advantage that they result in highly pure, sterile and versatile surfaces.

The sterilisation of implantable devices is a subject of great concern for the medical industry. Since ionising radiation is preferentially used for this purpose, attention must be paid to possible effects on the structural and mechanical properties of polymers (through chain scission or cross-linking). L. A. Pruitt from UC Berkeley has reviewed the specific behaviour of the different medical polymer classes to γ- and high-energy electron irradiation and environmental effects. The biocidal efficiency relies on free radical formation and on the ability to reduce DNA replication in any bacterial spore present in a medical device.

The latter point, radiation effects on living cells and tissues, is the subject of the final contribution in this volume. M. Scholz from the Gesellschaft für Schwerionenforschung (GSI) summarises the (damaging) biological effects of ion beam irradiation and the considerable differences with respect to conventional photon radiation. These studies are of particular importance for radiation protection and radiotherapy. The advantages of a tumor treatment by carbon ion beams (effectiveness, concentrated energy release, possibility to use the presence of positron emitting ^{10}C and ^{11}C isotopes for positron emission tomography) are also presented in a comprehensive way.

I hope that the combination in a single special volume of the Advances in Polymer Science of these highly complementary contributions is particularly helpful to scientists working in this rapidly developing area. I would also like to thank all the authors for their exemplary co-operation.

Lausanne, December 2002 H. H. Kausch

Advances in Polymer Science
Available Electronically

For all customers with a standing order for Advances in Polymer Science we offer the electronic form via SpringerLink free of charge. Please contact your librarian who can receive a password for free access to the full articles. By registration at:

http://www.springerlink.com/series/aps/reg_form.htm

If you do not have a standing order you can nevertheless browse through the table of contents of the volumes and the abstracts of each article at:

http://www.springerlink.com/series/aps/

There you will find also information about the

- Editorial Board
- Aims and Scope
- Instructions for Authors
- Sample Contribution

Contents

Recent Advances in Liquid Chromatography
Analysis of Synthetic Polymers
Taihyun Chang 1

Liquid Chromatography under Critical and Limiting Conditions:
A Survey of Experimental Systems for Synthetic Polymers
Tibor Macko, David Hunkeler 61

FTIR Microspectroscopy of Polymeric Systems
Rohit Bhargava, Shi-Qing Wang, Jack L. Koenig 137

Microwave Assisted Synthesis, Crosslinking, and Processing
of Polymeric Materials
Dariusz Bogdal, Piotr Penczek, Jan Pielichowski, Aleksander Prociak 193

Author Index Volumes 101–163 265

Subject Index 279

Recent Advances in Liquid Chromatography Analysis of Synthetic Polymers

Taihyun Chang

Department of Chemistry and Center for Integrated Molecular Systems, Pohang University of Science and Technology, Pohang, 790-784 Korea. *E-mail: tc@postech.edu*

Abstract Synthetic polymers are rarely homogeneous chemical species but have multivariate distributions in molecular weight, chemical composition, chain architecture, functionality, and so on. For a precise characterization of synthetic polymers, all the distributions need to be determined, which is a difficult task, if not impossible. Fortunately in most of the cases it is sufficient to analyze a limited number of molecular characteristics in order to obtain the information required for a given purpose. Nonetheless, it is still nontrivial if there exists distributions for more than one molecular characteristic. There have been continuing efforts to solve the problem. One approach is to find chromatographic methods sensitive to one molecular characteristic only. In favorable cases, the effect of all but one molecular characteristic can be suppressed to a negligible level. Various interaction chromatographic techniques as well as size exclusion chromatography are employed for the purpose. Also the multiple detection methods each sensitive to a specific molecular characteristic can provide additional information. Various detection methods developed recently such as FT-IR, FT-NMR, and mass spectrometry brought about significant progress in the characterization of complex polymers. This review presents the recent developments in the analysis of various heterogeneities in synthetic polymers by a variety of liquid chromatographic separation as well as detection methods.

Keywords Chromatography · Polymer characterization · Molecular weight · Chain architecture · Chemical composition · Functionality · Stereoregularity

1	Introduction	4
2	Chromatographic Separation Principles	5
3	Molecular Weight Distribution of Linear Polymers	12
4	Chain Architecture	18
4.1	Branched Polymers	18
4.2	Ring Polymers	22
5	Chemical Composition Distribution	22
5.1	Mixtures	25
5.2	Block Copolymers	28

© Springer-Verlag Berlin Heidelberg 2003

5.3	Random Copolymers	37
5.4	Functionality	46
6	**Microstructures**	50
7	**Outlook**	55
References		56

List of Symbols and Abbreviations

2D	Two dimension(al)
2D-LC	Two-dimensional liquid chromatography
-b-	-block-
BA	Butyl acrylate
C18	Octadecyl
C18AB	Polymeric phase octadecyl
CCD	Chemical composition distribution
c_i	Concentration of a solute in the interstitial space
c_m	Concentration of a solute in the mobile phase
c_p	Concentration of a solute in the pore space
c_s	Concentration of a solute in the stationary phase
DNA	Deoxyribonucleic acid
EA	Ethyl acrylate
ELSD	Evaporative light scattering detector
EPDM	Ethylene-propylene-diene rubber
ESI	Electron spray ionization
FAD	Full adsorption-desorption
FT-IR	Fourier transform-infrared spectroscopy
FT-MS	Fourier transform-mass spectrometry
FT-NMR	Fourier transform-nuclear magnetic resonance spectroscopy
-g-	-graft-
HEMA	2-Hydroxyethyl methacrylate
HPLC	High performance liquid chromatography
i-	Isotactic
IC	Interaction chromatography
IR	Infrared spectroscopy
K	Equilibrium constant
K_{IC}	Equilibrium constant at interaction chromatography
K_{SEC}	Equilibrium constant at size exclusion chromatography
k'	Capacity factor
LC	Liquid chromatography

LC-CAP	Liquid chromatography at the critical adsorption point
LCCC	Liquid chromatography at the critical condition
LC-LCA	Liquid chromatography under limiting conditions of adsorption
LC-LCD	Liquid chromatography under limiting conditions of desorption
LC-PEAT	Liquid chromatography at the point of exclusion-adsorption transition
LC-TEA	Liquid chromatography at the theta exclusion-adsorption condition
M	Molecular weight
MA	Methyl acrylate
MALDI-TOF MS	Matrix assisted laser desorption/ionization time of flight mass spectrometry
MALS	Multi-angle light scattering
MAn	Maleic anhydride
MMA	Methyl methacrylate
M_n	Number average molecular weight
M_w	Weight average molecular weight
MWD	Molecular weight distribution
NMR	Nuclear magnetic resonance spectroscopy
NP	Normal phase
NPLC	Normal phase liquid chromatography
PB	Polybutadiene
PBA	Poly(butyl acrylate)
PDMA	Poly(decyl methacrylate)
PDMS	Poly(dimethylsiloxane)
PE	Polyethylene
PEMA	Poly(ethyl methacrylate)
PEO	Poly(ethylene oxide)
PI	Polyisoprene
PLMA	Poly(lauryl methacrylate)
PMMA	Poly(methylmethacrylate)
PnBMA	Poly(n-butyl methacrylate)
PPO	Poly(propylene oxide)
PS	Polystyrene
PtBMA	Poly(*tert*-butyl methacrylate)
R	Gas constant
RP	Reversed phase
RPLC	Reversed phase liquid chromatography
S	Styrene
s-	Syndiotactic

SEC	Size exclusion chromatography
SI	Polystyrene-*block*-polyisoprene copolymer
T	Temperature
TGIC	Temperature gradient interaction chromatography
THF	Tetrahydrofuran
TLC	Thin layer chromatography
v/v	Volume/volume
V_i	Interstitial volume
V_m	Mobile phase volume
V_p	Pore volume
V_r	Retention volume
V_s	Stationary phase volume
ΔG^o	Standard Gibbs free energy change
ΔG^o_{IC}	Standard Gibbs free energy change in interaction chromatography
ΔG^o_{SEC}	Standard Gibbs free energy change in size exclusion chromatography
ΔH^o	Standard enthalpy change
ΔH^o_{IC}	Standard enthalpy change in interaction chromatography
ΔH^o_{SEC}	Standard enthalpy change in size exclusion chromatography
ΔS^o	Standard entropy change
ΔS^o_{IC}	Standard entropy change in interaction chromatography
ΔS^o_{SEC}	Standard entropy change in size exclusion chromatography
ν	Number average degree of polymerization

1
Introduction

Liquid chromatography (LC) is a powerful tool for the characterization of natural and synthetic macromolecules that are often heterogeneous in molecular weight, chain architecture, chemical composition, microstructure, and so on. Among numerous variations of LC methods, size exclusion chromatography (SEC) has dominated the area of the molecular weight distribution (MWD) analysis of synthetic polymers [1–3]. SEC has many advantages over other classical techniques in the characterization of molecular weight distribution of polydisperse polymers in speed, effort, required amount of sample, etc.

SEC fractionates polymer molecules by partition equilibrium of polymer chains between common solvent phases located at the interstitial space and the pores of the column packing materials, typically in the form of uniform size porous beads. Therefore the partition equilibrium is mainly governed by the conformational entropy difference of the polymer chains located in two different physical environments. The eluent is commonly chosen to minimize the enthalpic interaction between the polymeric solutes and the stationary phase. In results, SEC separates the polymer molecules in terms of the size of a polymer chain in the elution solvent. If there exists a simple relationship between the polymer chain size and its molecular weight such as for linear and chemically homogeneous polymers, SEC retention volume is well correlated with the molecular weight of polymers. However, the same cannot be said for nonlinear chain polymers or copolymers, in which a simple relationship between the molecular weight and the molecular size does not exist and SEC is not able to separate them according to their molecular weight.

For the same reason, SEC is not an efficient tool to separate the molecules in terms of chemical heterogeneity, such as chemical composition differences of copolymers, tacticity, and end-group difference. Interaction chromatography (IC) is suitable for the purpose since its separation mechanism is sensitive to the chemical nature of the molecules. In contrast to SEC, IC utilizes the enthalpic interaction, adsorption or partition of solute molecules to the stationary phase and IC has been employed frequently to separate polymers in terms of their chemical composition distribution (CCD) or functionality [4]. IC has also been used for the separation of polymers in terms of molecular weight [5, 6]. In general IC exhibits higher molecular weight resolution than SEC, but SEC is more universal in a sense that most of the thermodynamically good solvents can be used for the SEC separation unless there exists specific interaction between the polymer of interest and the porous packing materials. On the other hand, the enthalpic interaction strength in IC has to be controlled precisely to achieve a reproducible and high-resolution separation. Therefore the mobile and stationary phases have to be chosen for individual polymer system of interest and the gradient elution is often required [7].

Recent development in liquid chromatography analysis of complex polymers shows a clear trend to combine more than one LC separation mechanism/technique together with multiple detection techniques. It is a quite natural direction for the analysis of complex polymers with multivariate distributions in molecular characteristics. The coupled LC techniques have gained wide attention recently in the characterization of complex polymers and there are a number of monographs and reviews on this topic [4, 8–13]. In this chapter, recent advances in LC separation of polymers are reviewed. References are generally restricted to the works published after 1995 since most of the works prior to the mid-1990s have been well summarized already [8, 10, 11, 13].

2
Chromatographic Separation Principles

The retention in SEC, in which the solute partition takes place between the common eluent phases in two different environments, interstitial and pore space, is expressed as follows:

$$V_r = V_i + K_{SEC}V_p, \quad K_{SEC} = \frac{c_p}{c_i} \qquad (1)$$

where the subscripts p and i refer to the pore and interstitial spaces, respectively. The distribution constant K_{SEC}, the ratio of the solute concentration in pore space (c_p) to interstitial space (c_i), has a value between 0 (total exclusion) and 1 (total permeation). Like other equilibrium constants, K_{SEC} is related to the standard Gibbs free energy change (ΔG^o) of the partition process and ΔG^o can be further divided into the enthalpic and entropic contribution of the process:

$$K_{SEC} = \exp\left(-\frac{\Delta G^o_{SEC}}{RT}\right) = \exp\left(\frac{\Delta S^o_{SEC}}{R} - \frac{\Delta H^o_{SEC}}{RT}\right) \qquad (2)$$

In the SEC separation process, the separation condition is usually chosen to minimize the enthalpic interaction of polymer solutes with the packing materials. Therefore in the *ideal* SEC condition ($\Delta H^0=0$) K_{SEC} is a function of the conformational entropy loss ($\Delta S^o<0$) of polymer chains when transferred into the restricted pore space:

$$K_{SEC} \cong \exp\left(\frac{\Delta S^o_{SEC}}{R}\right) \qquad (3)$$

If this condition is fulfilled, SEC separates the polymers exclusively in terms of the size of polymer chain in the eluent relative to the size of the pores. This condition constitutes the background of the Cassasa theory of K_{SEC} [14] and the universal calibration method [15].

On the other hand, the retention in IC is generally expressed as follows.

$$V_r = V_m + K_{IC}V_s, \quad K_{IC} = \frac{c_s}{c_m} \qquad (4)$$

where K_{IC} is the distribution constant of an analyte between the mobile and stationary phase, and subscripts m and s stand for the mobile and stationary phase, respectively. In the IC separation process both ΔS^o and ΔH^o play roles to control the solute retention, which is generally expressed with the capacity factor (k'):

$$k' \equiv \frac{V_r - V_m}{V_m} = K_{IC}\frac{V_s}{V_m} = \exp\left(-\frac{\Delta G^o_{IC}}{RT}\right)\cdot\frac{V_s}{V_m} \qquad (5)$$

$$\ln k' = \frac{-\Delta H^o_{IC}}{RT} + \frac{\Delta S^o_{IC}}{R} + \ln\phi, \quad \phi \equiv \frac{V_s}{V_m} \qquad (6)$$

where ΔH^o and ΔS^o are the standard enthalpy and entropy changes associated with the solute sorption process to the stationary phase, which can be adsorption, partition or others depending on the nature of the chromatographic methods. In some literatures, it is argued that the ΔS^o in the ideal IC can be assumed to be zero, but it is highly unrealistic since ΔS^o involves the entropy change of various origins. The sorption of polymer chains dissolved in the mobile phase to the stationary phase involves the change of polymer chain conformation, redistribution of solvent molecules associated with polymer chains as well as with stationary phase, and so on. If porous packing materials were used, the conformational entropy change term due to the size exclusion process also contributes to ΔS^o. In this case, ΔH^o also becomes complicated since it should be dependent on the accessibility of polymer chains to the inner surface of the pores. There are some efforts to take all these contributions into account to devise a universal equation without great success.

Therefore both SEC and IC separation mechanisms operate in the chromatographic separation of polymers when porous packing materials is used. Which mechanism plays a dominant role is a matter of choice of the separation condition. Figure 1 shows the feature very well for the separation of polystyrene (PS) standards by use of a C18 silica column and CH_2Cl_2/CH_3CN mixture as stationary and mobile phase, respectively [5]. Figure 1A shows the solvent effect on the PS retention at the column temperature of 30.5 °C. For PS CH_2Cl_2 is a good solvent and CH_3CN is a poor solvent. The "good" and "poor" in this case stand for the solvent strength of the eluent and are not necessarily parallel to the thermodynamic quality of the solvent [16]. In other words, "good" solvent keeps polymers in the mobile phase more strongly and vice versa. In Fig. 1A, at high CH_2Cl_2 contents, high molecular weight PS elutes first before the elution of injection solvent at the elution volume of $V_i + V_p = V_m$, indicating that the size exclusion separation mechanism plays a dominant role. As the CH_2Cl_2 content decreases, the retention of PS gradually increases due to the decrease of solvent power. When the eluent composition reaches a certain value ($CH_2Cl_2/CH_3CN=57/43$, v/v) all the PS samples elute at the same elution time. Further decrease of the good solvent composition in the eluent has the PS samples elute in the IC regime (after the elution of the injection solvent). In the IC regime, the elution sequence is reversed and low molecular weight PS elutes first. Also the retention of PS increases very steeply with molecular weight. This behavior is explained by an empirical relationship known as Martin's rule [17]. It is based on the additivity of the free energy increments of structural elements of the solute molecules: ΔG^o in Eq. (5) is proportional to the degree of polymerization and the retention increases exponentially with molecular weight.

Figure 1B displays the similar effect with temperature variation at the fixed eluent composition ($CH_2Cl_2/CH_3CN=57/43$, v/v). The overall temperature dependence, i.e., decrease of retention with temperature, indicates that the interaction of polymer chains with C18 silica stationary phase is exothermic. Provided that the

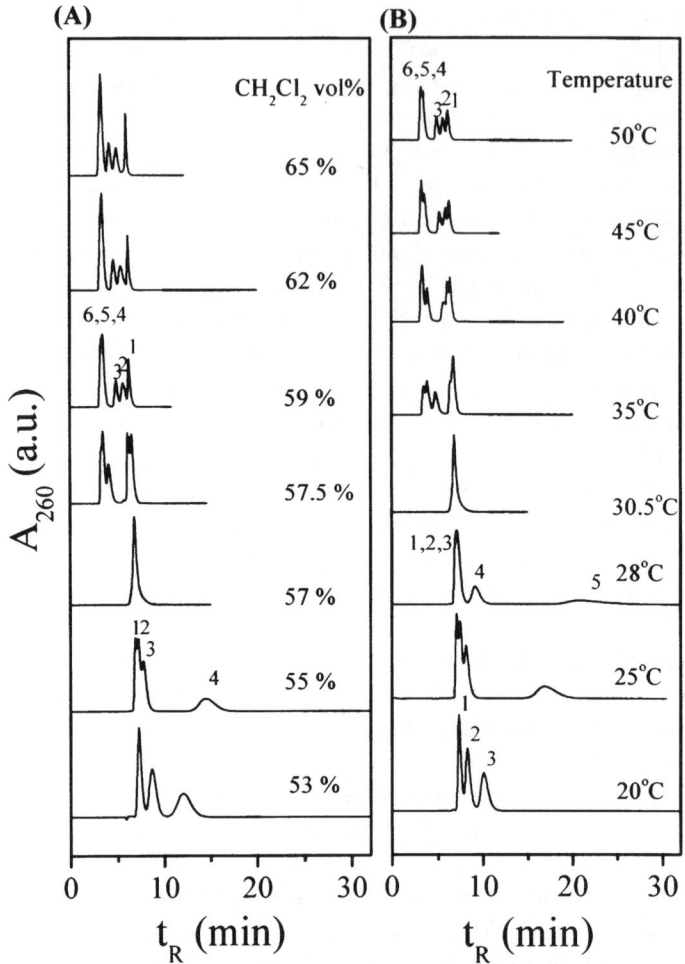

Fig. 1A,B. Chromatograms of six PS standards (M_w: (1) 2.5, (2) 12, (3) 29, (4) 165, (5) 502, (6) 1800 kg/mol). Injection samples were made by dissolving PS standards in respective eluents at the concentration of 1 mg/ml for each standards and the injection volume was 50 µl. Single RP column (Nucleosil C18; 250×4.6 mm, 100Å pore, 5µm particle) was used and the flow rate was 0.5 ml/min: **A** solvent composition effect – the chromatograms were obtained at a fixed column temperature of 30.5 °C with the mixed eluents of CH_2Cl_2/CH_3CN at different compositions as labeled in the plot; **B** temperature effect – the eluent was a mixture of CH_2Cl_2/CH_3CN at a fixed composition of 57/43 (v/v) for all chromatograms. The isothermal chromatograms were obtained at different column temperatures as labeled in the plot. Reproduced from [5] with permission

thermodynamic variables ΔH^o and ΔS^o do not change much in the temperature range shown in Fig. 1B, the temperature effect can be explained by the RT term in the denominator in Eq. (5). The retention variation with temperature for the separation in the SEC regime at high temperatures indicates that the separation condition is certainly not an *ideal* SEC condition. If the separation in the SEC regime at the high temperatures is a pure entropic process we do not expect such a significant retention change with temperature since there is no temperature factor in Eq. (3).

The separation condition of CH_2Cl_2/CH_3CN of 57/43 (v/v) and 30.5°C is a unique situation where the molecular weight resolution is completely lost. This is the point of enthalpy-entropy compensation that is known as the chromatographic critical condition [18]. At this point, PS samples of different molecular weights elute together. If this characteristics were maintained when different polymer chains or functional groups are attached to PS chains such as in block copolymers with PS block(s) or end functional PS chains, it would be possible to separate them in terms of the additional groups while keeping the PS chain portion from contributing to the retention. The term of "chromatographic invisibility" was proposed for this feature and the chromatographic separation utilizing this feature has gained wide popularity recently for the characterization of complex polymers. The study on this interesting feature started in the 1970s [18, 19] and has been supported by a number of other experimental studies [9–11] although there remains some controversy as to whether such a precise co-elution condition indeed exists [20–24]. Furthermore, other anomalies such as limited recovery for high molecular weight polymers as well as peak broadening and/or splitting have been reported at the critical condition [25, 26].

Nevertheless, the critical condition has been successfully employed for the chromatographic separation of the components in polymer blends [27–33], and for the separation of polymers with respect to the functional groups [33–41] as well as for the characterization of individual block of block copolymers [22, 35, 42–49]. This technique is variously termed as liquid chromatography at the critical condition (LCCC), liquid chromatography at the point of exclusion-adsorption transition (LC-PEAT), or liquid chromatography at the critical adsorption point (LC-CAP). Berek and coworkers further divided the method depending on the solvent strength of the injection sample solvent relative to the mobile phase; liquid chromatography at the critical adsorption point (LC-CAP), liquid chromatography under limiting conditions of adsorption (LC-LCA) and liquid chromatography under limiting conditions of desorption (LC-LCD) (Fig. 2) [50, 51].

Enthalpy and entropy compensation phenomena have been found not only in the chromatographic analysis of polymers but also in many kinds of kinetic and equilibrium processes for which a linear relationship exists between ΔH^o and ΔS^o [52, 53]. In the LCCC analysis of synthetic polymers, it is generally assumed that the retention of polymers is balanced by unfavorable entropic effect due to the size exclusion mechanism ($\Delta S^o < 0$) and the favorable enthalpic effect due to the solute-stationary phase interaction ($\Delta H^o < 0$) [54–56]. Consequently, it is expected that a polymer species at the LCCC condition elutes independent of its molecular weight

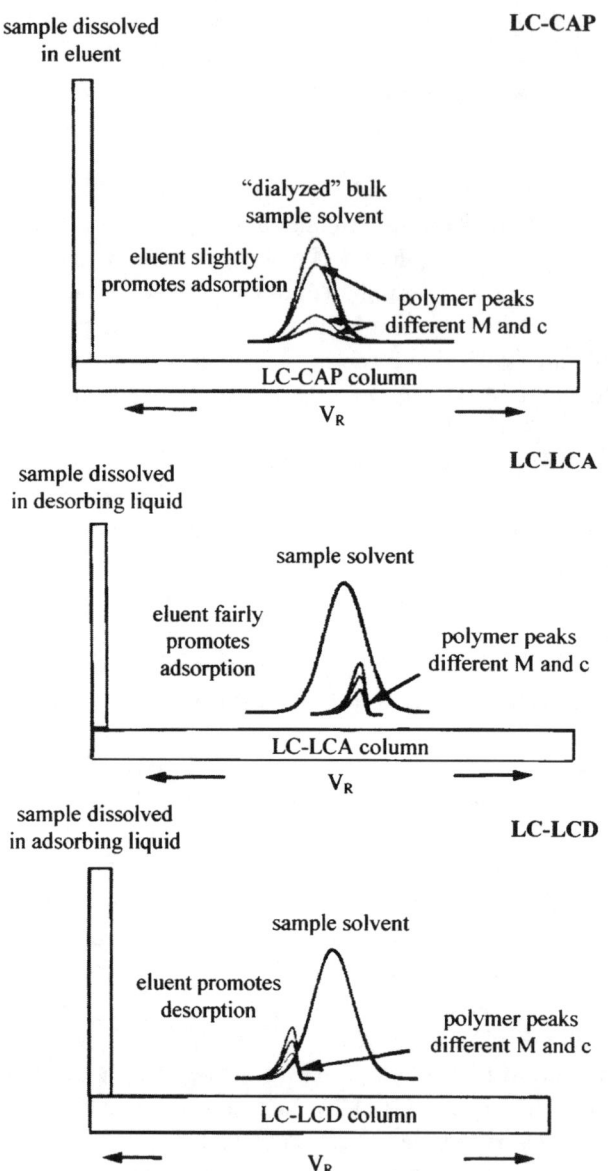

Fig. 2. Schematic representation of LC-CAP, LC-LCA, and LC-LCD. Reprinted from [50] with permission

at one single retention volume that roughly corresponds to the total mobile phase volume within the column(s). However, the enthalpy and entropy compensation point can also be found for a process of both positive ΔS^o and ΔH^o. For examples, in the reversed phase LC (RPLC) separation of ethylene oxide, enthalpy and entropy compensation was observed with the opposite entropic and enthalpic effects ($\Delta S^o > 0$ and $\Delta H^o > 0$), and the coelution retention volume was far apart from the total mobile phase volume [57]. In this case the enthalpic and entropic effect of poly(ethylene oxide) (PEO) sorption to the stationary phase is quite different from the usual LCCC separation condition. Such an enthalpy and entropy compensation phenomenon is often found in the hydrophobic interaction chromatography, a very useful LC technique in the separation of biopolymers, such as protein or nucleic acids [58, 59].

The exponential relationship between k' and ΔG^o in Eq. (5) indicates that the retention in IC varies much if ΔG^o of analytes spans a wide range, which in turn makes an isocratic elution difficult. The same problem exists in the IC analysis of synthetic homopolymers as can be seen in Fig. 1. According to the Martin's rule, K (also k') increases exponentially with the degree of polymerization of polymer molecules. Therefore the elution peak of synthetic polymers with a broad molecular weight distribution may become extremely broad and it appears as if isocratic elution of synthetic polymers is hardly feasible [6, 60]. In order for a polymer sample with a wide molecular weight distribution to be eluted in a reasonable experimental time period, it is necessary to change K during the elution. One way to control the retention is to change ΔG^o itself. Solvent gradient elution is a method to change ΔG^o, thus k', by changing the eluent composition to increase the solvent strength during the elution [7]. The solvent gradient HPLC fractionation of synthetic polymers works well, but it has a few drawbacks in that it is difficult to use many useful detection methods for the characterization of polymers such as differential refractometry, light scattering, and viscometry due to the background signal drift. Unlike the analysis of small molecules that are unique chemical species in general, synthetic polymers are collections of molecules having continuous distributions in various molecular characteristics. In order to characterize such synthetic polymers it is necessary to determine the distribution curve quantitatively with high reproducibility. The background signal drift makes the quantitative analysis difficult, if not impossible.

An alternative method to control the retention is to change the temperature as easily inferred from Eq. (5) and Fig. 1B. It is well known that temperature can alter the retention behavior of IC. In fact, temperature dependency of k' has been widely used to study the thermodynamics involved in the LC separation process. However, temperature has not been a popular variable to control the retention in practical IC separations since temperature is not as effective as solvent composition if the sample is a mixture of the components spanning a large ΔG^o range. It is mainly because the possible range of temperature variation is limited by the freezing and boiling of the mobile phase and/or precipitation of the polymeric solutes. However, temperature was found to permit a very efficient retention control in the MWD analysis of homopolymers, in particular for polymers with narrow molecular weight distribution [5].

In order to interpret chromatographic separation results correctly, one has to be aware of the "local polydispersity"". Local polydispersity is the presence of a variety of different types of molecules in the eluate eluted at the same retention volume. One source of local polydispersity is the axial dispersion taking place during the separation process in the column, which causes band broadening of the elution peak even for a homogeneous chemical species. The extent of the band broadening depends on the nature of the chromatographic separation mechanism as well as the various parameters in the operation of the instruments. Even if we assume "perfect resolution" that the axial dispersion effect is negligible, local polydispersity still exists in the chromatographic separation of complex polymers. For example, SEC separates on the basis of molecular size in solution and thus, for complex polymer molecules, such as copolymers or branched polymers, a variety of combinations of molecular weight, composition and chain architecture can have the same molecular size (Fig. 3). If the local polydispersity effect is not taken into account appropriately, highly misleading analyses may result [61].

3
Molecular Weight Distribution of Linear Polymers

As mentioned earlier, all synthetic polymers have finite MWD and the measurement of MWD is usually the first step in the characterization of polymers. SEC has been the most widely employed method for the MWD measurements of polymers. In the SEC analysis, the precision of the measurements depends on various factors in the operation of the instrument and the data reduction [1, 3]. Various efforts including round robin tests to establish a standardized procedure in the SEC analysis have been made [62, 63]. However, even for linear homopolymers there is an intrinsic limitation of SEC to measure the precise MWD due to the band-broadening problem, in particular for the polymers with narrow molecular weight distribution [64].

The MWD of polymers prepared by living anionic polymerization is a good example [65]. According to the early theoretical work of Flory, an "ideal" living anionic polymerization is expected to yield a polymer with Poisson distribution of chain lengths [66]:

$$w_i = \frac{i v^{i-1} e^{-v}}{(v+1)(i-1)!} \qquad (7)$$

where w_i stands for the weight fraction of the i-mer. The Poisson distribution has an asymptotic M_w/M_n (weight average molecular weight/number average molecular weight) value of $1+1/v$ where v is the number average degree of polymerization. The typical M_w/M_n value of such polymers measured by SEC is typically 1.01 or larger, where even 1.01 is significantly larger than the value anticipated from the Poisson distribution for high molecular weight polymers (Fig. 4). This is mainly

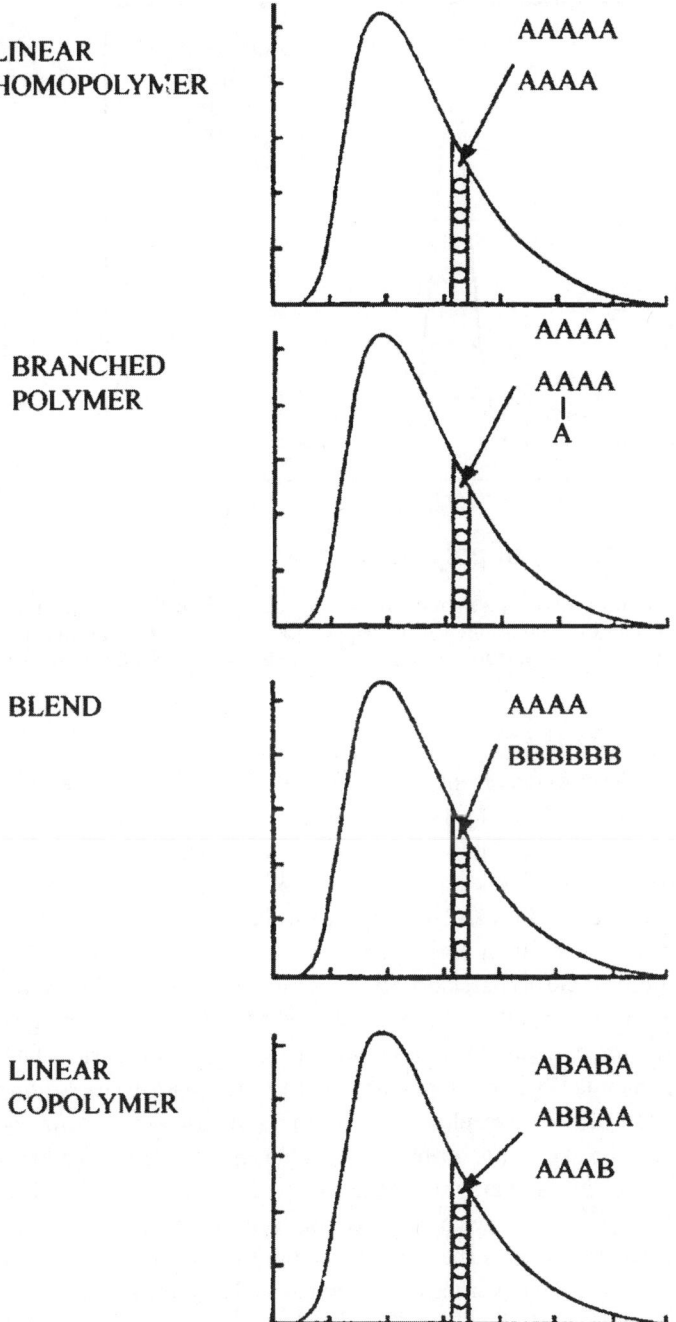

Fig. 3. Schematic representation of local polydispersity that may take place in the LC separation of polymers

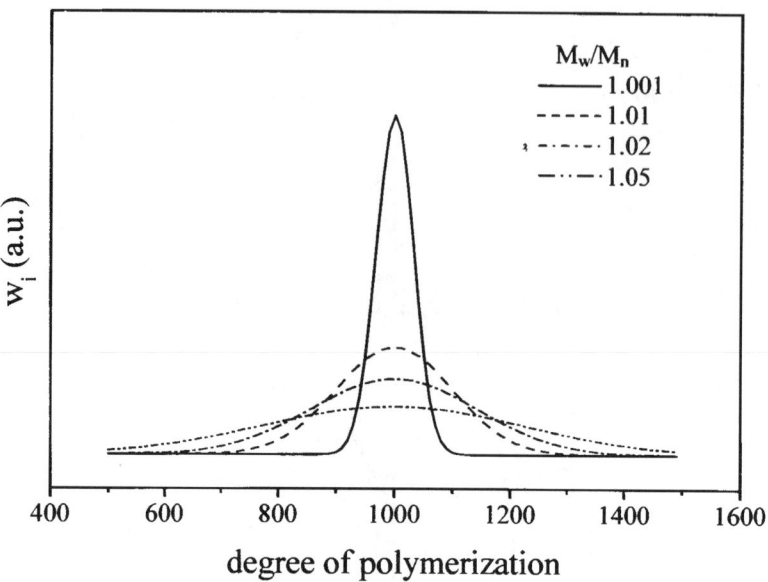

Fig. 4. Simulated distribution curves of weight fraction vs degree of polymerization having average degree of polymerization of 1000 and various M_w/M_n values. All the distribution curves were calculated assuming Gaussian distributions except for the Poisson distribution curve (M_w/M_n=1.001)

due to the band-broadening effect from which SEC suffers significantly. The widely used calibration method relative to standard samples cannot provide the correct MWD of such narrowly dispersed polymers [64]. The band-broadening problem was recognized from the early developmental stage of SEC. One way to resolve the problem is to correct the band broadening by deconvolution of the elution peak with appropriate broadening functions [67–70].

With the advances of new characterization techniques, the validity of the M_w/M_n values determined by SEC has been questioned: Shimpf et al. found much lower M_w/M_n values for PS made by anionic polymerization from thermal field flow fractionation analysis [71]. Development of light scattering detectors allows on-line determination of the molecular weight of the polymers eluted from the separation column. In principle, the on-line light scattering detection can allow a more precise analysis of MWD than the calibration method since it can provide correct weight average molecular weight for each slice across the elution peak despite the axial dispersion. However, the difficulty is that the analysis relies on the signals from two detectors, a light scattering detector as well as a concentration detector. Therefore the noise at the low detection signal portions of an elution peak limits the precision of the analysis and the possible error due to the interdetector delay volumes or interdetector broadening affects the analysis significantly [72,

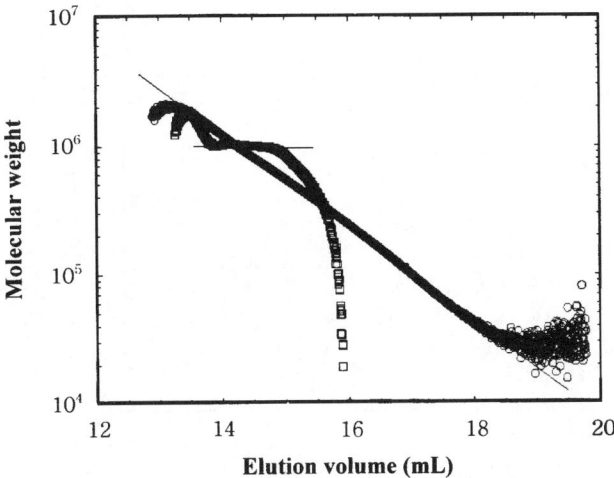

Fig. 5. Comparison of dependences of log M vs elution volume for a PS with narrow MWD (*open squares*) and a PS sample with broad MWD (*open circles*). Molecular weights are determined by on-line light scattering detection. The flat portion of the narrowly dispersed PS indicates that the fractions eluted at the elution volume range contain similar molecular weight polymers in contrast to the broad MWD sample. Reproduced from [74] with permission

73]. If the instrumental problems were correctly taken care of, light scattering detection clearly shows a narrower distribution of the polymers prepared by anionic polymerization (Fig. 5). Recently Netopilik et al. proposed a method to extract the M_w/M_n values of PS with narrow molecular weight distribution from SEC-light scattering detection results assuming a log-normal MWD [74].

Shortt [75] and Wyatt and Villalpando [76] used an interesting approach to measure M_w/M_n values for PS prepared by anionic polymerization. Instead of molecular weight, they measured the radius of gyration distribution by SEC coupled with multi-angle light scattering (MALS) detection. The absolute value of the radius of gyration can be determined from the angular dependency of the scattered light measured by MALS independently from any other detectors, which allows the measurement free from the complication of interdetector delay volumes or interdetector broadening effect. Assuming the Gaussian shape of the elution peak, they obtained much lower M_w/M_n values from the radius of gyration calibration curve.

A more promising experimental approach to measure MWD of polymers with narrow MWD is to use IC. IC has been employed mostly for copolymer separation with respect to chemical composition distribution, but IC has also been applied to separation according to molecular weight. The early IC separations were done by normal phase (NP) thin-layer chromatography and many of these separations suffered from relatively poor resolution, particularly for high molecular weight polymers [77–79]. The separation efficiency was greatly improved by employing the re-

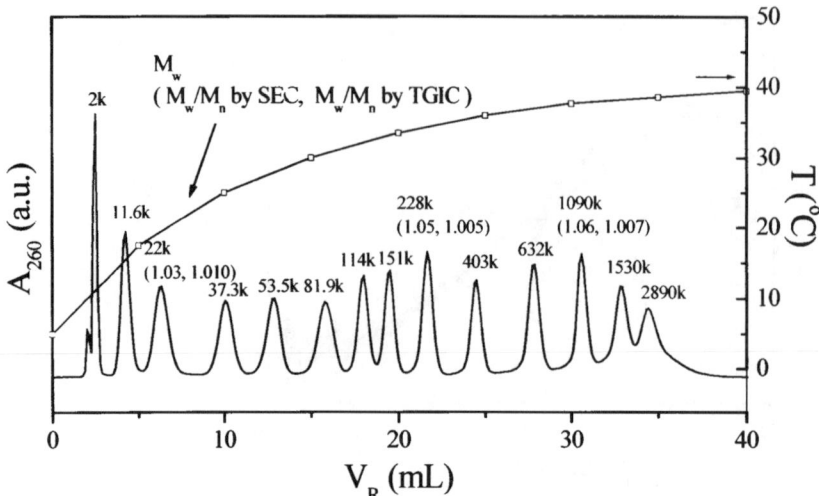

Fig. 6. TGIC separation of 14 standard PSs. Mobile phase was a mixture of CH_2Cl_2/CH_3CN at the volume ratio of 57/43. Injection sample concentration was 1 mg/ml of each PS standards and the injection volume was 50 µl. Single RP column (Nucleosil C18; 250×4.6 mm; 100 Å pore; 5 µm particle size) was used and the flow rate was 0.5 ml/min. Temperature program is shown in the plot. Each peak is labeled with the M_w values determined by TGIC and some of them are also marked with the M_w/M_n values measured by SEC and TGIC. Reproduced from [91] with permission

versed phase (RP) stationary phases. Armstrong and Bui were the first to separate PS successfully in terms of molecular weight by RP thin layer as well as column chromatography [80]. Subsequently a number of reports on the RPLC separation of high polymers, PS [81–85], poly(methyl methacrylate) (PMMA) [86], poly(ethylene oxide) (PEO) [57] have followed.

Although these works demonstrated the capability of IC to separate the synthetic polymers according to their molecular weight, MWD analysis using IC was first reported by Lee and Chang for PS standards [87]. They changed the column temperature to control the IC retention. IC shows a much higher resolution than SEC and the MWD of the PS standards was determined as much smaller than the values measured by SEC as shown in Fig. 6. Subsequently similar results were found for other polymers made by anionic polymerization such as PMMA [88] and polyisoprene (PI) [89]. Lochmüller et al. first demonstrated the possibility of IC retention control by varying the column temperature for the separation of PEO [57]. Chang and coworkers have utilized the method extensively for the precise characterization of complex polymers and named the technique as temperature gradient interaction chromatography (TGIC) [5, 90, 91]. Recently Bruheim et al. employed packed capillary liquid chromatography to reduce the TGIC analysis time. The low thermal mass of the RP packed capillary columns (0.32 mm internal diameter) al-

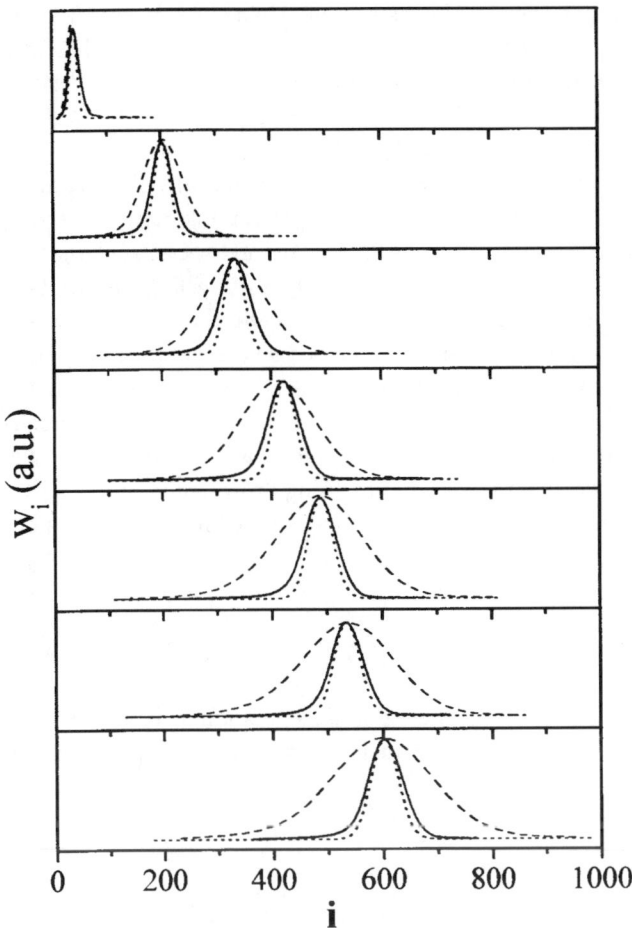

Fig. 7. Weight fraction (w_i) vs degree of polymerization (i) plot of SEC (*dashed lines*) and TGIC (*solid lines*) characterization results of a PS set synthesized under identical condition. Theoretical Poisson distributions matching the most probable value of MWD are also plotted with *dotted lines*. It clearly shows the band broadening of SEC elution peaks. Reproduced from [93] with permission

lowed rapid temperature programming for retention control. By use of temperature gradients as steep as 40 °C/min, rapid separation (11 min) of PS standards (2–400 kg/mol) was achieved [92].

Lee et al. [93] reported a systematic study to measure the precise MWD of a set of PS of varying molecular weights synthesized by anionic polymerization under identical conditions [94]. The polymerization time varying molecular weights and molecular weight distributions of the PS were analyzed by TGIC as well as SEC. Figure 7 shows the MWD of the set of PS measured by SEC and TGIC together with the Poisson distribution (Eq. 7). The instrumental band broadening of TGIC was found insignificant compared to the peak dispersion due to MWD. Therefore the MWD of the PS determined by TGIC was found to be close to the true MWD, which approaches the Poisson distribution in the late stage of the polymerization, in accordance with the early prediction of Flory [66].

Even though the MWD of the polymers prepared by anionic polymerization was found to have a much narrower distribution than commonly referred M_w/M_n values measured by SEC, they are still far from the "monodisperse" polymers as can be seen in Figs. 4, 6, and 7. It would be interesting to investigate further on how the finite MWD would affect the various properties using the fractionated samples of narrower MWD. For example, Busnel et al. reported on the band broadening in SEC separation using PS samples fractionated by TGIC. Chromatograms were fitted by exponentially modified Gaussian functions and mapping of band broadening was obtained for different column sets. Interpretation of the skewing of the chromatograms was proposed with a new model using Brownian motion properties inside the pores, which could explain why band broadening and tailing become so important near total exclusion volume [70].

4
Chain Architecture

In this section, recent efforts of the chromatographic separation with respect to the chain architecture are summarized, which include branched polymers and ring polymers.

4.1
Branched Polymers

Branching sometimes occurs from side reactions of the polymerization process or is made on purpose. Copolymerization of macromonomers and graft polymerization are good examples of the branching made on purpose. For short branching such as linear low-density polyethylene (PE) or graft copolymers with short grafted chains, the separation principle is often not much different from CCD analysis of copolymers. On line IR or NMR detection method coupled with SEC and/or IC and/or temperature rising elution fractionation can be used efficiently for the purpose. This will be discussed in more detail in the next section. For long chain branching, the branching frequency is low and the spectroscopic methods are often not sensitive enough to measure the branching density. For the analysis of randomly generated long chain branching, the most popular approach is to estimate the average number of branching according to the formula of Zimm and Stockmayer [95]:

$$g = \left(\frac{R_{br}^2}{R_{lin}^2} \right) \tag{8}$$

where g is the branching ratio defined as the ratio of radii of the branched (*br*) and linear (*lin*) polymer chain at the same molecular weight. For randomly branched polymers, g is related with the average number of branching per molecule, m:

$$g_3 = \left[\left(1+\frac{m}{7}\right)^{1/2} + \frac{4m}{9\pi}\right]^{-1/2} \qquad (9)$$

for 3 functional branching,

$$g_4 = \left[\left(1+\frac{m}{6}\right)^{1/2} + \frac{4m}{3\pi}\right]^{-1/2} \qquad (10)$$

for 4 functional branching, and

$$g = \frac{6f}{(f+1)(f+2)} \qquad (11)$$

for f random length arm star.

Therefore the problem boils down to the measurement of molecular weight dependence of the size (radius) distribution of polymer chains. SEC combined with light scattering and viscometry detection used to be the most popular method to study chain branching. Light scattering and viscometry detection measure the molecular weight and the size of the polymers in the SEC fraction, respectively [96]. Recently developed MALS detector can provide the information on the radius of gyration and molecular weight directly and SEC/MALS combination can be used for the purpose [97].

Although the SEC combined with the detection methods of molecular weight and size has been widely used for the branching analysis, the analysis scheme is based on the assumption that the SEC eluate at one elution time is homogeneous; i.e., a fraction of the SEC elution peak contains the polymer species of the same molecular weight as well as the number of the branching. This is generally not true since SEC only fractionates polymers according to the hydrodynamic volume and different polymer species can elute at the same retention volume. Therefore it is important to keep the possible local polydispersity problem in mind in the SEC analysis [61, 72, 98].

Unlike SEC, IC utilizes the interaction between polymer segments and the stationary phase and IC is able to separate branched polymers with better sensitivity to molecular weight (not chain size) than SEC. However, IC has not been used to characterize the randomly branched polymers since the system is too complicated to allow precise characterization. On the other hand, IC has shown a great potential in characterizing model branched polymers with well-defined structures. Such

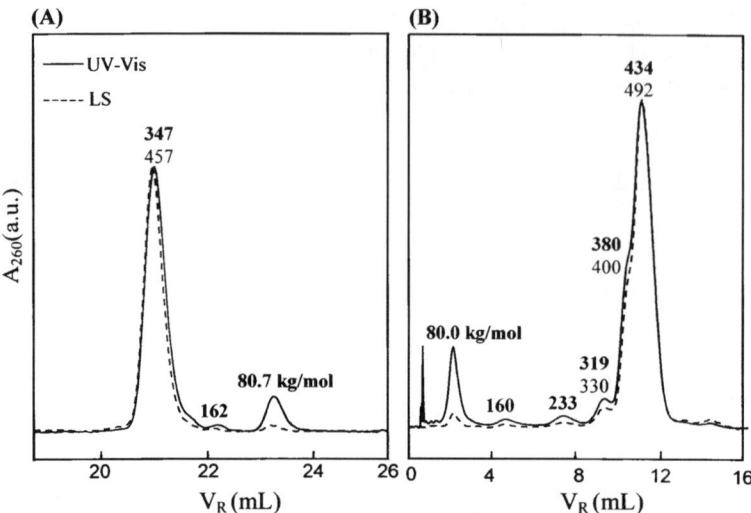

Fig. 8 A SEC chromatograms; **B** TGIC chromatograms of a six-arm star shaped PSs recorded by a UV/Vis (*solid line*) and a light scattering (*dotted line*) detector. In the SEC analysis, four- to six-arm star PSs are not resolved at all while one and two arm species are resolved. On the other hand, in TGIC analysis all six species having one to six arms are resolved. Each *peak* is labeled with M_w measured by light scattering detection (*plain letters*) and by calibration relative to linear PS standards (*bold letters*). For the SEC analysis four columns (Polymer lab. two mixed C, one mixed D, and one mixed E) were used with THF eluent. For the TGIC analysis, one C18 bonded silica column (Nucleosil C18, 250×2.1 mm, 100 Å pore, 5 μm particle size) was used and the eluent was a mixture of CH_2Cl_2/CH_3CN (57/43, v/v) at a flow rate of 0.1 ml/min. Reproduced from [100] with permission

polymers are usually prepared by a controlled polymerization method such as anionic polymerization [99]. Figure 8 shows a good comparison of SEC (A) and TGIC (B) separation of a six-arm star PS [100]. This star polymer was obtained by linking living polymer anions (arm polymer) to a hexavalent linking agent, 1,2-bis(trichlorosilyl)ethane [101]. The SEC chromatogram in Fig. 8A shows a well-resolved elution peak of the excess arm polymer and a trace of two-arm polymers, but three- to six-arm star polymers are not resolved. This is because the hydrodynamic volume of the star polymers with a uniform arm molecular weight does not change much as the number of arms increases. The molecular weights of the major peak measured by two methods, calibration relative to linear PS and light scattering detection, clearly demonstrate the effect. While the light scattering analysis yields the molecular weight close to six times the arm molecular weight indicating that the major component is the six-arm star, the molecular weight from the calibration method shows a much lower value equivalent to the molecular weight of linear chains having the same chain size as the six-arm star polymer.

On the other hand, TGIC separation in Fig. 8B shows a far higher resolution and all the species having a different number of arms are resolved. Furthermore

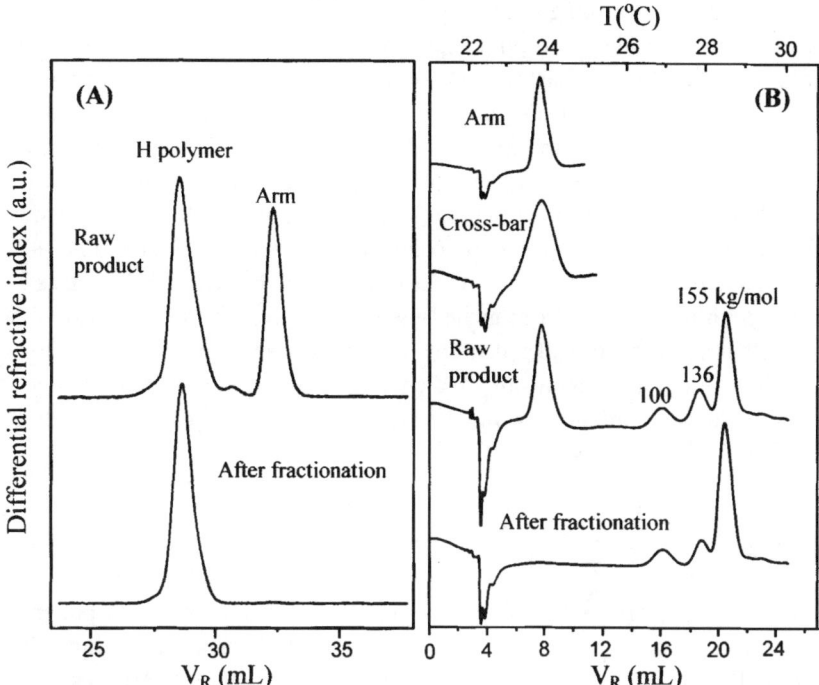

Fig. 9. A SEC chromatogram of H-shaped PB before (*top*) and after fractionation (*bottom*). Four styragel columns with a porosity range from 10^5 to 500 Å were used with THF eluent. **B** TGIC chromatograms of different stages during the H-shaped PB synthesis. *From the top*, arm, cross bar, H-shaped PB before and after the fractionation. One RP column (Nucleosil C18, 250×4.6 mm, 100 Å, 5 µm particle size) was used and the eluent was 1,4-dioxane at a flow rate of 0.5 ml/min. In contrast to the SEC chromatogram, TGIC chromatogram shows the existence of side products, two missing arms (100 kg/mol), one missing arm (136 kg/mol), and higher molecular weight coupled cross-bar products. Reproduced from [103] with permission

the molecular weight determined by light scattering detection and calibration methods are in much better agreement. It is due to the fact that IC utilizes the interaction of the polymer chains with the stationary phase and the interaction strength depends on molecular weight, but not significantly on chain architecture. As the number of arms increases, that chain architecture starts to affect the interaction strength and the resolution deteriorates, but it is apparent that IC is more sensitive to the molecular weight than SEC in resolving branched polymers. Lee et al. reported on the linking reaction kinetics of the star polymers by TGIC and they could measure the linking reaction rate constants of the living PS anions to a hexavalent linking agent, 1,2-bis(trichlorosilyl)ethane [102].

Perny et al. compared the analysis results of an H-shaped polymer by SEC and TGIC, which provided an opportunity to examine critically the purity of branched polymers prepared by anionic polymerization [103]. The H-shaped polybutadiene

(PB) was prepared by linking linear polybutadienyl anions (arm) to a tetravalent chlorosilane functional (two at both chain ends) polybutadiene (cross bar). After purification by a classical fractionation procedure the H-shaped PB appears to have a high structural uniformity judging from the conventional analysis method including SEC, membrane osmometry, and NMR. Figure 9A shows the SEC chromatogram of the H-shaped polymers before and after the fractionation. TGIC analysis, however, revealed the presence of a large amount (~30%) of various side products, missing arm H-polymer and coupled cross bars as displayed in Fig. 9B. This is a good example to demonstrate the difficulty in the synthesis as well as in the characterization of model branched polymers. Although SEC coupled with various detection methods still constitutes a majority of such analyses, satisfactory SEC analysis results are usually not good enough to endorse the structural intactness. More critical examination with IC is highly desirable for the characterization of such branched polymers.

4.2
Ring Polymers

Ring polymers are of interest due to the topological influence of the polymer chains on their physical properties [104]. Various physical properties of ring polymers have been predicted theoretically as well as by computer simulation studies and have been examined experimentally. Anionic polymerization has been used most widely to obtain high molecular weight ring polymers with narrow molecular weight distribution. The basic strategy is to form precursor polymers with two carbanion chain ends and to have them react intramolecularly, under extreme dilution, with a difunctional electrophile to close the ring. However, side reactions produce linear precursor polymers and intermolecular reactions simultaneously produce dimeric and higher molecular weight linear polymers. Therefore, it is difficult to obtain pure ring polymers directly and further fractionation has been necessary in order to obtain ring polymers with high purity.

The hydrodynamic volume of ring polymers is smaller than their linear precursors of the equivalent molecular weight, but the difference is not large enough to be fully resolved by SEC. For example, Fig. 10 shows SEC chromatograms of ring PS of various molecular weights and their linear precursors [105]. The ring PS were prepared by linking two chain ends of the linear precursor PS prepared by anionic polymerization of styrene using Na-naphthalene as an initiator and already extensively fractionated from linear contaminants by fractional precipitation [106]. The overlapped SEC elution peaks clearly indicate that the separation by SEC cannot be complete.

Skvortsov and Gorbunov formulated the theoretical basis for the LCCC separation of ring polymers from linear contaminants [107, 108]. Recently several research groups have examined the theoretical prediction [105, 109–111]. For exam-

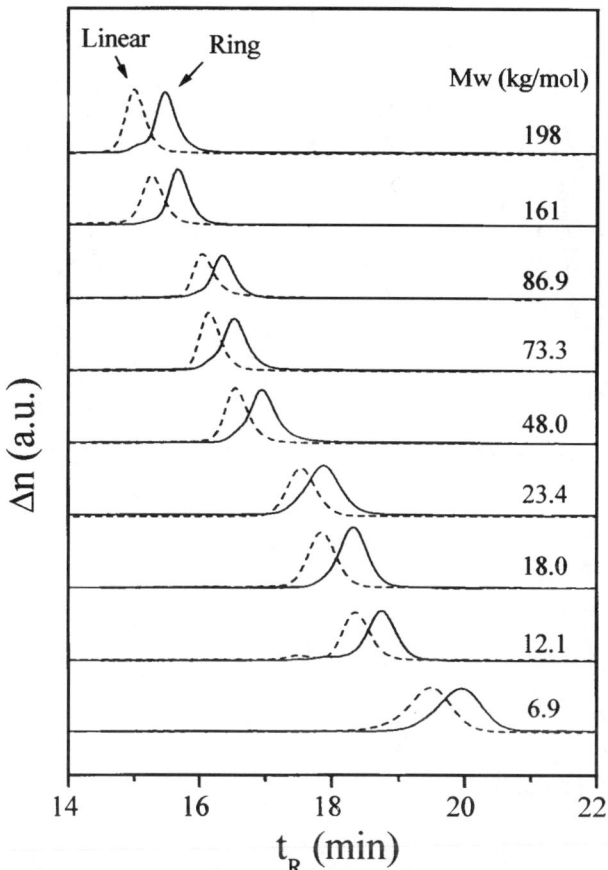

Fig. 10. SEC chromatograms of ring PSs (*solid line*) and corresponding linear precursors (*dashed line*). Linear polymers elute earlier than rings but the elution peaks are partially overlapped. Some elution peaks of ring polymers show small shoulders at lower retention time indicating the contamination of the linear precursors. Two SEC columns (Polymer Lab. mixed C) were used with THF eluent at a flow rate of 1.0 ml/min. Reproduced from [105] with permission

ple, Lee et al. showed that at the critical condition of linear PS, the same set of ring PS shown in Fig. 10 could be fully resolved as displayed in Fig. 11 using C18 bonded silica stationary phase and CH_2Cl_2/CH_3CN mixed eluent system [105]. The chromatograms show that the ring PS still contains a significant amount of contaminants and the majority of the contaminants is the linear chains. Subsequent MALDI-TOF MS analysis clarified the structure of the contaminants [112]. More recently Lee at al. examined the LCCC separation behavior of the ring polymers in more detail [20]. They found that the retention of the ring polymers did not follow the theoretical prediction of Skvortsov and Gorbunov quantitatively and possible

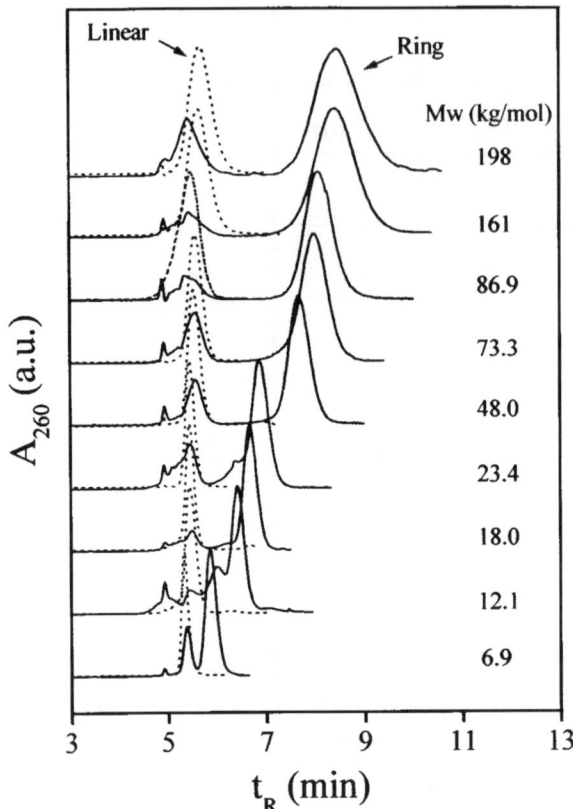

Fig. 11. LCCC chromatograms of ring PSs (*solid line*) and corresponding linear precursors (*dotted line*) at the chromatographic critical condition of linear PS. Linear precursors elute at about the same retention time (5.4 min) independent of the molecular weight while the retention of ring polymer increases with molecular weight. The *small peaks* appearing near t_R= 5 min are the injection solvent peaks. The elution peak of ring polymers is completely separated from the linear precursors down to the baseline. The chromatograms of ring PSs show that most of the samples contain contaminants mostly the linear precursors. Column: Nucleosil C18AB; 250×4.6 mm; 100 Å pore; 5 µm particle size at 43 °C. Eluent: CH_2Cl_2/CH_3CN (57/43, v/v) at a flow rate of 0.5 ml/min. Reproduced from [105] with permission

reasons including the difficulty to find a rigorous chromatographic critical condition were discussed.

5
Chemical Composition Distribution

Rigorously speaking, no synthetic polymer is homogeneous either in molecular weight or in chemical composition. Even a homopolymer would differ in initiator, terminal groups, tacticity, regioregularity, etc. Copolymers contain more than one kind of monomer and also have chemical composition distribution (CCD). Copolymers can be further divided into random or statistical copolymers, block copolymers, graft copolymers, etc. depending on the monomer sequence and chain architecture. Therefore copolymers may have heterogeneity in a number of molecular characteristics and their characterization is much more complicated than homopolymers. In addition, chain terminal groups or functional groups can contribute to the molecular heterogeneity. The contribution of the chain ends is more important for low molecular weight polymers.

Characterization of polymer mixtures is also of interest due to the wide use of polymer blend systems. Mixtures of homopolymers are relatively a simple form of chemical heterogeneity compared to copolymers. Even in this case, precise characterization is often non-trivial since many of polymer blend systems contain various additives in addition to polymer resins. In this section, recent progress on the characterization of synthetic polymers having chemical heterogeneity is reviewed. For the sake of convenience, the content is divided into mixtures, block copolymers, random copolymers, and functionality distribution.

5.1
Mixtures

Application of conventional SEC to the characterization of polymer mixtures is straightforward if the polymer chain sizes of the components in a mixture are different enough to be resolved as separate peaks. Otherwise, the species having similar chain size would elute together and the eluate has local heterogeneity not only in molecular weight but also in composition (Fig. 3). There have been several approaches to get round the problem. One approach is to use multiple detectors and the chromatograms recorded by different detectors can be decomposed into the individual contribution of each component [98, 113–116]. This method can be applied with relative ease but the precision of the method is not high unless the components of a mixture show large differences in sensitivity in the detection methods.

One of the popular methods used for the separation of polymer mixtures is the liquid chromatography at the chromatographic critical condition (LCCC). At the critical condition of one component, the other component, which is not under critical condition, can be separated in SEC mode [27–31, 33]. It is necessary to choose a proper combination of stationary and mobile phase considering the interaction strength to elute the component to be analyzed in the SEC mode. For example, Pasch et al. showed in the analysis of mixtures of methacrylate-based polymer blends that, at the critical point of PMMA, less polar methacrylates can be analyzed in the SEC mode using silica gel as the stationary phase and a methyl ethyl ketone/cyclohexane mixture as the eluent. For the analysis of polar polymethacr-

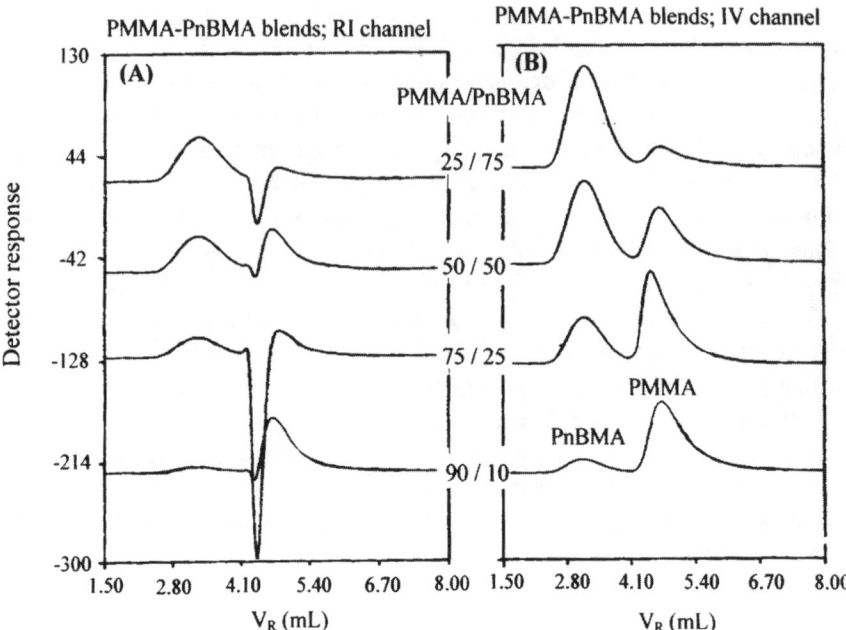

Fig. 12A,B. LCCC chromatograms of blends of PMMA and PnBMA under the critical condition of PMMA. Column: LiChrospher 300 Å + 1000 Å, mobile phase: methyl ethyl ketone/cyclohexane (72/28, v/v), detector: **A** differential refractometer; **B** capillary viscometer. The large injection solvent peak is suppressed in the chromatograms recorded by the viscosity detector. Reproduced from [30] with permission

ylates at the critical point of poly(decyl methacrylate) (PDMA), an RP stationary phase and THF/CH_3CN mixture are a useful combination [29]. The isocratic nature of the LCCC elution allowed employing viscometry detection for the on-line determination of the molecular weight [30, 31]. In contrast to the refractive index detection, where a strong solvent peak interferes with the elution peak of the polymer under critical condition in most cases, the viscosity signal is not affected by the injection solvent peak (Fig. 12). Yun et al. reported the LCCC separation of PS homopolymer from poly(S-co-MA) copolymers by use of enhanced-fluidity liquid mobile phases (THF/CO_2) with packed-capillary NPLC. The critical condition for PS was approached by changing the concentration of CO_2 in the mixture combined with temperature and pressure variation. Long packed capillaries could be used in this application because the enhanced-fluidity mobile phases have low viscosities [33].

Another interesting coupled LC method for the characterization of polymer blends utilizes dual SEC/IC separation mechanisms simultaneously [47, 89, 117–119]. In this method, one component is separated by the SEC mechanism while the other is separated by the IC mechanism simultaneously. Figure 13 shows an ex-

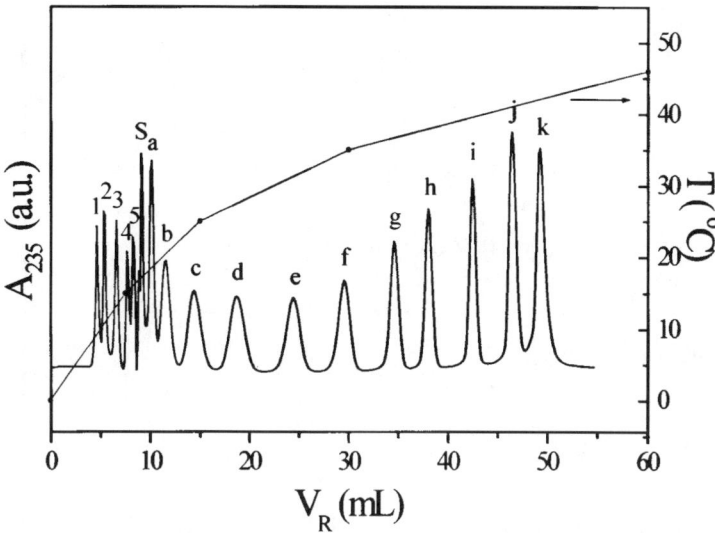

Fig. 13. Chromatogram of a mixture of 11 PS (a–k, in the increasing order of molecular weight, 1.7–2890 kg/mol) and 5 PMMA standards (1–5, in the decreasing order of molecular weight, 1500–2.0 kg/mol) separated simultaneously by dual TGIC and SEC mechanism, respectively. S is the injection solvent peak. The eluent was 57/43 (v/v) mixture of CH_2Cl_2/CH_3CN and three RP columns having different pore sizes (Nucleosil C18; 100, 500, 1000 Å pore size each; 250×4.6 mm). The temperature of the circulating fluid is programmed to change by five segment linear ramps from 5°C to 45°C as shown in the figure. Reproduced from [117] with permission

ample of the SEC/IC dual mechanism separation of PS/PMMA binary mixture [117]. The eluent was 57/43 (v/v) mixture of CH_2Cl_2/CH_3CN and three C18 bonded silica columns having different pore sizes were used to enhance the resolution of the SEC separation. The solvent power is strong for PMMA to be separated by the size exclusion mechanism while PS is separated by the interaction mechanism at the separation condition. In order to control the IC retention of PS, column temperature was varied during the elution. The retention of PMMA is not affected much by the temperature change since the size exclusion separation mechanism is mainly an entropic effect and not sensitive to temperature. Chang and coworkers also reported successful separations by RP-TGIC of PS/PI and PI/PMMA using C18 silica column [21, 89]. For these polymer pairs less polar polymers eluted in the IC regime. The elution sequence can be reversed under NPLC condition [118]. More recently, Molander et al. applied the method to a mixture of high molecular weight semicrystalline polypropylene and polyphenylene ether grafted to maleic anhydride (MAn) using a capillary column packed with large-pore silica. Rapid temperature programming from 110 to 170°C was required in order to resolve the materials eluting PP by SEC mechanism and polyphenylene ether by NP-TGIC mechanism [119].

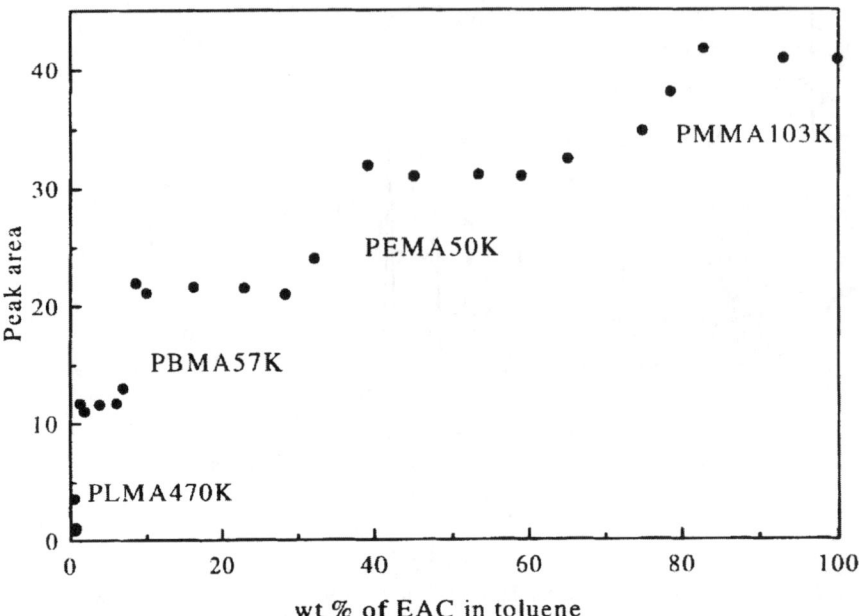

Fig. 14. Dynamic integral desorption isotherm for a polymer blend containing PS (92 kg/mol), PLMA (470 kg/mol), PnBMA (57 kg/mol), PEMA (50 kg/mol), PMMA (103 kg/mol), and PEO (45 kg/mol). The polymers (0.01 mg each) were pre-adsorbed in a FAD column (Develosil, 50×4.6 mm) using toluene eluent and desorbed with the increasing content of ethyl acetate (EAC). Reproduced from [126] with permission

The other method to separate polymer mixtures utilizes the full adsorption and desorption of particular components by switching eluent [120–127]. Under appropriate choice of a sorbent and eluent system, one component of the polymer mixture is separated in terms of molecular weight in the conventional SEC mode while the other components are fully retained on the surface of the sorbent. In the next step the eluent is changed so that one of the previously retained polymers is desorbed and analyzed in the SEC mode [32, 123]. In the favorable case a blend of more than two components can be fractionated in sequence [124, 126]. Figure 14 shows an example of the separation of a four components blend.

5.2
Block Copolymers

Block copolymer is a type of copolymers, which has a chain structure of homopolymer blocks connected in series by a chemical bond. These block copolymers are commonly synthesized by either end coupling of two different polymers or growing different chains in a sequential manner. The precise analysis of diblock copolymers requires a characterization of two-dimensional distribution in both composition and molecular weight. For multiblock copolymers it becomes more complicated. However, it is obvious that the analysis of block copolymers is simpler than random copolymers that can be regarded as multiblock copolymers with numerous block sizes connected in a random sequence.

Major effort has been made on LCCC separation at the critical condition of one block. It is based on the assumption that a block can be made "chromatographically invisible" at its critical condition and the retention of the block copolymer is determined solely by the other block that is not under critical condition [55, 128, 129]. This assumption is only valid if the retention of the homopolymer of the "invisible" block is indeed independent of molecular weight at the critical condition and if the retention of the "visible" block is not affected by the presence of the "invisible" block. If these conditions were satisfied, LCCC would undoubtedly be a powerful tool for the characterization of selected blocks within block copolymers. The applications of LCCC for the characterization of block copolymers made so far can be divided into two categories.

One is to elute the block copolymers in the IC regime (thus eluting after the total mobile phase volume) if the "visible" block is more interactive with the stationary phase than the "invisible" block. Since the elution condition needs to be fixed at the critical condition for the "invisible" block, the application is limited to end group analysis or to rather short visible block lengths due to the exponential increase of the retention with the chain length. The analysis of end groups with this method will be discussed later. There are not many reports on the block copolymer analysis in this LCCC mode. Lee et al. characterized both blocks of a low molecular weight poly(ethylene oxide)-*block*-poly(L-lactide) (PEO-*b*-PLLA) at the critical condition of PEO [46]. As shown in Fig. 15, at the critical condition of PEO, the block copolymers were separated well according to the number of lactide units in the IC regime. At the off-critical condition, PEO block also contributes to the retention and MWD of PEO causes significant peak broadening. MALDI-TOF mass spectrum of each peaks revealed the MWD of PEO. PLLA-*b*-PEO-*b*-PLLA triblock copolymers were also characterized at the critical condition of PEO [45]. The dependence of the LCCC retention of the diblock and triblock copolymers on the degree of polymerization of PLLA block(s) follows Martin's rule very well. Unlike the case of the diblock copolymer, however, splitting of the elution peaks containing the same number of lactide units was found. The peak splitting was ascribed to the different length distributions of PLLA blocks at the two ends of the PEO block (Fig. 16). This result indicated that the interaction of the triblock copolymers with the stationary phase is affected by the distribution of the interacting blocks at the two ends of the center PEO block, in addition to the total number of lactide units in the triblock copolymer.

The other LCCC separation mode is to elute block copolymers in the SEC regime (eluting before injection solvent peak) if the "visible" block is less interactive

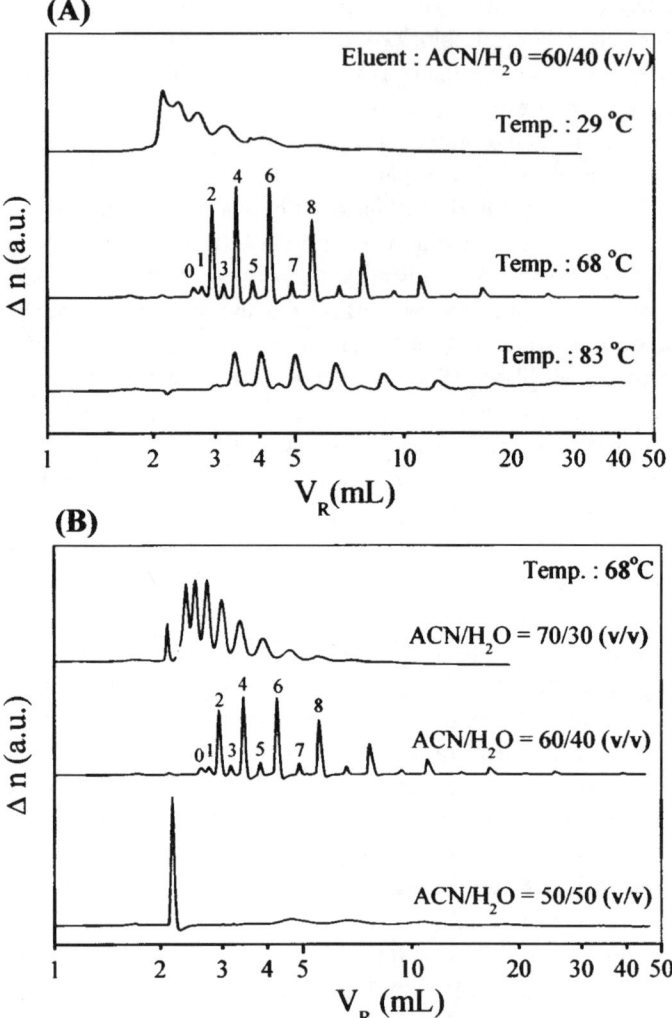

Fig. 15A,B. Change of RPLC chromatograms of the PEO-*b*-PLLA with: **A** the temperature; **B** the eluent composition near the critical condition of poly(ethylene glycol). At the critical condition (middle chromatograms in A and B) the separation takes place according to the number of L-lactide units and the elution peaks are sharp while the resolution is poor at the off-critical conditions. The numbers indicate the number of L-lactide units in the block copolymer. Column: Luna C18; 100 Å; 250×4.6 mm. Eluent: mixture of CH_3CN (ACN) and H_2O. Reproduced from [46] with permission

with the stationary phase. Most of the LCCC applications to the characterization of block copolymers have been made in this mode and the determination of the molecular weight distribution of the "visible" block has been made using the

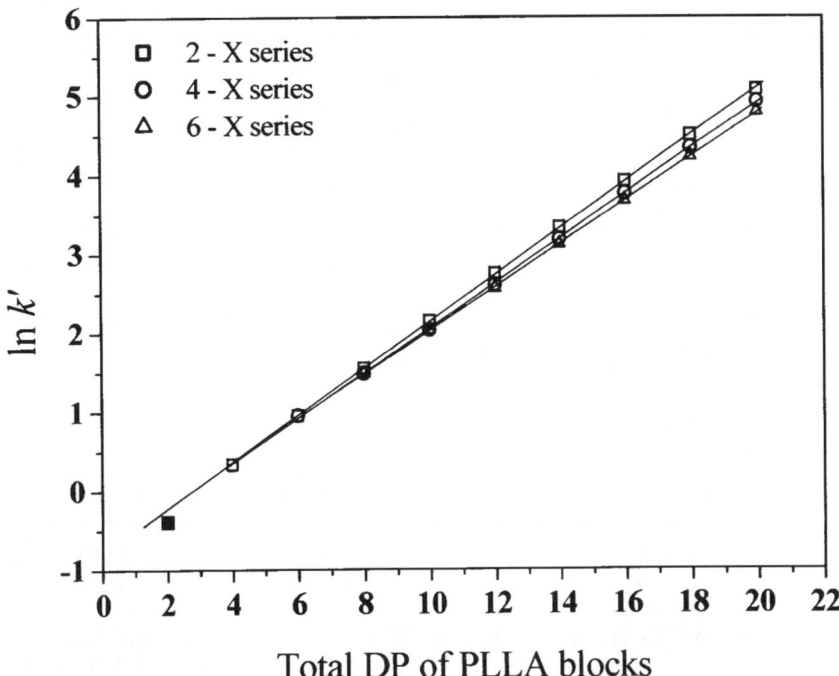

Fig. 16. Dependence of capacity factor on total degree of polymerization of PLLA blocks in PLLA-b-PEO-b-PLLA triblock copolymer at the critical condition of PEO block. *Squares, circles and triangles* are assigned to each series of the triblock copolymers with two, four, and six L-lactide units at one end and a varying block length at the other end of the PEO block, respectively. Column: Luna C18; 250×4.6 mm; 100 Å at 54 °C. Eluent: mixture of CH_3CN/H_2O (54/46, v/v). Reproduced from [45] with permission

standard calibration method commonly used in SEC [22, 35, 42–44, 47–49]. Gankina et al. demonstrated the possibility of LCCC analysis of block copolymers by TLC [130] and Zimina et al. applied LCCC to characterize PS-b-PMMA and PS-b-PBMA diblock copolymers [131, 132]. Subsequently Pasch and coworkers have done extensive works on the LCCC characterization of various diblock copolymers [10, 11, 49, 133–135].

All of these works appear to support the applicability of LCCC for the characterization of selected block of block copolymers, but the precision of the method was not tested rigorously. A 10% error in the accuracy of an individual block molecular weight would not be uncommon in the characterization of block copolymers with conventional characterization methods. Therefore the apparent consistency found with several independent block copolymers would not constitute the validity of the method. In order to test the method rigorously, it is highly desirable to use a custom-made set of block copolymers.

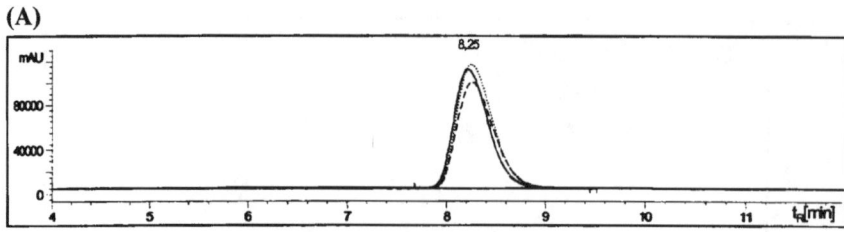

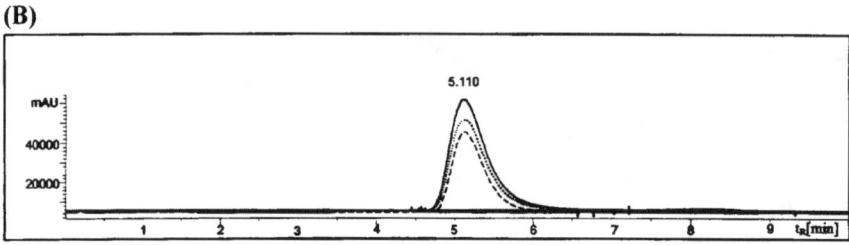

Fig. 17 A LCCC chromatograms of PtBMA-*b*-PMMA. Precursor PMMA (M_n=31.0 kg/mol) with different PtBMA block lengths (M_n=0, 46.3, 85.3 kg/mol) at the critical condition of PtBMA. Column: two Nucleosil C18; 250×4 mm; 300 Å and 1,000 Å. Eluent: THF/CH$_3$CN (49.5/50.5, w/w). **B** LCCC chromatograms of PMMA-block-PtBMA. Precursor PtBMA (M_n= 73.0 kg/mol) with different PMMA block lengths (M_n=29.6, 108, 167 kg/mol) at the critical condition of PMMA. Column: two Nucleosil; 150×4 mm, 100 Å and 250×4 mm, 300 Å. Eluent: THF/*n*-hexane (82/18, w/w). Reproduced from [44] with permission

Recently Falkenhagen et al. and Lee et al. reported rather contradictory results on LCCC studies with custom-made sets of block copolymers. These block copolymer sets were prepared by anionic polymerization keeping one block length constant while varying the other block length so that a rigorous examination was possible. Also they used an RP- and an NP-LCCC methods so that they could examine both blocks of block copolymers. Falkenhagen et al. reported that LCCC provided a accurate molecular weight of the visible block of PMMA-*b*-PtBMA (Fig. 17) [44], while Lee et al. reported that a block in PS-*b*-PI was not completely "invisible" at the critical condition of its homopolymer, and the retention of block copolymers is affected to some extent by the length of the "invisible" block under its chromatographic critical condition (Fig. 18) [22]. The retention of the "visible" block eluting in the SEC regime gradually increases and the apparent molecular weight of the block decreases with the increase of the "invisible" block length.

Lee at al. further extended the study using a single solvent, 1,4-dioxane to establish the LCCC condition for PI with the C18 bonded silica stationary phases [21]. Employment of single solvent for LCCC experiments is highly desirable. It improves the reproducibility of the LCCC experiments since the polymer retention changes very sensitively with eluent composition in mixed solvents. Also it elimi-

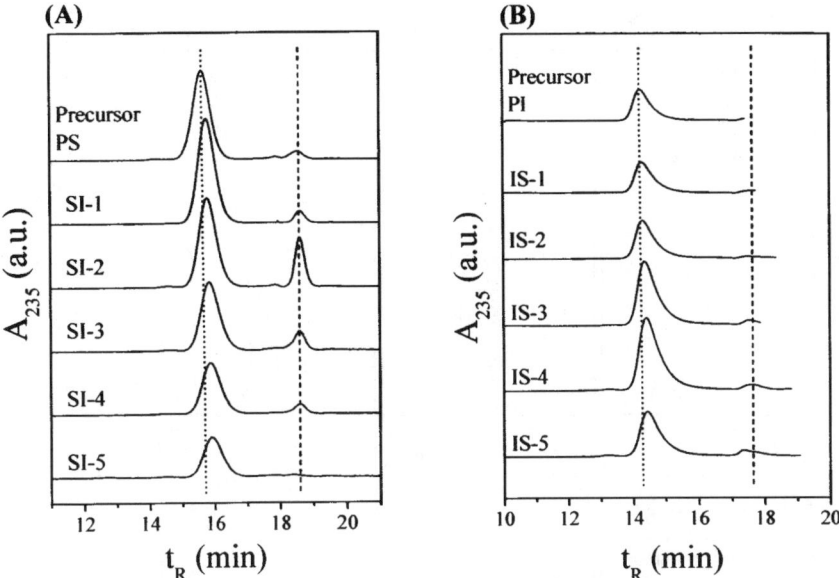

Fig. 18A,B. LCCC chromatograms of PS-*b*-PI samples under the critical condition of the block of varying molecular weight: **A** fixed PS block series (M_w (PS)=12.0 kg/mol, M_w (PI)=3.0, 6.0, 11.1, 21.4, 34.2 kg/mol from SI-1 to SI-5) at the critical condition of PI. Column: three RP columns (Nucleosil C18; 100Å, 500 Å, and 1000 Å; 250×4.6 mm) at 47°C. Eluent: CH_2Cl_2/CH_3CN (78/22, v/v); **B** fixed PI block series (M_w (PI)=12.5 kg/mol; M_w (PS)=3.3, 5.9, 13.5, 26.6, 38.1 kg/mol from IS-1 to IS-5) at the critical condition of PS. Column: three NP columns (Nucleosil; 100 Å, 500 Å, and 1000 Å; 250×4.6 mm) at 7°C. Eluent: THF/isooctane (50/50, v/v). The *vertical dashed line* indicates the elution time of critical component for each case. *Vertical dotted lines* are drawn for visual aid to compare the retention time of the block copolymers. Reproduced from [22] with permission

nates the preferential sorption of a component in the mixed eluent, which may cause various complications [136]. As displayed in Fig. 19A, the critical condition of PI was unambiguously established at 47.7°C at least for the PI samples investigated. A small change of temperature clearly shifts the elution peaks from the co-elution point. The PS precursor and the two PS-*b*-PIs having the same PS block length are supposed to co-elute at the critical condition of PI and may show some deviation at the off-critical condition. However, as shown in Fig. 19B, the elution behavior of block copolymers hardly changes. Also a rigorous chromatographic invisibility was not achieved and the retention of the block copolymers was affected by the length of the PI block under the critical condition.

Combining these results it can be concluded that the LCCC experiments permit a reasonable estimation of the individual block length of diblock copolymers but the feasibility of the precise characterization of block copolymers remains questionable. A more systematic study with a variety of block copolymers and chroma-

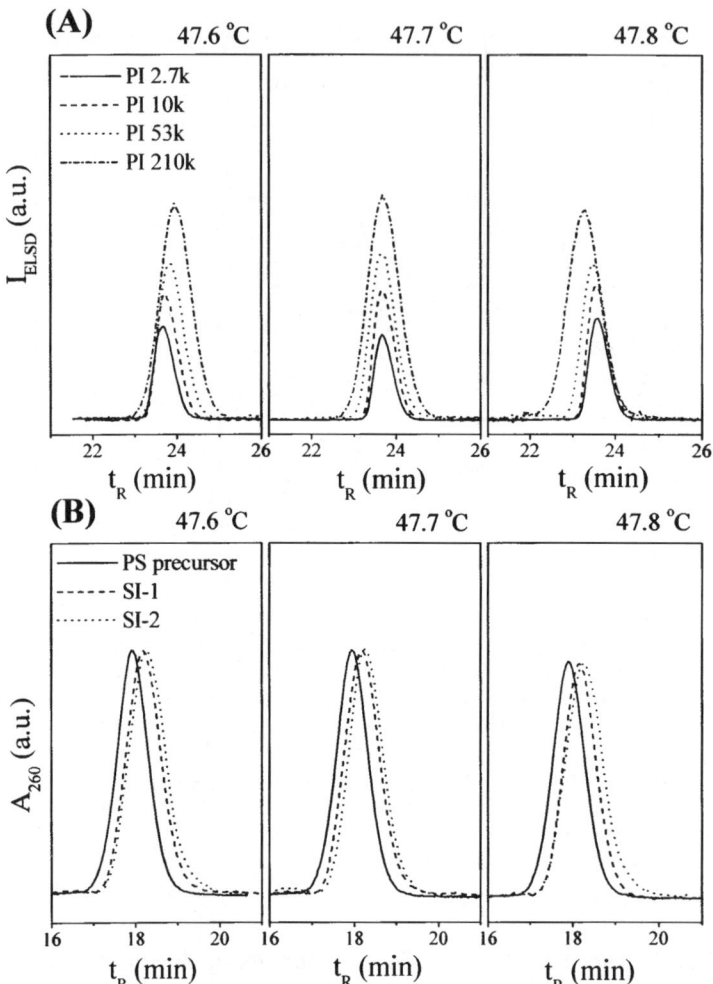

Fig. 19. A Overlapped chromatograms of four PI obtained at three different temperatures near the critical condition of PI. Column: three RP columns (Nucleosil C18; 100 Å, 300 Å, and 500 Å; 250×4.6 mm), eluent: 1,4-dioxane. Amplitudes of the elution peaks are rescaled for visual aid. **B** Chromatograms of the PS precursor (M_w=12.0 kg/mol) and two PS-*block*-PI with fixed PS block length at 12.0 kg/mol (M_w (PI)=6.0 and 21.4 kg/mol for SI-1 and SI-2, respectively) at the same temperatures as in A. Reproduced from [47] with permission

tographic conditions is called for to clarify whether a rigorous characterization of the individual block length is indeed feasible. In any event, one needs to be aware of the possible problems associated with the LCCC analysis of block copolymers. The problem may be related with the difficulty to achieve an unambiguous critical condition [20, 23–25].

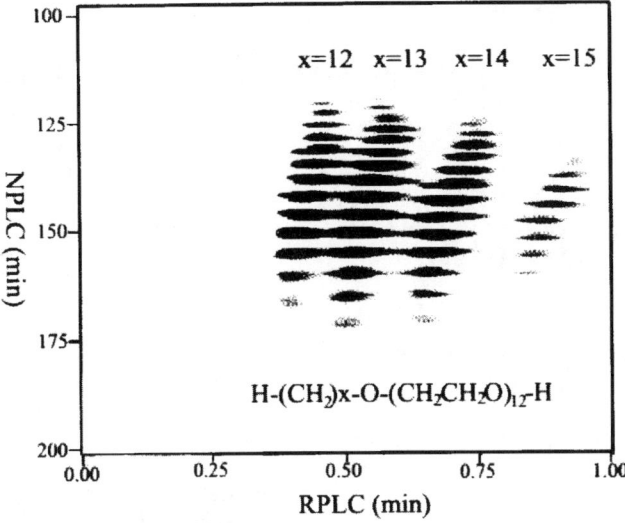

Fig. 20. 2D-LC (NPLC and RPLC) chromatogram of an alcohol ethoxylate (Neodol 25–12) with the corresponding chemical structure. For the first dimension NPLC was used. Column: Zorbax silica, eluent: H_2O/CH_3CN; 100/20–20/80 concave gradient in 300 min with a flow rate of 0.05 ml/min. The second dimension was RPLC. Column: Pecosphere C18; 100 Å; 33×4.6 mm. Eluent: CH_3OH/H_2O (95/5) with a flow rate of 1.5 ml/min. Reproduced from [137] with permission

Another promising method is 2D-LC, a combination of adequately selected two different LC separation conditions. If the separation condition of one dimension separates one block in the IC mode while the other block does not show a strong interaction with the stationary phase, preferably weak enough to elute in the SEC mode (like simultaneous SEC/IC separation condition for mixtures), the interacting block can be separated by IC mode nearly independent of the non-interacting block length and vice versa. Employing the 2D-LC method, Murphy et al. and Jandera et al. separated individual blocks of alcohol ethoxylate and PEO-*b*-PPO, respectively (Fig. 20) [137–139]. Due to the amphiphilic nature of the block copolymer, the combination of NPLC and RPLC worked well.

Recently Trathnigg et al. characterized fatty acid polyglycol esters using 2D-LC with LCCC as the first and RPLC as the second dimension [140]. Fractions from LCCC are transferred to RPLC using the full adsorption-desorption (FAD) technique [125], by which they are focused and reconcentrated before injection into the second dimension (Fig. 21). This is achieved by increasing the water content of the mobile phase behind the LCCC column. Monoester oligomers of up to 20 oxyethylene units can be resolved to the baseline.

Such 2D-LC separations had been applied mainly for low molecular weight block copolymers with a high contrast in polarity of two blocks. Recently Park et

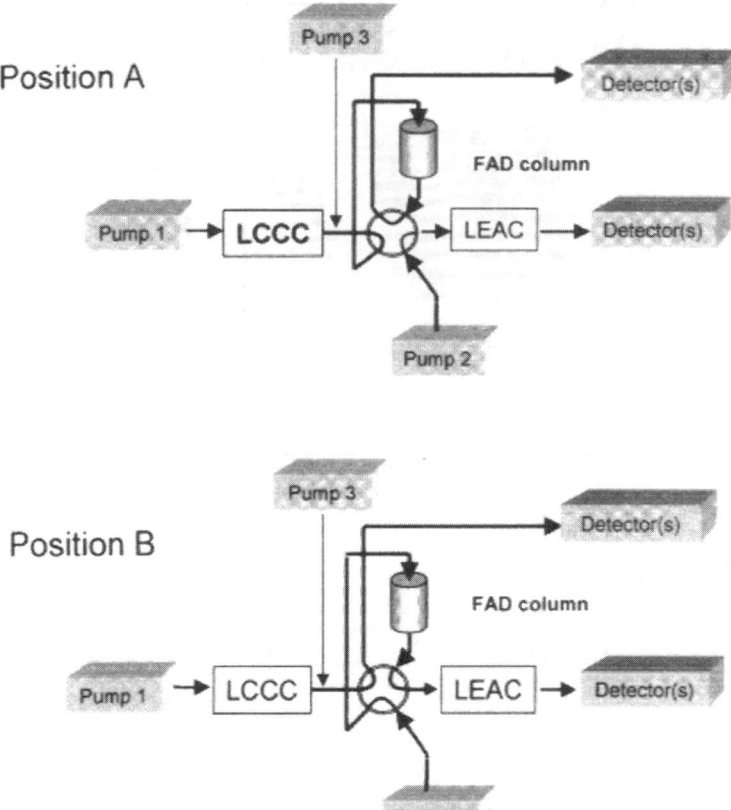

Fig. 21. Experimental set-up for 2D-LC with a combination of LCCC as the first and RPLC as the second dimension, and a full adsorption-desorption (FAD) column for focusing of fractions. Reproduced from [140] with permission

al. successfully carried out a 2D-NPLC/RPLC fractionation for PS-b-PI diblock copolymers prepared by anionic polymerization into fractions, which have narrower distributions in molecular weight as well as in chemical composition [141]. The polarity difference of two blocks of PS and PI is subtler than the previous examples but the 2D-NPLC/RPLC fractionation can fractionate individual blocks of PS-b-PI well. The working principle of the separation method was confirmed for a low molecular weight PS-b-PI (Fig. 22). With the aid of MALDI-TOF MS, it was confirmed that the separation method could resolve each *mer* of the PS-b-PI. They extended the application to a high molecular weight PS-b-PI (24 kg/mol) and succeeded to further fractionate the block copolymer (Fig. 23). Significant variation was observed in average molecular weight as well as in composition of the fractionated samples. As displayed in Fig. 24 these variations were large enough to show

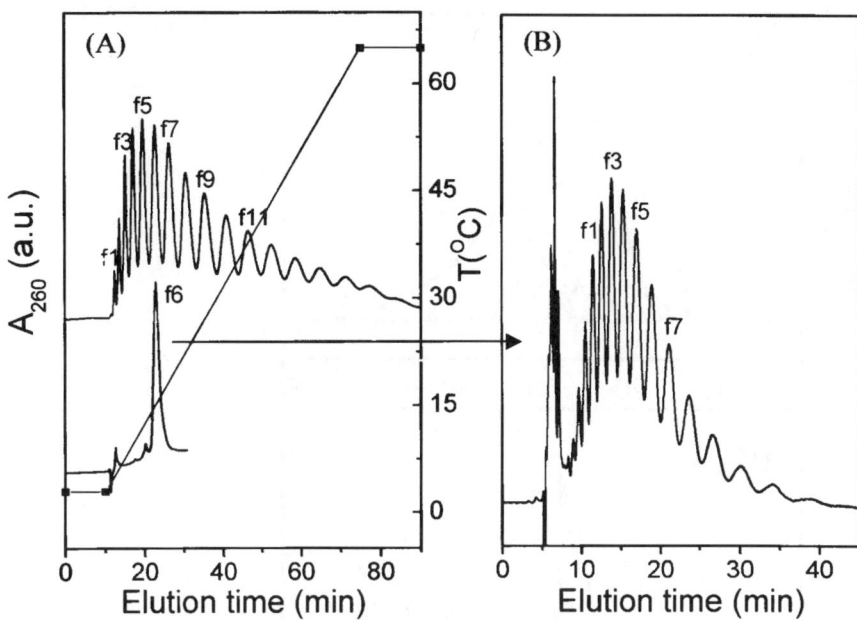

Fig. 22A,B. 2D-LC separation of a low molecular weight PS-*b*-PI (2.4 kg/mol, 30.4% PI content): **A** NP-TGIC chromatograms of the PS-*b*-PI (*top*) and a fraction, f6 (*bottom*), which separates PS-*b*-PI in terms of the PS block length only. Temperature program is also shown in the plot; **B** RPLC chromatogram of the NP-TGIC fraction, f6, which separates in terms of the PI block length only. It elutes as a single peak in NP-TGIC, but shows multiple peaks in RPLC indicating that it contains homogeneous PS block length (eight monomer units) but various PI block lengths. Reproduced from [141] with permission

different morphologies for the fractions taken from the same mother block copolymer [141].

5.3
Random Copolymers

Characterization of random copolymers needs to measure at least two distributions: molecular weight and composition, which are not necessarily independent of each other. A good example of the bivariate distribution is the high conversion copolymers that often experience a significant variation in composition as a function of monomer conversion and CCD may vary with the molecular weights. The ideal method to characterize such copolymers is to measure the 2-D map of the bivariate distributions, i.e., along the horizontal and vertical axes in Fig. 25. However, none of the known chromatographic techniques rigorously works in this way and a chromatographic separation is commonly executed along a not well-defined pathway that depends on both molecular weight and composition. Several meth-

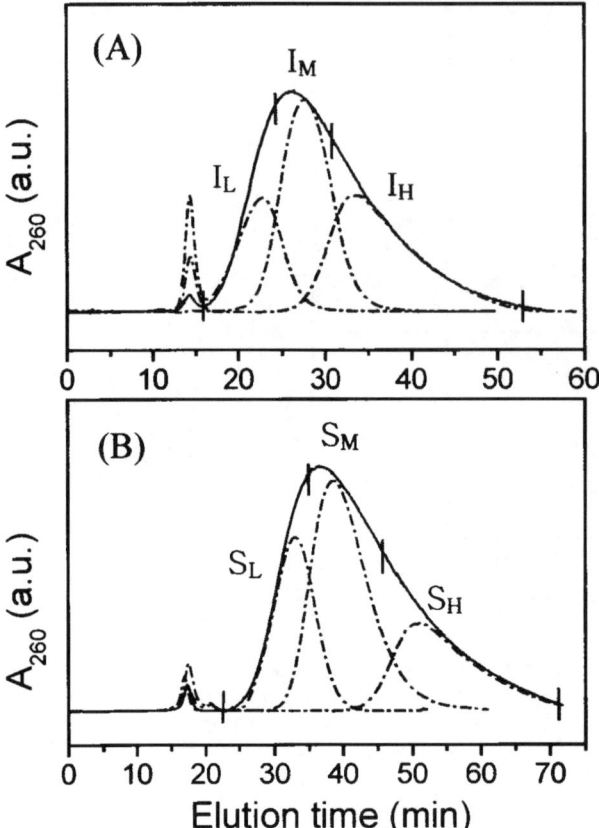

Fig. 23A,B. 2D-LC fractionation of high molecular weight PS-*b*-PI (24 kg/mol, 34.8 wt% PI content); **A** RPLC chromatograms of the mother PS-*b*-PI (*solid line*) and its three fractions (*dash-dot line*); low (I_L), middle (I_M), and high (I_H) molecular weight portion of the PI block; **B** NPLC chromatograms of the middle fraction of RPLC separation (*solid line*) and its three fractions (*dash-dot line*); low (S_L), middle (S_M), and high (S_H) molecular weight portion of the PS block. RPLC fractionates the PI block while NPLC fractionates the PS block. The range of the fraction collected is indicated with *small vertical bars*. Reproduced from [141] with permission

ods have been proposed to determine the bivariate distributions of random copolymers.

One approach is the SEC fractionation under the assumption that SEC separates random copolymers in terms of molecular weight and the analysis of the fractions using multiple detection methods to measure molecular weight as well as composition. For example, Montaudo and Montaudo fractionated two copolymers, poly(MMA-*co*-BA) obtained at high and low conversions by SEC and measured the molecular weight and composition of the fractions by MALDI-TOF MS

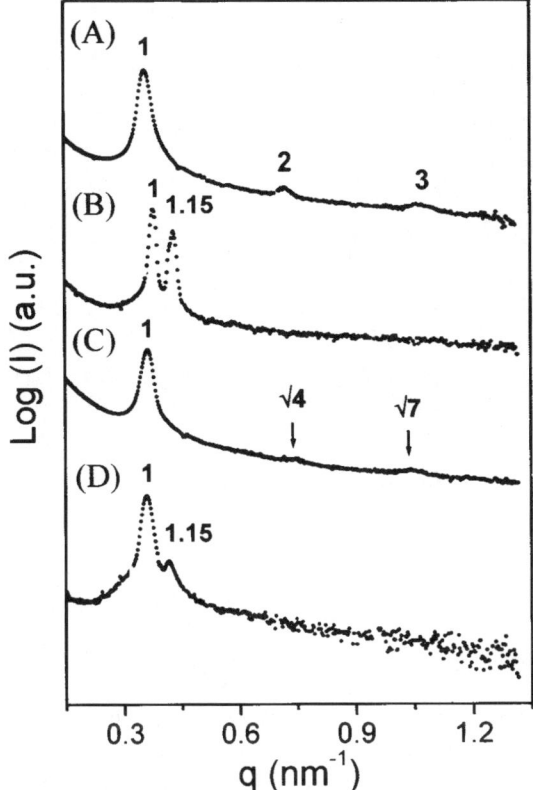

Fig. 24A–D. Small angle X-ray scattering profiles of the mother PS-b-PI and three fractions: **A** $S_L I_H$ fraction showing lamellar morphology; **B** $S_M I_M$ fraction showing gyroid morphology; **C** $S_H I_L$ fraction showing hexagonal cylinder morphology; **D** mother PS-b-PI showing gyroid morphology. The number on the peak positions represents the scattering vector of the peak position relative to the first order scattering maximum. Reproduced from [141] with permission

and NMR, respectively [142]. Bivariate distribution maps were derived by combining SEC/NMR and SEC/MALDI-TOF MS data assuming the Stockmayer's theoretical distribution of comonomers to estimate the composition distribution of each fraction (Fig. 26). Recently Montaudo extended the application to poly(S-co-MAn) as well as poly(S-co-MMA) [143].

Mrkvickova investigated PMMA grafted with poly(dimethylsiloxane) (PDMS) by SEC coupled with refractive index and low-angle laser light scattering detectors [144, 145]. Using toluene and THF as isorefractive index eluents for PMMA and PDMS, respectively, a variation in chemical composition and molecular weight of individual copolymer blocks as a function of hydrodynamic volume were measured. Krämer et al. characterized poly(S-co-EA) by SEC coupled with NMR [146].

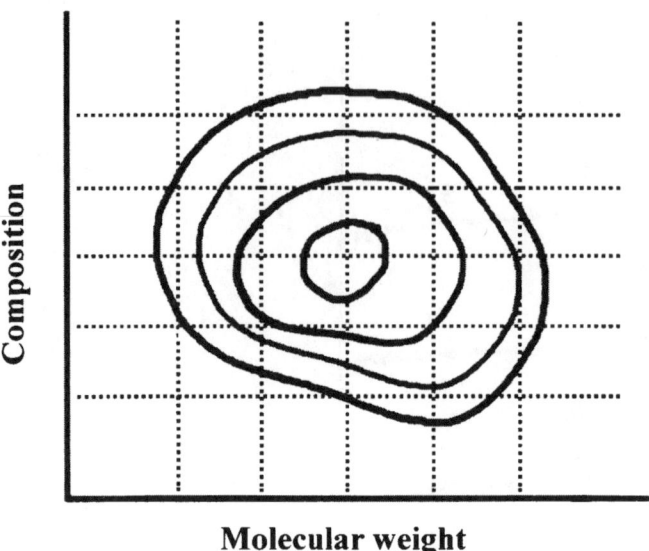

Fig. 25. A combined molecular weight–chemical composition distribution. Ideal 2-D separation would be along the horizontal and vertical axes to obtain both molecular weight and chemical composition distributions

Using on-line coupled SEC-^{1}H NMR, the chemical composition of eluates was continuously monitored. It was found that the ethyl acrylate-rich copolymers exhibited a broader molar mass and chemical composition distribution than styrene-rich copolymers. The results indicated that the block character of the copolymers with respect to ethyl acrylate units increased with increasing molar mass.

SEC separations of copolymers inevitably suffer from the local polydispersity to variable extent depending on systems. On the other hand, IC retention is known to depend on chemical composition more strongly than molecular weight and 2D-LC analysis of random copolymer has been extensively carried out by coupling IC and SEC under the premise that IC and SEC separates exclusively in terms of chemical composition and molecular weight, respectively. In an ideal case, once IC separates a copolymer in terms of composition exclusively, subsequent SEC analysis does not suffer from the local polydispersity problem except for the instrumental broadening [147].

Berek proposed that this near independence of IC retention of copolymers on molecular weight could be ascribed to the "peak compression" effect in the solvent gradient IC elution [148]. The peak compression effect stands for the focusing effect of the elution peak in the solvent gradient elution LC. It is due to the eluent composition gradient along the column and the tendency of polymer chains moving faster than the eluent due to the size exclusion effect. Since it is a common practice that the solvent gradient is applied in the increasing solvent power, a copoly-

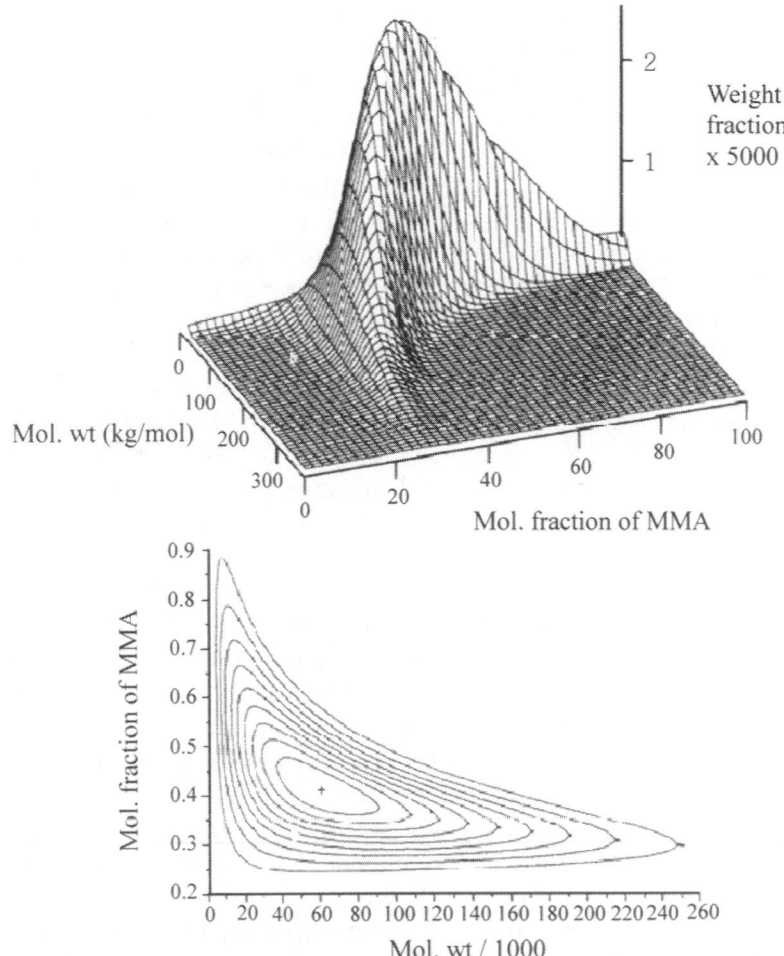

Fig. 26. *Top*: bivariate distribution of chain sizes and compositions for poly(MMA-*co*-BA) obtained at high conversions. *Bottom*: corresponding contour map. The random copolymer was fractionated by SEC and the average molecular weight and average composition of each fractions were determined by MALDI-TOF MS and NMR, respectively. The mol fraction of MMA was calculated according to the Stockmayer's theoretical prediction. Reproduced from [142] with permission

mer species with a given composition travels along the column with a velocity corresponding to that of the eluent composition exhibiting a solvent power just strong enough to prevent the sorption of the polymer chains to the stationary phase (Fig. 2). Under specific circumstances, retention of polymers at this "barrier" depends almost exclusively on the copolymer composition and not on its molar mass. This proposition was based on the mechanism of the liquid chromatography at the limiting conditions of desorption (LC-LCD) [148]. Degoulet et al. also proposed a simple analytical model for the self-focusing process of elution peaks to predict optimal conditions of operation and pointed out the uselessness of long columns [149]. In any event the molecular weight independence of the solvent gradient LC is only valid when the sorption and desorption of polymer chains exclusively depends on its composition and not on molecular weight.

The first separation of random copolymers by gradient LC according to CCD was reported by Teramachi et al. for poly(S-co-MA) [150]. The subsequent applications of the gradient LC for the analysis of copolymers are well summarized in the monograph of Glöckner [4]. Many recent efforts have employed a greater variety of detection methods and correlated the chromatography analysis results with theoretical predictions and/or other physical properties of the polymers. For examples, Braun et al. separated poly(MMA-co-EA) and poly(MMA-co-BA) with respect to chemical composition by gradient NPLC [151]. They found that the molecular weight did not affect the separation and that for high-conversion samples the chemical heterogeneity increases with increasing content of ethyl acrylate. The HPLC separation result was correlated to theoretical predictions and to the glass transition temperatures of the samples, which show a broadening with increasing chemical heterogeneity. Krämer et al. successfully analyzed poly(S-co-EA) with respect to CCD by gradient RPLC coupled with on-line ^{1}H NMR detection.[152] Information on the chemical composition and the chemical heterogeneity of copolymers with high conversion was obtained. The experiments have been carried out using conventional HPLC grade solvents and no deuterium lock. The results have been correlated with a HPLC procedure based on calibration with narrow-distributed copolymer standards (Fig. 27).

Murphy et al. reported on the use of LC-particle beam mass spectrometry for the quantitative analysis of copolymer composition of several high molecular weight poly(MMA-co-BA) by monitoring the low-mass fragments formed by thermal decomposition and electron impact ionization when using a particle beam interface [153]. The fragment ions produced are proportional to the comonomers present and are quantitatively related to the copolymer composition. Schoonbrood et al. examined poly(S-co-HEMA) prepared with free-radical bulk and emulsion copolymerization at 50°C with gradient LC and with ^{1}H NMR [154]. By using the CCD obtained from LC and comparing the hydroxyl proton signal from NMR measurements of emulsion copolymers to that of a bulk copolymer with the same overall composition, it could be shown that styrene and 2-hydroxyethyl methacrylate copolymerize during the greater part of the emulsion polymerization.

The analysis strategy of graft copolymers is not much different from random copolymers. Schunk and Long examined graft copolymers prepared by radical polymerization of methyl methacrylate with a macromonomer of PDMS [155]. Separation emphasizing chemical composition heterogeneity was performed by gradient NPLC combining precipitation and adsorption retention. Comparison of

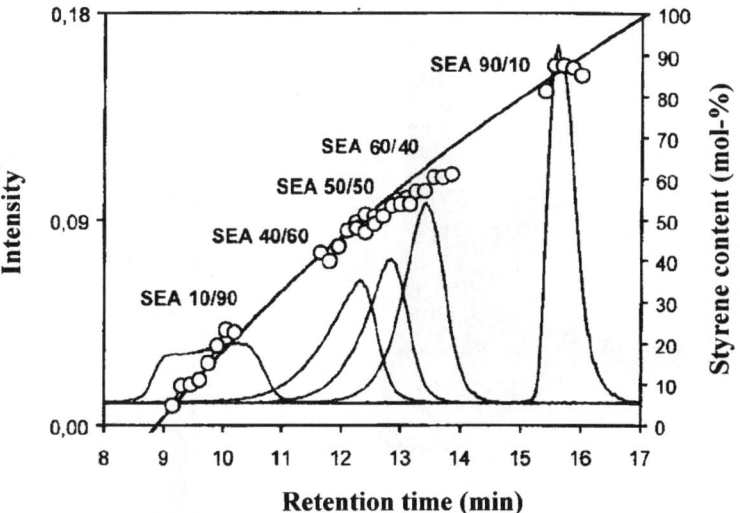

Fig. 27. RPLC chromatograms of poly(S-*co*-EA)s (SEA) of different composition with high conversion. Styrene contents obtained by HPLC calibration (*lines*) and on-line HPLC/^{1}H NMR (*circles*) experiments match each other very well. Column: Nucleosil C18; 100 Å; 250×4.6 mm. Eluent: THF/CH$_3$CN, gradient elution from 10/90 to 100/0 (v/v) in 25 min. Reproduced from [152] with permission

FT-IR and evaporative light-scattering detection indicated decreasing PDMS macromonomer incorporation corresponding to increasing retention time for a graft copolymer. More detailed information was obtained by multidetector SEC of composition fractions from gradient elution HPLC.

Adrian et al. applied 2D-LC, LCCC in the first dimension and SEC in the second dimension, to investigate the grafting of butyl acrylate onto PS-*b*-PB [156]. Separating the graft backbone and the graft product by LCCC at the critical condition of PBA, a 2-D map of the chemical composition and molar mass was obtained with the aid of quasi on-line FT-IR detection. The graft products could be separated into the component of graft copolymer, backbone PS-*b*-PB and PBA homopolymer. It was also found that the grafting reaction results in the formation of a complex product due to the partial degradation of the backbone during the grafting reaction (Fig. 28).

Siewing et al. investigated the grafting reaction of methyl methacrylate onto ethylene-propylene-diene rubber (EPDM), which is a complex mixture of nongrafted EPDM, the graft copolymer EPDM-*g*-PMMA, and the PMMA homopolymer [157]. Using LCCC at the critical condition of PMMA, the PMMA homopolymer could be isolated, but a separation of the graft copolymer and EPDM was not possible. The complete separation of all components was achieved by gradient LC. Combining gradient HPLC and SEC in a fully automated 2D-LC setup, the com-

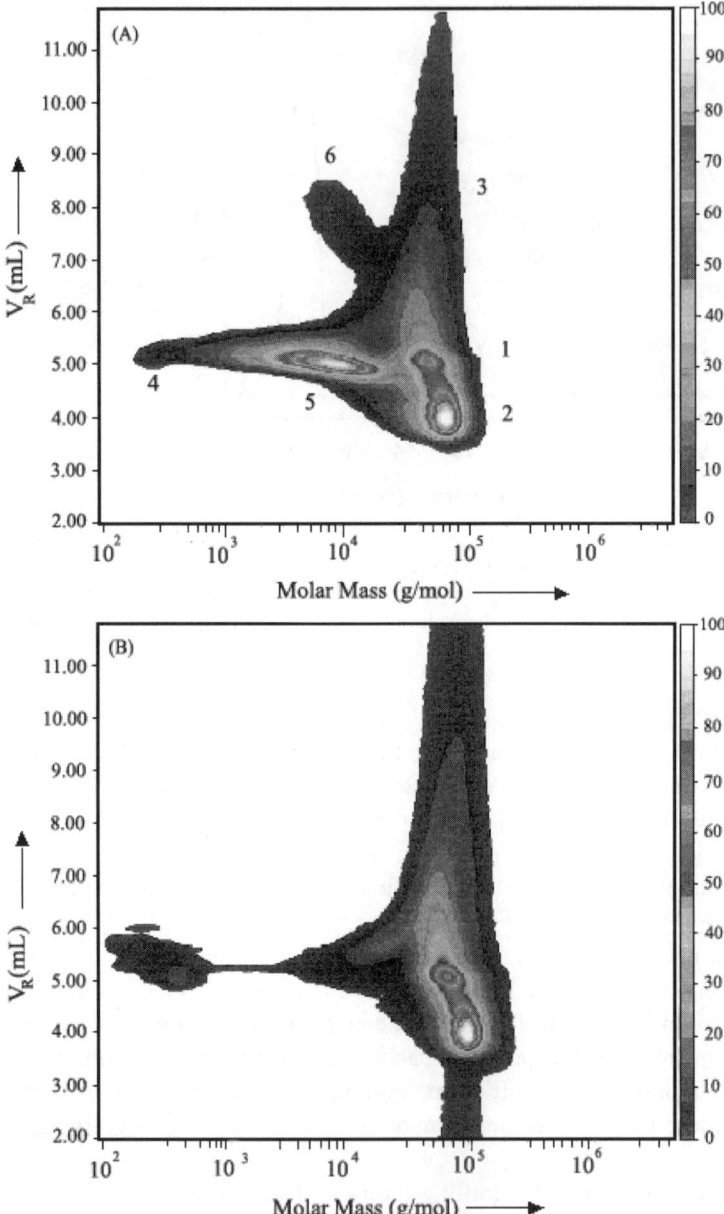

Fig. 28A,B. Contour plot of the 2D separation of the (PS-*block*-PB)-graft-PBA. 1 – ungrafted PS-*b*-PB; 2 – graft polymer; 3 – graft and block copolymers with degraded PB blocks; 4 – additive; 5 – PBA; 6 – BA-maleic acid copolymer. The first dimension was LCCC for PBA. Column: Knauer Si 300+1000 Å. Eluent: THF/cyclohexane 15.5/84.5 (v/v). The second dimension was SEC (PL mixed D and THF). Detection: **A** ELSD; **B** UV. Reproduced from [156] with permission

plex distributions of chemical composition and molar mass could be fingerprinted simultaneously.

Tanaka et al. used NPLC and RPLC to determine CCD of PMMA-g-PS with different compositions and conversions synthesized from the copolymerization of methyl methacrylate and ω-p-vinylbenzyl polystyrene macromonomer [158]. The good agreement between CCD obtained by both HPLC modes showed that the molecular weight effect on the CCD is negligible. As the macromonomer content increases, the CCD becomes sharper. These results are in accordance with the theoretical predictions. As the conversion increases at the same feed composition, the CCD becomes broader towards the low macromonomer content side, which is in contrast to the CCD of the samples obtained previously from ω-methacryloyl polystyrene macromonomer [159]. They also showed that the CCD is broadened as the graft length increases, in copolymer samples with a similar composition, in accordance with the theoretical prediction [159].

There also have been a few reports on the IC characterization of random copolymers made by step reaction polymerization. For example, Rissler used gradient capillary RPLC to characterize commercial polyesters. A good resolution was obtained for low molecular weight samples with respect to MWD, CCD and functionality type distribution [160]. Philipsen et al. studied the MMD, CCD, and the functionality distribution of low molecular weight aromatic copolyester by gradient elution LC [161]. By a combination of SEC and NPLC they found a significant differences in MWD/CCD between strongly resembling copolyesters that can only be the cause of the relative importance of reaction kinetics in step reaction copolymers. This makes the assumption that a predictable, theoretical statistically determined CCD is formed in all cases questionable.

Brun has extended the concept of LCCC to the cases of statistical copolymers as well as porous stationary phases with heterogeneous surfaces (viz., surfaces with both inert and active groups) [162]. The theory predicted that a statistical copolymer with narrow CCD always possesses a single adsorption-desorption transition point and behaves like a homopolymer with a single energy of interaction between the effective monomer units and the active groups at the surface. If copolymer has a broad CCD, each compositionally homogeneous fraction has its own adsorption-desorption threshold.

The molecular-statistical theory of polymer solutions in confined media was also applied to the conventional chromatographic theory of gradient elution [163]. This approach leads to the prediction of the special mode of interactive polymer chromatography: gradient elution at critical point of adsorption. It was demonstrated theoretically and experimentally that under appropriate conditions elution of each compositionally homogeneous fraction of copolymer occurs at the critical mobile phase composition. This critical mobile phase composition depends only on the local structure of the copolymer chain and is independent of its molecular weight. As a consequence, gradient elution produces the chemical composition distribution of the copolymer. The theory was experimentally tested by the isocratic and gradient NPLC elution of chlorinated polyethylene with various chlorine contents and molecular weights.

Bartkowiak and Hunkeler separated poly(S-*co*-MMA) according to their chemical composition by LCCC [164]. They also showed the possibility to couple with SEC for the simultaneous identification of copolymer composition and molecular weight distributions. Sauzedde and Hunkeler tested the effect of molecular weight and composition in LCCC retention with two series of poly(S-*co*-MMA) [165]. The first series varies with respect to the chemical composition in the copolymer, at a constant molar mass, while the second has molecular weight variation at a fixed composition. The copolymers were fractionated using SEC and LCCC. The influence of the molecular weight on the retention was negligible in the studied range compared to composition effect.

Berek et al. evaluated the applicability of the FAD approach for separation of random copolymers, which was proven to be powerful in the fractionation of polymer blends [166]. Dynamic integral desorption isotherms for a series of PMMA homopolymers and poly(S-*co*-MMA) were measured. Nonporous silica FAD column packing and various adsorption-promoting and desorption- promoting liquids were used. The desorption isotherms from FAD columns strongly depended on both (co)polymer molar mass and copolymer chemical composition. The interference of both parameters prevents the direct use of FAD for fractionation of the copolymers. Nguyen et al. characterized graft copolymers obtained from radical dispersion copolymerization of methacryloyl-terminated PEO macromonomer and styrene by FAD/SEC and gradient LC [127]. FAD/SEC and gradient LC allowed separation of homo-PS from graft-copolymer and determination of both its amount and molar mass. The results indicate that there are at least two or maybe three polymerization loci, namely the continuous phase, the particle surface layer, and the particle core. The graft copolymers are produced mainly in the continuous phase while PS or copolymer rich in styrene units is formed mostly in the core of monomer-swollen particles.

5.4
Functionality

Functionality refers to the moieties chemically distinct from the repeating units of polymer chains. It can be the terminal groups at chain ends or the functional groups deliberately added to a polymer chain. Determination of functionality type distribution of reactive oligomers (telechelics) is an old problem in polymer characterization. The classical method is the combination of SEC separation and functionality sensitive detection. Recent advances of the detection methods such as FT-NMR and MALDI-TOF MS have brought about big improvements in the precision of functionality characterization [12].

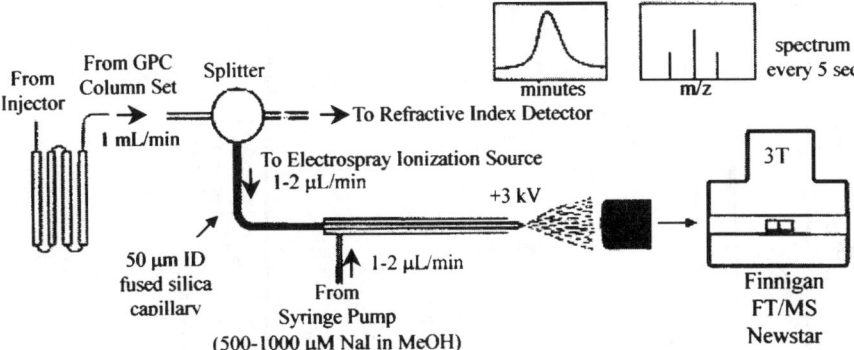

Fig. 29. Experimental setup for SEC coupled to Fourier transform mass spectrometry via an electron spray ionization source. Reproduced from [169] with permission

Pasch and Gores investigated by SEC with MALDI-TOF MS detection of the functional heterogeneity of PMMAs prepared by group-transfer polymerization, which contain oligomers with cyclic end-groups in addition to the expected linear polymers [167]. For technical products it is shown that oligomers with cyclic end-groups may be formed in significant amounts and, therefore, have to be accounted for in structural characterization.

The SEC fractionation/MALDI-TOF MS detection procedure was also applied to poly(bisphenol-A carbonate) samples [168]. The MALDI-TOF mass spectra of the SEC fractions allow not only the detection of linear and cyclic oligomers contained in these samples, but also the simultaneous determination of their average molar masses. From the average mass data of the SEC fractions, they could construct SEC calibration plots of the cyclic and linear oligomers.

Aaserud et al. reported an on-line coupling of SEC to Fourier transform mass spectrometry (FT-MS) using a modified commercial electron spray ionization (ESI) interface (Fig. 29) [169]. They analyzed a glycidyl methacrylate/butyl methacrylate copolymer with a broad molecular weight distribution, where fractionation and high resolving power were required for adequate characterization. The SEC/ESI/FT-MS also allowed for an unequivocal end-group determination and characterization of a secondary distribution due to the formation of cyclic reaction products.

In many cases functionality does not affect the hydrodynamic volume of the whole polymer chain significantly and SEC can separate them in terms of molecular weight. In order to separate the polymers in terms of functionality, however, IC has to be used. Pasch and Hiller separated a technical oligo(ethylene oxide) by isocratic RPLC with respect to degree of polymerization and functional end groups, and analyzed the chemical structure by on-line ^{1}H-NMR detection [170]. The experiments was conducted under conditions that are common for HPLC

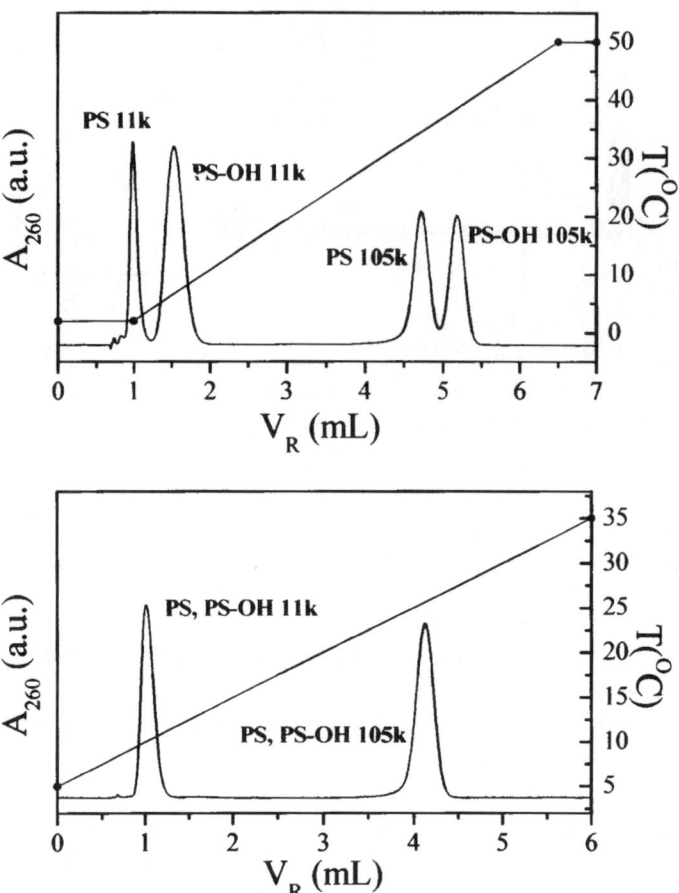

Fig. 30. TGIC separation of PS samples with different end groups (hydrogen terminated vs hydroxy terminated) by NP-TGIC. *Top* – column: Nucleosil; 100 Å; 250×2.1 mm, eluent: isooctane/THF=55/45 (v/v) and RP-TGIC. *Bottom* – column: Nucleosil C18; 100 Å; 250×2.1 mm, eluent: CH_2Cl_2/CH_3CN=57/43 (v/v)). Flow rates are 0.1 ml/min. Temperature programs are also drawn in each figure. Reproduced from [118] with permission

separations; a sufficiently high flow rate, moderate sample concentration, and on-flow detection. Lee et al. demonstrate very high sensitivity of NP-TGIC on the polar end group [118]. They showed that hydroxy group terminated PS could be fully resolved from normally terminated PS of identical molecular weight up to 100 kg/mol, which was impossible either with SEC or with RP-TGIC (Fig. 30).

As mentioned briefly in Sect. 5.2, the analysis of functionality has been carried out extensively by LCCC method [9]. Adrian et al. analyzed bisphenol A based epoxy resins by SEC, LCCC and MALDI-TOF MS with respect to molar mass, functionality, and branching [34]. LCCC yielded quantitative data on different

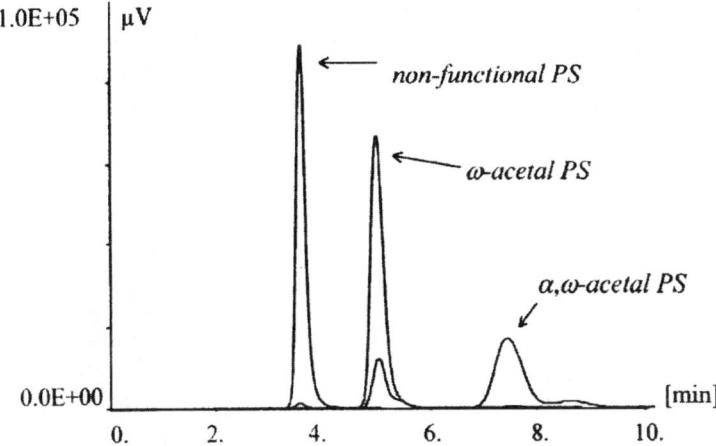

Fig. 31. LCCC chromatograms of PS standards (M_w=13.7 kg/mol), ω-acetal-PS (M_w= 4.0 kg/mol) and α, ω-acetal-PS (M_w=5.1 kg/mol) at the critical condition for PS in CH_2Cl_2/hexane (63.5/36.5, w/w). Flow rate was 1 ml/min. Column: Nucleosil amino; 100 Å; 250×4.6 mm, Detection: UV (λ=261 nm) or RI, T=25°C. Reproduced from [35] with permission

functionality fractions, including fully epoxydized, glycidyl-diol and bisdiol oligomers, which were efficiently identified by MALDI-TOF MS. In addition, these measurements yielded information on the number of branching points per oligomer molecule. Keil et al. characterized commercial PPO with respect to functionality and molar mass by coupled SEC/LCCC and MALDI-TOF MS [40]. The analysis of technical PPO triols revealed the presence of diols and mono-ols in addition to the triols.

Baran et al. fractionated end functional PS and PEO as well as their block copolymers by LCCC [35]. Various functional PS (α- and/or α, ω-alcohol, acetal, aldehyde and carboxy) could be readily separated from non-functional PSs under various LCCC conditions (Fig. 31).

Evreinov et al. fractionated four types of branched polyesters based on adipic acid, diethylene glycol and trimethylolpropane, pentaerythritol, or glycerol, or on diethylene glycol, dimethyl adipate, and trimethylolpropane with respect to functionality types (the number and type of end groups) using LCCC [36]. They also reported on fractionation of oligoesters based on adipic acid and two diols (ethylene- and butylene glycols) with respect to the number of end functional groups using LCCC [37].

Gancheva et al. employed LCCC to investigate the cationic polymerization of a cyclic acetal initiated by mono and bifunctional initiators [38, 39]. Termination with 9-anthrylmethyl improved the detection of the end functionality and the separation into functionality fractions. Mengerink et al. separated and identified the linear and cyclic polyamide-6 by LCCC and MALDI-TOF MS, respectively [41]. The highest cyclic structure present and detected was the cyclic pentacontamer. Quantification with evaporative light-scattering detection of the components separated by LCCC using a universal calibration curve or an iterative procedure was developed.

6
Microstructures

Tacticity refers to the stereoregularity in the pendant group of the chain backbone, which is another kind of heterogeneity to affect the physical properties of synthetic polymers a great deal, in particular for the crystallization behavior of polymers. Unlike the monomer sequence in copolymers, which is dictated by the reactivity ratio of the monomers, the microstructures are developed more randomly during the polymerization. Therefore it is often assumed that the microstructure is developed in a random manner and the average ads composition is used as the parameter to represent the tacticity. Provided that the average ads composition is of only interest, NMR is the most powerful technique to measure the average microstructure of such polymers. However, there are many examples of polymerization in which microstructures are developed by non-Bernoullian or non-Markovian processes [171].

One way to probe the tacticity heterogeneity is to fractionate the polymer in question according to the tacticity distribution or the molecular weight distribution first and measure the remaining molecular characteristics by other means. However, all the heterogeneities in synthetic polymer systems usually occur together and it is difficult to isolate the effect of single origin. If one assumes that the molecular size does not change with tacticity, SEC is an effective method to separate the polymer in terms of molecular weight. For example, Krämer et al. characterized by SEC coupled with NMR the microstructure of poly(S-*co*-EA) prepared by radical polymerization. Using ^{13}C NMR, the compositional styrene- and ethyl acrylate-centered triads was determined and compared to theoretical values. Good agreement between experimental and calculated triad concentrations was obtained for styrene-rich copolymers. Ethyl acrylate rich copolymers exhibit significant deviations from the calculated values in particular for the ethyl acrylate homo-triads [146].

With the recent development of the polymerization techniques to prepare highly stereoregular polymers, there have been various efforts to isolate the tacticity distribution from other heterogeneity [172]. Different tactic polymers have slightly different hydrodynamic volumes but not large enough for SEC to distinguish the difference effectively [173, 174]. On the other hands, IC is able to separate the stereoregular polymers in terms of tacticity. When one considers small molecules, the polymers with different tacticity are diastereomers. Therefore it is not surprising

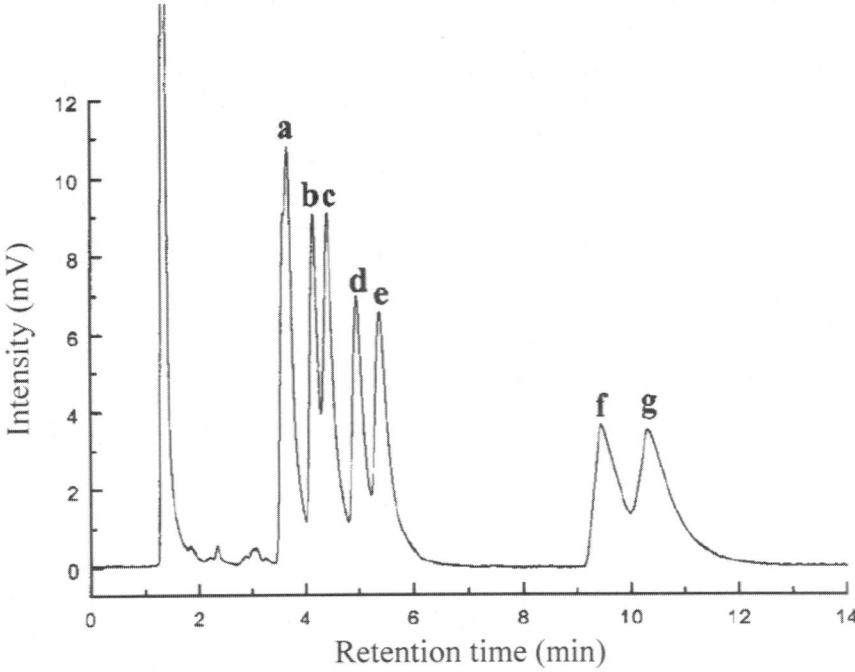

Fig. 32. Diastereomer separation of oligostyrene (n=4) with *sec*-butyl end group on carbon-clad zirconia column (two 50×4.6 mm, 3 μm particle size) using a mobile phase of 100% CH_3CN. Reproduced from [180] with permission

that IC can separate them. For example, there have been a number of reports to separate oligostyrenes according to the number of repeating units and the solvent selective fine resolution of the stereoisomers of individual oligomers using solvent and temperature gradient HPLC [92, 175–180]. The separation efficiency of the diastereomers depends on the choice of mobile and stationary phase very sensitively. A recent report by Sweeney et al. showed a full resolution of all the diastereomers of PS tetramer by use of carbon clad zirconia stationary phase [180]. Figure 32 shows the full separation of PS tetramers.

However, such diastereomeric separation becomes more and more difficult as the molecular weight increases since the chromatographic retention is determined by both tacticity and molecular weight, and synthetic polymers have finite distributions in both. Therefore the chromatographic separation of polymers is restricted to highly stereoregular polymers. Inagaki and coworkers were the first to report the separation of PMMA according to the tacticity by thin layer chromatography [181, 182]. Sato et al. showed that stereoregular PMMA eluted in the sequence of isotactic, atactic and syndiotactic PMMA in the gradient elution of a mixture of

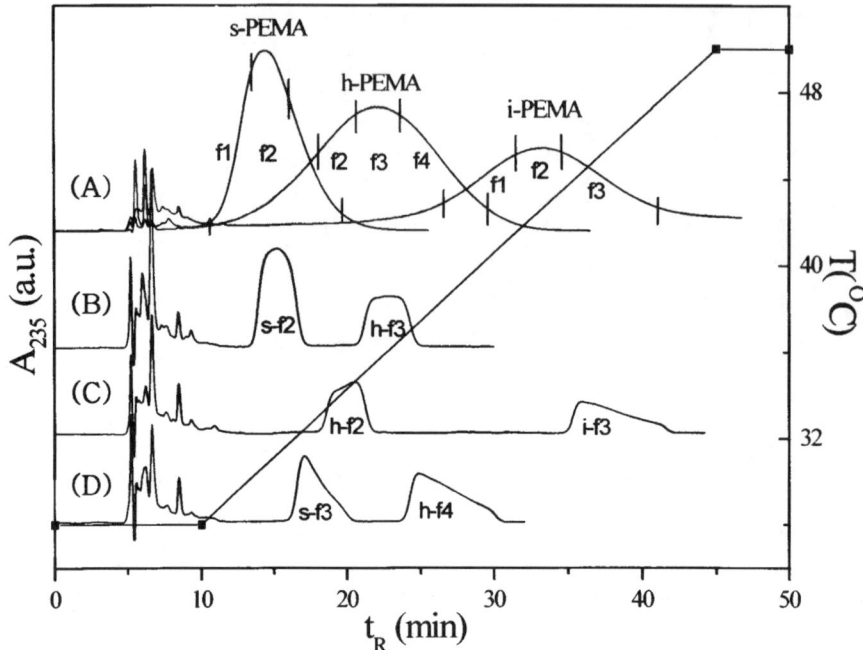

Fig. 33. TGIC chromatograms of three PEMA samples with differing tacticity. (A) (s-, h-, and i-PEMA: rr triad content=0, 53, and 91%), and TGIC chromatograms of fractions of similar molecular weight but of different tacticity (B~D). Column: LUNA C18; 100 Å; 250×4.6 mm, Eluent: CH_2Cl_2/CH_3CN (30/70, v/v). Temperature program is also shown in the plot. Reproduced from [186] with permission

dichloromethane and *n*-hexane (or ethyl ether or benzene) when a cross-linked polyacrylonitrile gel was used as the stationary phase [183]. The elution sequence was reversed when styrene gel and a mixture of CH_2Cl_2/CH_3NO_2 were used as stationary and mobile phase, respectively.

LCCC was also applied successfully to separate stereoregular PMMA [174, 184] and poly(ethyl methacrylate) (PEMA) [173, 185]. At the critical condition of syndiotactic PEMA (aminopropyl bonded silica and mixed eluent of acetone/cyclohexane or THF/cyclohexane), isotactic and atactic PEMAs eluted in the SEC regime. Recently Cho et al. employed TGIC to show a very high resolution fractionation of stereoregular PEMAs [186]. They used C18 silica stationary phase and a mixture of CH_2Cl_2 and CH_3CN as the mobile phase. They were able to show unambiguously that the retention of PEMA depends on both molecular weight and tacticity and the equivalent molecular weight fractions having very narrow MWD (fractionated by TGIC) could be fully resolved by the difference in tacticity (Fig. 33).

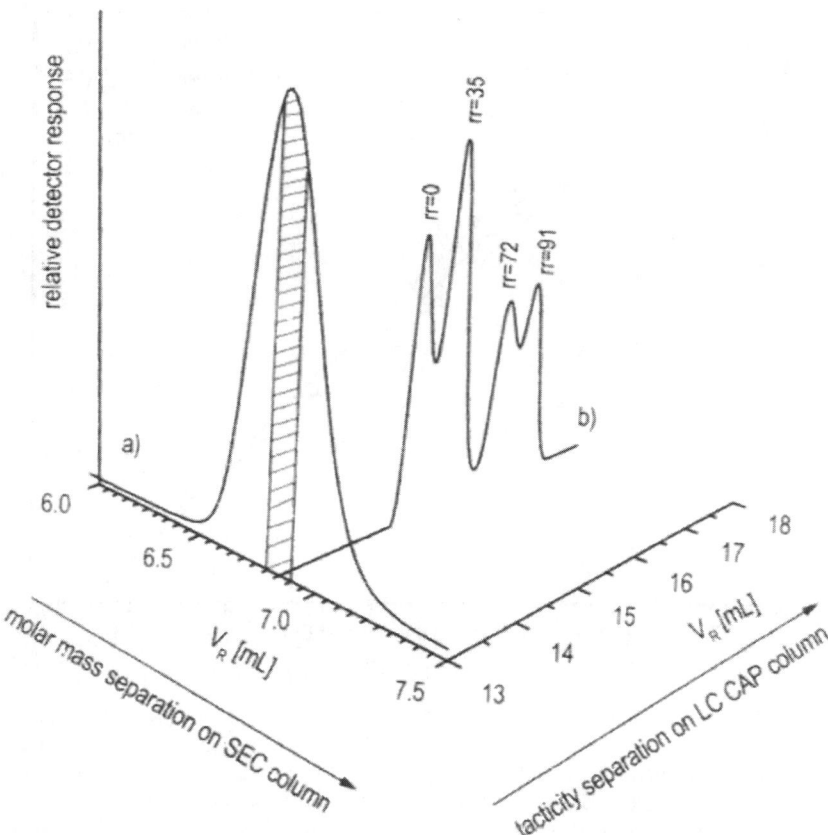

Fig. 34. SEC/LCCC separation of PEMA sample composed of four constituents with similar molar mass (M_w=12, 8.2, 8.8, and 13.6 kg/mol) but differing in their tacticity (rr content 0, 35, 72, 91%). A 100-μl sample of the SEC fraction (eluted in the range 6.8–6.9 ml) was subjected into the LCCC column for tacticity separation. A common eluent, THF/cyclohexane mixture (64/36 w/w) was used for both SEC and LCCC separation at 35°C. Column: PL-gel mixed D for SEC and Develosil SG-NH_2 for LCCC. Reproduced from [173] with permission

For the full characterization of such stereoregular polymers, the difficulty is again in finding the separation mode to resolve the polymers in one parameter without being affected by the others. The work of Janco et al. is close to the concept, in which they demonstrated a simultaneous discrimination of stereoregular PEMAs according to their molecular weight and tacticity by 2D coupling of SEC and LCCC (Fig. 34) [173]. Narrow molecular weight fractions of PEMA eluted from the SEC column were on-line forwarded into the LCCC column (aminopropyl bonded silica) to be separated according to their tacticity.

The 2D analysis was based on the assumption that the SEC retention is not affected by tacticity, which is not entirely true although the effect is not large. For

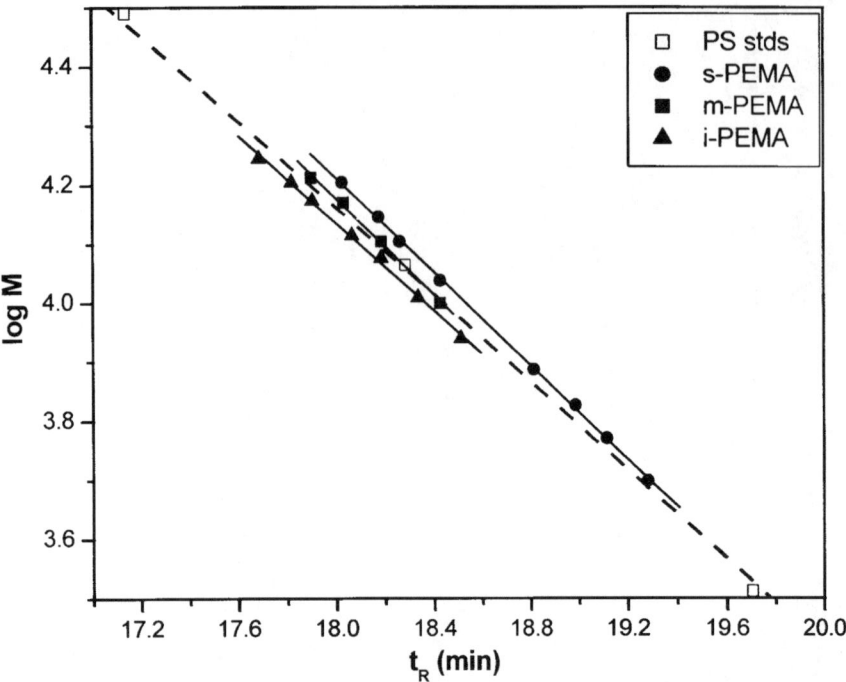

Fig. 35. SEC calibration curves of different stereoregular PEMA samples (*filled circles*: syndiotactic-PEMA, *filled squares*: moderately syndiotactic-PEMA, *filled triangles*: isotactic-PEMA) whose rr triad content are 0, 53, and 91%, respectively. *Dashed line* is a calibration curve obtained from three PS standards (*open squares*), 30.9, 11.6, and 3.3 kg/mol. Column: two mixed bed columns (Polymer Lab., Mixed C), Eluent: THF at a flow rate of 0.8 ml/min, column temperature: 40°C. Reproduced from [190] with permission

stereoregular PMMA samples, it has been known that i-PMMA is more expanded than s-PMMA in the unperturbed state and in several good solvents [187, 188]. However, there are few works to discriminate the stereoregular polymers by SEC. For example, Ute et al. fractionated individual "mers" of stereoregular PMMA by supercritical fluid chromatography and compared the SEC retention of 25mer and 50mer of isotactic and syndiotactic PMMA. They found that isotactic 50mer eluted a little faster than syndiotactic 50mer, but the difference was negligible for 25 mer [189]. Recently Cho et al. carried out a more systematic study, in which they fractionated the anionically polymerized PEMA further by RPLC and determined the accurate molecular weights of the fractions by MALDI-TOF MS. As displayed in Fig. 35, SEC calibration curves obtained with the fractionated PEMAs show a small but finite difference between the different stereoregular PEMAs. SEC retention of the same molecular weight PEMA increases with the rr triad content PE-

MA. Comparing the molecular weight of PEMA (at around 10 kg/mol) eluted at the same retention time, isotactic PEMA was about 20% smaller than syndiotactic PEMA and the deviation appears to increase with molecular weight. Such a magnitude of deviation may be tolerable for the analyses of most synthetic stereoregular polymers, but it is always important to be aware of the accuracy of an analysis method [190].

7
Outlook

Recent development in chromatographic separation as well as in detection methods has made it possible to investigate far more detailed molecular characteristics of complex polymer systems than ever. SEC is still undoubtedly the most popular tool for the separation of polymers, but IC and LCCC have enlarged their application area in a remarkable speed. In addition, coupling of more than one chromatographic separation mechanism greatly enhances the separation power. Introduction of efficient detection techniques such as light scattering, viscometry, mass spectrometry, density detector, evaporative light scattering, FT-IR, FT-NMR and the development of convenient interfaces to the detectors has made the chromatographic separation techniques more powerful. Yet, the power of those useful detection methods has not been fully exploited let alone the high cost of the devices. For example, the high resolution of MALDI-TOF MS is still limited to relatively low molecular weight polymers and the mass spectrum cannot provide a precise distribution for the samples with broad MWD. Also the MALDI process is not effective for highly hydrophobic polymers such as polyolefins. The sensitivity and resolution of the interfaced FT-IR detection is still not so satisfactory as to be used as a routine tool. ELSD shows excellent sensitivity for the polymers lacking chromophores, but it has a problem in quantitative analysis due to the elution peak distortion.

Despite the many progresses made recently, any single or multidimensional LC separation is only the first step in molecular characterization of complex polymers. Possible local polydispersity of the fractions obtained in the course of separation must be understood properly and the detector data have to be processed accordingly. Therefore it is important to understand the limitations of each separation/detection method as well as the advantages. Furthermore, at the moment no universal approach does exist and customized procedure must be developed to solve each particular complex polymer system of interest.

At the same time, analysis speed needs to be improved. The characterization techniques of biopolymers have made an astonishingly fast progress, which enabled to complete the sequencing of human DNA. Employment of more variety of LC instrumentation, such as monolithic column, capillary as well as open tubular column, supercritical fluid mobile phase, capillary electrophoresis, electrochromatography would gain more attention in the analysis of synthetic polymers in the future. In addition, the coupling with the separation techniques other than classical HPLC techniques appears to be promising. Field flow fractionation techniques show high resolution and would supplement classical HPLC methods more in the future [191]. Phase fluctuation chromatography has shown good potential for large-scale fractionation [192–195]. Coupling of LC methods with these separation techniques have been attempted and yielded promising results [196, 197].

Undoubtedly the future development of polymer characterization will depend on the demand from the synthesis and application area. Requirements on more tailored properties will increase the need of more complex polymers as well as more precisely controlled synthesis, which in turn need high precision characterization. High resolution, selectivity, sensitivity, quantitativeness and speed would remain as the key words in the future development in chromatography characterization of complex polymers.

Acknowledgements The help of Soojin Park, Kyoon Kwon and Donghyun Cho in preparing the plots and proof reading is gratefully acknowledged. This work was supported in part by KOSEF (Center for Integrated Molecular Systems).

References

1. Mori S, Barth HG (1999) Size exclusion chromatography. Springer, Berlin Heidelberg New York
2. Potschka M, Dubin PL (eds) (1996) Strategies in size exclusion chromatography. ACS
3. Yau WW, Kirkland JJ, Bly DD (1979) Modern size-exclusion liquid chromatography, practice of gel permeation and gel filtration chromatography. Wiley, New York
4. Glöckner G (1992) Gradient HPLC of copolymers and chromatographic cross-fractionation. Springer, Berlin Heidelberg New York
5. Chang T, Lee HC, Lee W, Park S, Ko C (1999) Macromol Chem Phys 200:2188
6. Lochmüller CH, Jiang C, Liu QC, Antonucci V, Elomaa M (1996) Crit Rev Anal Chem 26:29
7. Snyder LR, Stadalius MA, Quarry MA (1983) Anal Chem 55:1413A
8. Berek D (2000) Prog Polym Sci 25:873
9. Entelis SG, Evreinov VV, Gorshkov AV (1986) Adv Polym Sci 76:129
10. Pasch H (1997) Adv Polym Sci 128:1
11. Pasch H, Trathnigg B (1997) HPLC of polymers. Springer, Berlin Heidelberg New York
12. Pasch H (2000) Adv Polym Sci 150:1
13. Trathnigg B (1995) Prog Polym Sci 20:615
14. Casassa EF (1967) J Polym Sci Part B 5:773
15. Grubisic Z, Rempp P, Benoit H (1967) J Polym Sci Part B 5:753
16. Snyder LR, Kirkland JJ (1979) Introduction to modern liquid chromatography. Wiley-Interscience, New York
17. Martin AJP (1949) Biochim Soc Symp 3:4
18. Belenkii BG, Gankina ES, Tennikov MB, Vilenchik LZ (1976) Doklady Akad Nauk SSSR 231:1147
19. Belenkii BG, Gankina ES, Tennikov MB, Vilenchik LZ (1978) J Chromatogr 147:99
20. Lee W, Lee H, Lee HC, Cho D, Chang T, Gorbunov AA, Roovers J (2002) Macromolecules 35:529

21. Lee W, Park S, Chang T (2001) Anal Chem 73:3884
22. Lee W, Cho D, Chang T, Hanley KJ, Lodge TP (2001) Macromolecules 34:2353
23. Cifra P, Bleha T (2000) Polymer 41:1003
24. Philipsen HJA, Klumperman B, vanHerk AM, German AL (1996) J Chromatogr A 727:13
25. Berek D (1996) Macromol Symp 110:33
26. Berek D, Janco M, Meira GR (1998) J Polym Sci Polym Chem 36:1363
27. Dudorina AV, Gorshkov AV, Filatova NN, Evreinov VV, Entelis SG (1995) Vysokomol Soedin 37:1957
28. Esser KE, Braun D, Pasch H (1999) Angew Makromol Chem 271:61
29. Pasch H, Rode K, Chaumien N (1996) Polymer 37:4079
30. Pasch H, Rode K (1996) Macromol Chem Phys 197:2691
31. Pasch H, Rode K (1998) Polymer 39:6377
32. Janco M, Berek D, Onen A, Fischer CH, Yagci Y, Schnabel W (1997) Polym Bull 38:681
33. Yun H, Olesik SV, Marti EH (1998) Anal Chem 70:3298
34. Adrian J, Braun D, Rode K, Pasch H (1999) Angew Makromol Chem 267:73
35. Baran K, Laugier S, Cramail H (2001) J Chromatogr B 753:139
36. Evreinov VV, Filatova NN, Gorshkov AV, Entelis SG (1995) Vysokomol Soedin 37:2076
37. Evreinov VV, Filatova NN, Gorshkov AV, Entelis SG (1997) Vysokomol Soedin Ser B 39:907
38. Gancheva VB, Vladimirov NG, Velichkova RS (1996) Macromol Chem Phys 197:1757
39. Gancheva VB, Vladimirov NG, Velichkova RS (1996) Macromol Chem Phys 197:1771
40. Keil C, Esser E, Pasch H (2001) Macromol Mater Eng 286:161
41. Mengerink Y, Peters R, deKoster CG, van der Wal S, Claessen HA, Cramers CA (2001) J Chromatogr A 914:131
42. Braun D, Esser E, Pasch H (1998) Int J Polym Anal Charact 4:501
43. Chang T, Lee W, Park S, Cho D (2001) Am Lab 33:24
44. Falkenhagen J, Much H, Stauf W, Muller AHE (2000) Macromolecules 33:3687
45. Lee H, Chang T, Lee D, Shim MS, Ji H, Nonidez WK, Mays JW (2001) Anal Chem 73:1726
46. Lee H, Lee W, Chang T, Choi S, Lee D, Ji H, Nonidez WK, Mays JW (1999) Macromolecules 32:4143
47. Lee W, Park S, Chang T (2001) Anal Chem 73:3884
48. Murgasova R, Capek I, Lathova E, Berek D, Florian S (1998) Eur Polym J 34:659
49. Pasch H, Esser E, Kloninger C, Iatrou H, Hadjichristidis N (2001) Macromol Chem Phys 202:1424
50. Berek D (1998) Macromolecules 31:8517
51. Berek D (2001) Mater Res Innov 4:365
52. Liu L, Guo QX (2001) Chem Rev 101:673
53. Ranatunga R, Vitha MF, Carr PW (2002) J Chromatogr A 946:47
54. Kosmas M, Kokkinos I, Bokaris EP (2001) Macromolecules 34:7537
55. Guttman CM, DiMarzio EA, Douglas JF (1996) Macromolecules 29:5723
56. Skvortsov AM, Gorbunov AA, Berek D, Trathnigg B (1998) Polymer 39:423
57. Lochmüller CH, Moebus MA, Liu QC, Jiang C, Elomaa M (1996) J Chromatogr Sci 34:69
58. Vailaya A, Horvath C (1996) Ind Eng Chem Res 35:2964
59. Queiroz JA, Tomaz CT, Cabral JMS (2001) J Biotechnol 87:143
60. Stadalius MA, Quarry MA, Mourey TH, Snyder LR (1986) J Chromatogr 358:17
61. Balke ST (1991) Characterization of complex polymers by size exclusion chromatography and high-performance liquid chromatography. In: Barth HG, Mays JW (eds) Modern methods of polymer characterization. Wiley, New York
62. Robert E, Fichter J, Godin N, Boscher Y (1997) Int J Polym Anal Charact 3:351
63. Mori S (1998) Int J Polym Anal Charact 4:531
64. Johnson AF, Mohsin MA, Meszena ZG, Graves-Morris P (1999) J Macromol Sci-Rev Macromol Chem Phys C39:527
65. Hsieh HL, Quirk RP (1996) Anionic polymerization: principles and practical applications. Marcel Dekker, New Tork

66. Flory PJ (1940) J Am Chem Soc 62:1561
67. Tung LH (1966) J Appl Polym Phys 10:375
68. Tung LH, Moore JC, Knight GW (1966) J Appl Polym Phys 10:1261
69. Hamielec AE (1970) J Appl Polym Phys 1970:1519
70. Busnel JP, Foucault F, Denis L, Lee W, Chang T (2001) J Chromatogr A 930:61
71. Schimpf ME, Myers MN, Giddings JC (1987) J Appl Polym Sci 33:117
72. Netopilik M (1998) J Chromatogr A 793:21
73. Netopilik M (1998) J Chromatogr A 809:1
74. Netopilik M, Podzimek S, Kratochvil P (2001) J Chromatogr A 922:25
75. Shortt DW (1994) J Chromatogr A 686:11
76. Wyatt PJ, Villalpando DN, Alden P (1997) J Liq Chromatogr Relat Technol 20:2169
77. Otocka EP, Hellman MY (1970) Macromolecules 3:362
78. Kamiyama F, Matsuda H, Inagaki H (1970) Polym J 1:518
79. Belenkii BG, Gankina ES (1970) J Chromatogr 53:3
80. Armstrong DW, Bui KH (1982) Anal Chem 54:706
81. Lochmüller CH, McGranaghan MB (1989) Anal Chem 61:2449
82. Alhedai A, Boehm RE, Matire DE (1990) Chromatographia 29:313
83. Shalliker RA, Kavanagh PE, Russell IM (1991) J Chromatogr 543:157
84. Northrop DM, Martire DE, Scott RPW (1992) Anal Chem 64:16
85. Shalliker RA, Kavanagh PE, Russell IM (1994) J Chromatogr 664:221
86. Lochmüller CH, Jiang C, Elomaa M (1995) J Chromatogr Sci 33:561
87. Lee HC, Chang T (1996) Polymer 37:5747
88. Lee W, Lee HC, Chang T, Kim SB (1998) Macromolecules 31:344
89. Lee W, Lee HC, Park T, Chang T, Chae KH (2000) Macromol Chem Phys 201:320
90. Lee W, Lee HC, Park T, Chang T, Chang JY (1999) Polymer 40:7227
91. Chang T, Lee W, Lee HC, Cho D, Park S (2002) Am Lab 34:39
92. Bruheim I, Molander P, Theodorsen M, Ommundsen E, Lundanes E, Greibrokk T (2001) Chromatographia 53:S266
93. Lee W, Lee H, Cha J, Chang T, Hanley KJ, Lodge TP (2000) Macromolecules 33:5111
94. Kwon K, Lee W, Cho D, Chang T (1999) Korea Polym J 7:321
95. Zimm BH, Stockmayer WH (1949) J Chem Phys 17:1301
96. Rudin A (1991) Measurement of long-chain branch frequency in synthetic polymers. In: Barth HG, Mays JW (eds) Modern methods of polymer characterization. Wiley, New York
97. Podzimek S (2002) Am Lab 34:38
98. Balke ST, Mourey TH (2001) J Appl Polym Sci 81:370
99. Hadjichristidis N, Pitsikalis M, Pispas S, Iatrou H (2001) Chem Rev 101:3747
100. Lee HC, Chang T, Harville S, Mays JW (1998) Macromolecules 31:690
101. Frater DJ, Mays JW, Jackson C, Sioula S, Efstradiadis V, Hadjichristidis N (1997) J Polym Sci Polym Phys 35:587
102. Lee HC, Lee W, Chang T, Yoon JS, Frater DJ, Mays JW (1998) Macromolecules 31:4114
103. Perny S, Allgaier J, Cho D, Lee W, Chang T (2001) Macromolecules 34:5408
104. Semlyen JA (1996) Large ring molecules. Wiley, New York
105. Lee HC, Lee H, Lee W, Chang T, Roovers J (2000) Macromolecules 33:8119
106. Roovers J, Toporowski PM (1983) Macromolecules 16:843
107. Skvortsov AM, Gorbunov AA (1986) Vysokomol Soedin 28:1686
108. Gorbunov AA, Skvortsov AM (1995) Adv Colloid Interface Sci 62:31
109. Lepoittevin B, Dourges MA, Masure M, Hemery P, Baran K, Cramail H (2000) Macromolecules 33:8218
110. Pasch H, Deffieux A, Henze I, Schappacher M, Riquelurbet L (1996) Macromolecules 29:8776
111. Blagodatskikh IV, Gorshkov AV (1997) Vysokomol Soedin Ser A 39:1681
112. Cho D, Park S, Kwon K, Chang T (2001) Macromolecules 34:7570
113. Mourey TH, Balke ST (1998) J Appl Polym Sci 70:831

114. Lee HC, Ree M, Chang T (1995) Polymer 36:2215
115. Trathnigg B, Feichtenhofer S, Kollroser M (1997) J Chromatogr A 786:75
116. Thitiratsakul R, Balke ST, Mourey TH, Schunk TC (1998) Int J Polym Anal Charact 4:357
117. Lee HC, Chang T (1996) Macromolecules 29:7294
118. Lee W, Cho D, Chun BO, Chang T, Ree M (2001) J Chromatogr A 910:51
119. Molander P, Greibrokk T, Iveland A, Ommundsen E (2001) J Sep Sci 24:136
120. Berek D, Nguyen SH (1998) Macromolecules 31:8243
121. Janco M, Berek D, Prudskova T (1995) Polymer 36:3295
122. Janco M, Prudskova T, Berek D (1995) J Appl Polym Sci 55:393
123. Janco M, Prudskova T, Berek D (1997) Int J Polym Anal Charact 3:319
124. Nguyen SH, Berek D (1998) Chromatographia 48:65
125. Nguyen SH, Berek D, Chiantore O (1998) Polymer 39:5127
126. Nguyen SH, Berek D (1999) Colloid Polym Sci 277:318
127. Nguyen SH, Berek D, Capek I, Chiantore O (2000) J Polym Sci Polym Chem 38:2284
128. Skvortsov AM, Gorbunov AA (1979) Polym Sci USSR 21:371
129. Gorbunov AA, Skvortsov AM (1988) Vysokomol Soedin Ser A 30:453
130. Gankina E, Belenkii B, Malakhova I, Melenevskaya E, Zgonnik V (1991) J Planar Chromatogr 4:199
131. Zimina TM, Kever YY, Melenevskaya YY, Zgonnik VN, Belen'kii BG (1991) Polym Sci USSR 33:1250
132. Zimina TM, Kever JJ, Melenevskaya EY, Fell AF (1992) J Chromatogr 593:233
133. Pasch H, Brinkmann C, Much H, Just U (1992) J Chromatogr 623:315
134. Pasch H, Brinkmann C, Gallot Y (1993) Polymer 34:4100
135. Pasch H, Gallot Y, Trathnigg B (1993) Polymer 34:4988
136. Macko T, Hunkeler D, Berek D (2002) Macromolecules 35:1797
137. Murphy RE, Schure MR, Foley JP (1998) Anal Chem 70:4353
138. Jandera P, Holcapek M, Kolarova L (2000) J Chromatogr A 869:65
139. Jandera P, Holcapek M, Kolarova L (2001) Int J Polym Anal Charact 6:261
140. Trathnigg B, Rappel C, Raml R, Gorbunov A (2002) J Chromatogr A 953:89
141. Park S, Cho D, Ryu J, Kwon K, Lee W, Chang T (2002) Macromolecules 35:5974
142. Montaudo MS, Montaudo G (1999) Macromolecules 32:7015
143. Montaudo MS (2002) Polymer 43:1587
144. Mrkvickova L (1997) Macromolecules 30:5175
145. Mrkvickova L (1999) J Liq Chromatogr Relat Technol 22:205
146. Krämer I, Pasch H, Handel H, Albert K (1999) Macromol Chem Phys 200:1734
147. Klumperman B, Cools P, Philipsen H, Staal W (1996) Macromol Symp 110:1
148. Berek D (1999) Macromolecules 32:3671
149. Degoulet C, Perrinaud R, Ajdari A, Prost J, Benoit H, Bourrel M (2001) Macromolecules 34:2667
150. Teramachi S, Hasegawa A, Shima Y, Akatsuka M, Nakajima M (1979) Macromolecules 12:992
151. Braun D, Kramer I, Pasch H (2000) Macromol Chem Phys 201:1048
152. Krämer I, Hiller W, Pasch H (2000) Macromol Chem Phys 201:1662
153. Murphy RE, Schure MR, Foley JP (1998) J Chromatogr A 824:181
154. Schoonbrood HAS, Aerdts AM, German AL, Vandervelden GPM (1995) Macromolecules 28:5518
155. Schunk TC, Long TE (1995) J Chromatogr A 692:221
156. Adrian J, Esser E, Hellmann G, Pasch H (2000) Polymer 41:2439
157. Siewing A, Schierholz J, Braun D, Hellman G, Pasch H (2001) Macromol Chem Phys 202:2890
158. Tanaka S, Uno M, Teramachi S, Tsukahara Y (1995) Polymer 36:2219
159. Teramachi S, Sato S, Shimura H, Watanabe S, Tsukahara Y (1995) Macromolecules 28:6183
160. Rissler K (2000) J Chromatogr A 871:243

161. Philipsen HJA, Wubbe FPC, Klumperman B, German AL (1999) J Appl Polym Sci 72:183
162. Brun Y (1999) J Liq Chromatogr Relat Technol 22:3027
163. Brun Y (1999) J Liq Chromatogr Relat Technol 22:3067
164. Bartkowiak A, Hunkeler D (2000) Int J Polym Anal Charact 5:475
165. Sauzedde F, Hunkeler D (2001) Int J Polym Anal Charact 6:295
166. Berek D, Nguyen SH, Pavlinec J (2000) J Appl Polym Sci 75:857
167. Pasch H, Gores F (1995) Polymer 36:1999
168. Puglisi C, Samperi F, Carroccio S, Montaudo G (1999) Rapid Commun Mass Spectrom 13:2260
169. Aaserud DJ, Prokai L, Simonsick WJ (1999) Anal Chem 71:4793
170. Pasch H, Hiller W (1996) Macromolecules 29:6556
171. Cheng HN, Kasehagen LJ (1993) Macromolecules 26:4774
172. Hatada K, Nishiura T, Kitayama T, Ute K (1997) Macromol Symp 118:135
173. Janco M, Hirano T, Kitayama T, Hatada K, Berek D (2000) Macromolecules 33:1710
174. Berek D, Janco M, Kitayama T, Hatada K (1994) Polym Bull 32:629
175. Lewis JJ, Rogers LB, Pauls RE (1983) J Chromatogr 264:339
176. Mourey TH, Smith GA, Snyder LR (1984) Anal Chem 56:1773
177. Sweeney AP, Wyllie SG, Shalliker RA (2001) J Liq Chromatogr Relat Technol 24:2559
178. Sweeney AP, Wong V, Shalliker RA (2001) Chromatographia 54:24
179. Pasch H, Hiller W, Haner R (1998) Polymer 39:1515
180. Sweeney AP, Wormell P, Shalliker RA (2002) Macromol Chem Phys 203:375
181. Inagaki H, Miyamoto T, Kamiyama F (1969) Polym Lett 7:329
182. Miyamoto T, Inagaki H (1970) Polym J 1:46
183. Sato H, Sasaki M, Ogino K (1989) Polym J 21:965
184. Berek D, Janco M, Hatada K, Kitayama T, Fujimoto N (1997) Polym J 29:1029
185. Kitayama T, Janco M, Ute K, Niimi R, Hatada K, Berek D (2000) Anal Chem 72:1518
186. Cho D, Park S, Chang T, Ute K, Fukuda I, Kitayama T (2002) Anal Chem 74:1928
187. Sawatari N, Konishi T, Yoshizaki T, Yamakawa H (1995) Macromolecules 28:1089
188. Cowie JMG (1972) Solutions of stereoregular polymers. In: Huglin MB (ed) Light scattering from polymer solutions. Academic Press, London
189. Ute K, Miyatake N, Osugi Y, Hatada K (1993) Polym J 25:1153
190. Cho D, Park I, Chang T, Ute K, Fukuda I, Kitayama T (2002) Macromolecules 35:6067
191. Colfen H, Antonietti M (2000) Adv Polym Sci 150:59
192. Fujiwara T, Kimura Y, Teraoka I (2001) Polymer 42:1067
193. Lee D, Teraoka I, Fujiwara T, Kimura Y (2001) Macromolecules 34:4949
194. Zheng HF, Teraoka I, Berek D (2001) Macromol Chem Phys 202:765
195. Xu YM, Teraoka I (1998) Macromolecules 31:4143
196. Vanasten AC, Vandam RJ, Kok WT, Tijssen R, Poppe H (1995) J Chromatogr 703:245
197. Venema E, deLeeuw P, Kraak JC, Poppe H, Tijssen R (1997) J Chromatogr A 765:135

Editor: K.-S. Lee
Received: November 2002

Adv Polym Sci (2003) 163: 61–136
DOI 10.1007/b11053

Liquid Chromatography under Critical and Limiting Conditions: A Survey of Experimental Systems for Synthetic Polymers

Tibor Macko[1] and David Hunkeler[2]

Laboratory of Polyelectrolytes and BioMacromolecules,
Swiss Federal Institute of Technology, Ecublens, CH-1015 Lausanne, Switzerland
E-mail: TMacko@dki-tu-darmstadt.de, David.Hunkeler@aquaplustech.ch

[1] Present address: German Institute for Polymeric Materials (Deutsches Kunststoff-Institut), Schlossgartenstr. 6, D-64289 Darmstadt, Germany
[2] Present address: AQUA+TECH Specialties S.A., Chemin du chalet du bac 4, CH-1283 La Plaine, Geneva CP 28, Switzerland

Abstract At the interface between the entropic size exclusion separation and the enthalpy-dominated liquid adsorption chromatography it is possible, experimentally, to identify conditions where polymer samples elute independent of their molar mass. These "critical conditions" function according to theory by compensating the entropy and enthalpy of the separation. For a series of common and specialty polymers, the mobile phase compositions and temperature which provide the molar mass independent elution behavior, for a given stationary phase, have been extracted from literature and summarized herein. This collection may help to select, or forecast, suitable LC systems, when an application of liquid chromatography under critical conditions for a polymer is required. Correlations between properties of solvents, sorbents and polymers, such as solubility parameters, eluotropic strength and Mark-Houwink constants have been extracted from the collected data. Specifically, solubity parameters of critical mobile phases corresponding to a pair polymer-sorbent are in a majority of cases very similar. The elution strength of the first component of a critical binary eluent correlates linearly with the volume percent of the second component, especially, when an identical silica gel is applied. However, the critical conditions are independent of the thermodynamic quality of the solvent. Under limiting conditions, the mobile phase may be even a precipitant or a strong adsorption promoting liquid for an injected polymer. Possibilities for increase of the upper molar mass separation limit are outlined and influence of sample solvent on elution behavior is described. Applications of liquid chromatography under conditions of enthalpy-entropy compensation for separation of homo- and copolymers are also, briefly, summarized.

Keywords Liquid chromatography · critical conditions · limiting conditions · synthetic polymers

1	Introduction	64
2	Summary of Systems under Conditions of Entropy-Enthalpy Compensation	66
2.1	Critical Conditions for Common Polymers (Table 1)	67
2.2	Critical Conditions for Specialty Polymers (Table 2)	89
2.3	Limiting Conditions (Table 3)	106
3	Correlations	112
3.1	Polymer-Sorbent versus Solubility Parameter of Mobile Phase	112

© Springer-Verlag Berlin Heidelberg 2003

3.2   Polymer-Sorbent versus Elution Strength and the Polarity Index
       of Mobile Phase 114
3.3   Polymer-Sorbent and Thermodynamic Quality of Mobile Phase . . . 115
3.4   The upper Molar Mass Limit versus the Nature of Mobile Phase . . . 117

4   **Mixed Mobile Phases** 118

5   **Single Mobile Phases** 119

6   **Polymer Peak and Solvent Peak under Critical Conditions** 120

7   **Limiting Conditions** 122

8   **Comparison of Molar Mass Calculated from Separation under
       Critical Conditions** 124

9   **Detection** 126

10   **Applications of Liquid chromatography under Critical and
       Limiting Conditions on Polymer Separation** 128

11   **Conclusions** 130

       References 131

List of Symbols and Abbreviations

a	Mark-Houwink exponent
adsorli	solvent supporting adsorption of a polymer on a sorbent
A	adsorption of polymer supporting liquid, adsorli
ACN	acetonitrile
CR	critical range of molar masses (where retention does not depend on molar mass)
D	desorption of polymer supporting liquid, desorli
DAD	diode array detector
DCM	dichloromethane
desorli	solvent supporting desorption of a polymer on a sorbent
ELSD	evaporative light scattering detector
EPDM	ethylene-propylene-dimer
ε	elution strength of mobile phase
FTIR	Fourier transform infrared
Inj. in	polymer was dissolved and injected in the given solvent
K	Mark-Houwink constant
LAC	liquid adsorption chromatography

LALLS	low angle laser light scattering
LC	liquid chromatography
LCA	limiting conditions of adsorption
LCCC	liquid chromatography at critical condition
LCD	limiting conditions of desorption
LCS	limiting conditions of solubility
MALDI-TOF-MS	matrix-assisted laser desorption/ionization mass spectrometry
MEK	methylethylketone
N	nonsolvent for polymer
near crit.	almost molar mass independent elution
NH_2	amino group bonded on sorbent surface
NMR	nuclear magnetic resonance
P'	polarity index'
PAM	polyacrylamide
PB	polybutadiene
PBGA	poly(butylene glycol adipate)
PBTF	polybutylenetereftalate
PDEGA	poly(diethylene glycol adipate)
PDMA	poly(decyl methacrylate)
PDMS	polydimetylsiloxane
PE	polyethylene
PEMA	poly(ethyl methacrylate)
PEO	poly(etylene oxide)
PEG	poly(etylene glycol)
PFS	polyphenylensulfone
PI	polyisoprene
PMMA	poly(methyl methacrylate)
PnBMA	poly(n-butylmethacrylate)
PPG	poly(propylene glycol)
PPGA	poly(propylene glycol adipate)
PPOA	poly(oxypropylene adipate)
PPOG	poly(oxypropylene glycol)
PS	polystyrene
PS/DVB	polystyrene/divinylbenzene
PTHF	polytetrahydrofuran
PtBMA	poly(t-butyl methacrylate)
PVAc	poly(vinyl acetate)
PVC	poly(vinyl chloride)
PVP	poly(vinyl pyrrolidone)
P2VP	poly(2-vinyl pyridine)
RI	refractometric detector
S	solvent for polymer

SEC	size exclusion chromatography
σ	solubility parameter
SFC	supercritical fluid chromatography
THF	tetrahydrofuran
TGIC	temperature gradient interaction chromatography
TLC	thin layer chromatography
US	molar mass range of all injected polymers
UV	ultraviolet detector
VIS	viscosity detector

1
Introduction

The chromatographic separation of polymers by **liquid chromatography under critical conditions (LCCC)**, also referred to as liquid chromatography (LC) at the critical point of adsorption, LC in the critical range or LC at the point of exclusion-adsorption transition, has attracted significant attention within polymer community. Russian scientists using TLC [1-3] and later LC [4,5] have been the first experimentally identify critical conditions. At the critical conditions polymers of a given kind are eluted independently from their molar mass (for example, Fig. 1 [6]).

This implies that these polymers are not separated and elute at constant elution volume. Pioneers of the critical conditions have recognized the universal character of this phenomenon in chromatography of polymers and they have experimentally demonstrated that critical conditions represent a natural connection between the

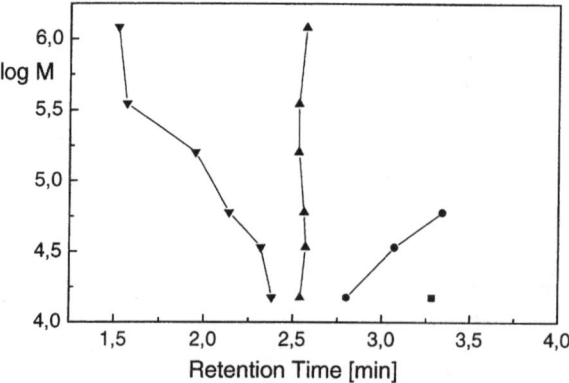

Fig. 1. Plots of log M vs. retention time for polystyrene standards at various temperatures. Polystyrene is eluted under critical conditions at a temperature of 95°C, at 115°C in exclusion mode, and at 85°C in adsorption mode. Mobile phase: Dimethylformamide. Column packing: Nucleosil C18. Symbols: 115°C (inverted triangles), 95°C (triangles), 85°C (circles), 70°C (squares) [6]

adsorption and exclusion modes of polymer separation [1,2,4,5]. They have found its "critical" sensitivity to variation of experimental variables, such as mobile phase composition, temperature, pressure and column packing as well as to chemical differences of macromolecules [1-5,7]. Scientists from St. Petersburg have also elucidated the theoretical foundation of critical phenomenon [5,8-11] for review see [12]. A second group of Russian investigators, Gorshkov, Evreinov, Entelis et al., has elaborated applications of the critical conditions, which were partly summarized in [13,14]. Independently, over the same period, experimental observations of transitions from exclusion to adsorption mode in LC polymers were reviewed by Klein and Treichel [15]. Balke [16], as well, has reported near critical elution for polystyrene. The first serie of quantitative measurements under critical conditions and spectrum of applications of LCCC were elaborated, as it will be later shown, by German investigators such as Pasch et al., Krüger, Much and Schulz. Recently it was experimentally confirmed that molar mass independent retention of polymers is possible to reach also in supercritical fluid chromatography [17].

According to the generally accepted theory [12], polymers in the course of movement through the column packing can experience an entropy-enthalpy balance. Therefore, the critical conditions can be also referred as **conditions of entropy-enthalpy compensation (CEEC)**. The term "conditions" seems to be more appropriate than "point" since, indeed, the critical behavior of a polymer is a collection of specific data, such as mobile phase composition, temperature, pressure and both sorption and structural characteristics of column packings. A change of only one of these parameters may dramatically change retention of the polymer. The term CEEC can include, in addition to adsorption, partition and solubility also other mechanisms, for example, ionic effects, which can also be compensated through enthalpy-entropy balance.

Some reviews regarding the CEEC of polymers have been published by Tennikov, Belenkii, Gankina and Nefedov [18–20] as well as by Evreinov, Entelis, Gorshkov on the separation of oligomers [13] and functional polymers [14]. The theoretical development has also recently been presented by Gorbunov and Skvorcov [12]. An analysis of the problems and queries of the critical conditions was carried out by Berek [21]. Complementing this, a practical review with a number of separation examples was written by Pasch and Trathingg [22]. The usefulness of LCCC was emphasized in a review about interaction chromatography of polymers [23]. An article discussing the place of the critical conditions between other chromatographic methods for polymer characterization has recently appeared [24]. Another, on the use of the critical conditions in combination with other separation methods, has been recently written by Kilz and Pasch [25]. The presents manuscript will not repeat the aforementioned efforts. Rather, experimental data where CEEC were found and utilized, are collected. Such a complete survey of data has not yet been published and enables the comparison of data. Thus, the goal of this review is to facilitate the selection of a suitable LC system under critical conditions for a polymer of interest. Moreover, the authors have utilized the collected data to correlate some characteristics of polymer-sorbent-eluent systems under CEEC.

2
Summary of LC Systems under Conditions of Entropy-Enthalpy Compensation

A molar mass independent retention of synthetic polymers has been experimentally identified by a number of investigators using isocratic elution. LC systems for polymers which are produced industrially in large quantities (commodities) are summarized in Table 1. Lower volume polymers, and prepared in laboratories (specialty polymers) are listed in Table 2. At LC, TLC and SFC analysis summarized in Tables 1 and 2, polymer samples were dissolved and injected in the used mobile phase. A transition from a separation governed by entropy, to separation directed mainly by enthalpy, including molar mass independent separations, may be reached also when polymers are dissolved and injected only in one component of the corresponding mixed mobile phase, as was first demonstrated by Hunkeler et al. [26]. This approach, which employs CEEC, is named limiting conditions [27] and will be discussed separately later. The corresponding systems are summarized in Table 3.

It should be noticed that not all experimental parameters have been published in the literature. Specifically, the molar mass of samples, the dimensions of column, the pore diameter of column packing, the flow rate, pressure, temperature or injected both concentration and volume are ofter not stated and are, therefore, missing in selected parts of Tables 1-3. The relationship between polymer and solvent (a component of the mixed mobile phase is a solvent or non solvent for a polymer) and the relationship between solvent and sorbent (*adsorli*, i.e., adsorption of polymer supporting liquid; *desorli*, i.e., a desorption of polymer supporting liquid), when defined in the literature, is noted in Tables 1-3. Measurements, which require other techniques as LC, are denoted by abbreviation (TLC or SFC). The abbreviations employed have been defined separately in list of symbols and abbreviations.

Table 1. Critical Conditions for Common Polymers (Polymers Injected in the Mobile Phase.)

Polymer	Mobile phase	Stationary Phase	Conditions	Analytical application	Investigators
Molar masses Critical range Used samples	Solvent / Nonsolvent Adsorli / Desorli	Particle diameter Pore size Column dimension	Temperature Pressure Flow rate Inj. volume Detector	(LC, TLC or SFC separation and characterization) and/or notices	Reference
PB Polybutadiene CR: 1kD-30kD	MEK / cyclohexane, 98 / 2 vol.	Nucleosil C18, 5 μm, 300°A+1000°A 25x0.4 cm	25°C RI, UV	Di and tri-block (PS-PB-PS) copolymers of styrene/butadiene. PB is soluble in eluent only up to 30kD; block copolymers are soluble.	Braun D, Esser E, Pasch H [28]
				Blends of styrene / butadiene and butyl rubber.	Esser KE, Braun D, Pasch H [29]
				Analysis of miktoarm star polymers. PS eluted in SEC mode, PI eluted in adsorption mode.	Pasch H, Esser E, Kloniger C, Iatrou H, Hadjichristidis N [30]
PB CR: 1.2kD-12kD	Hexane / toluene near crit. 84.4 / 15.6 vol.	Silica gel Si-100	24°C 1 ml/min 10 μl RI		Gorshkov AV, Evreinov VV, Entelis SG [31], Gorshkov AV, Evreinov VV, Entelis SG [32]
	Hexane / DCM 76 / 24 vol.				Gorshkov AV, Evreinov VV, Entelis SG [33], Gorshkov AV, Evreinov VV, Entelis SG [32]
	Heptane / MEK 99.5 / 0.5 vol.		24°C 1 ml/min RI	Mono and difunctional PB with terminal hydroxyl groups at critical conditions of nonfunctional PB.	Gorshkov AV, Evreinov VV, Entelis SG [33]

Table 1. (continued)

Polymer	Mobile phase	Stationary Phase	Conditions	Analytical application	Investigators
Molar masses Critical range Used samples	Solvent / Nonsolvent Adsorli / Desorli	Particle diameter Pore size Column dimension	Temperature Pressure Flow rate Inj. volume Detector	(LC, TLC or SFC separation and characterization) and/or notices	Reference
PB with terminal hydroxyls US: 2kD – 72kD	Heptane / dioxane 94.2 / 5.8 vol. 95.8 / 4.2 vol. 96.5 / 3.5 vol. 96.9 / 3.1 vol.	Lichrospher Si, 500°A	0.1ml/min 10°C 25°C 40°C 60°C UV	Zero, mono and difunctional molecules of oligodienes with terminal hydroxyl groups. OH groups transfered into naphthylurethane (UV-chromophor).	Estrin YaI, Kasumova LT [34]
	Hexane / dioxane 94.5 / 5.5 vol.	Lichrospher NH_2 + $AgCl$, 500°A	50°C, 0.1 ml/min UV		
PB CR: 2kD-30kD	n-heptane / dioxane 97.3 / 2.7 vol.	Nucleosil NH_2, 1000°A, 7 µm, 25×0.4 cm	60°C 1 ml/min 20µl RI, UV	Determination of number of hydroxyl groups in PB. Two-dimensional separation. The samples derivatized into phenyl urethane groups to make the OH groups detectable with UV.	Pulda J, Reisova A [35]
	n-heptane / dioxane 95 / 5 vol.		60°C 1 ml/min 20 µl RI		Pulda J [36]
PB CR: 0.9kD-9.3kD	THF / water 92 / 8 vol. S / N	Novapak C18	30°C 0.5 ml/min 25 µl ELSD		Cools PJCH, Herk AM, German AL, Staal WJ [37]

Polymer	Mobile phase	Stationary phase	Conditions	Notes	Reference
PI Polyisoprene	1,4-dioxane	Silica gel C18		TGIC of PI between 20–50°C.	Chang T, Lee HC, Lee W, Park S, Ko C [38]
PI CR: 2.7kD–108kD	1,4-dioxane	Nucleosil C18, 100°A+300°A +5000°A 3×(25×0.46) cm	47.7°C 1 ml/min 20 ul ELSD		Lee W, Park S, Chang T [39]
PI CR: 0.6kD–100kD	Isooctane / THF 99.12 / 0.88 vol.	Silica gel YMC-Pack SIL AP, 200°A, 5 μm, 25×0.46 cm	36°C 0.5 ml/min RI, ELSD		Czichocki G, Heger R, Goedel W, Much H [40]
	Isooctane/(0.5 vol.% acetic acid in methyl-tert. butyl-ether) 99.35 / 0.65 vol.	Silica gel YMC-Pack SIL AP, 200°A, 5 μm, 25×0.46 cm			
PI CR: 3.8kD–42kD	Hexane / isopropyl-alcohol Near crit. 91 / 1 vol. S/N	Lichrosorb Si NH$_2$, 25×0.46 cm	30°C 2 ml/min 20 μl RI	Separation of functionless, mono and bifunctional PI with terminal hydroxyl groups.	Pokorny S, Janca J, Mrkvickova L, Tureckova O, Trekoval J [41]
	Hexane / dichloro-ethane Near crit. 80 / 20 vol.	Lichrosorb Si NH$_2$, 25×0.46 cm			
PI US: 0.9kD–199kD CR: 2kD–148kD	DCM / ACN 78 / 22 vol.	Nucleosil C18, 100°A+500°A +1000°A 3×(25×0.46) cm	47°C UV, ELSD	Block copolymers PI-PS. Molar masses of PS determined are systematically smaller (deviation 0–24%) as molar mass of PS precursor.	Lee W, Cho DH, Chun BO, Chang T, Ree M [42]
PI CR: 1.5kD–1000kD	MEK / cyclohexane 92 / 8 vol.	Nucleosil C18, 300°A+1000°A, 2×(25×0.4) cm	30°C RI	Analysis of miktoarm star polymers. PB and PS eluted in SEC mode.	Pasch H, Esser E, Kloniger C, Iatrou H, Hadjichristidis N [30]

Table 1. (continued)

Polymer Molar masses Critical range Used samples	Mobile phase Solvent / Nonsolvent Adsorli / Desorli	Stationary Phase Particle diameter Pore size Column dimension	Conditions Temperature Pressure Flow rate Inj. volume Detector	Analytical application (LC, TLC or SFC separation and characterization) and/or notices	Investigators Reference
PBA Polybutylacrylate CR: 6kD-100kD	THF / cyclohexane 15.5 / 84.5 vol.	Silica gel Si-300°A + 1000°A, 10 µm, 2×(25×0.4) cm	50µl UV, ELSD	Grafting of PBA onto poly (styrene-block-butadiene). Two-dimensional separation. It is assumed that the separation of PBA is directed mainly by the styrene/butadiene ratio. PS is eluted in adsorption mode, PB in SEC mode.	Adrian J, Esser E, Hellmann G, Pasch H [43]
PMMA	Etylacetate near critical	Silica gel Porasil E+D (125 + 60 nm), 2×(105×0.78) cm	25°C 1.43 ml injected RI		Spychaj T [44]
PMMA, atactic US: 9.4kD-1350kD	ACN	Silica gel C18		TGIC of PMMA at 10–60°C.	Chang T, Lee HC, Lee W, Park S, Ko C [38]
PMMA, atactic CR: 13kD-613kD	ACN	Nucleosil C18, 5 µm, 30 nm, 25×0.6 cm	66°C 1 ml/min 50µl ELSD	Atactic PMMA eluted at critical conditions, syndiotactic PMMA in SEC mode, isotactic PMMA in adsorption mode.	Macko T, Hunkeler D, Berek D [6]
PMMA	ACN / DCM 42.8 / 57.2 vol. S / S, D / A	Lichrospher SI-300, 5 µm, 300°A, 35×0.05 cm		PMMA is totally adsorbed from DCM on silica gel.	Zimina TM, Kever JJ, Melenevskaya EY, Fell AF [45]

Polymer	Mobile phase	Stationary phase	Conditions	Application	Reference
PMMA	ACN / DCM / THF 48 / 48 / 4 vol. D / A / D S / S / S	Silica gel S5X, 5 µm, 300°A, 25×0.2 cm	27°C 0.2 ml/min 1 µl DAD	Block copolymers PS-PMMA.	Zimina TM, Fell AF, Castledine JB [46]
PMMA CR: 60kD-400kD	Methanol / chloroform 20 / 80 vol. N / S	Silica gel S5X, 5 µm, 300A, 25×0.2 cm		Zone splitting and considerable peak tailing under near critical conditions.	Zimina TM, Kever JJ, Melenevskaya EY, Fell AF [45]
PMMA CR: 6.5kD-160kD	MEK / cyclohexane 70 / 30 vol. S / N	Silica gel Lichrospher Si-300°A+1000°A, 10µm, 2×(20×0.4) cm	25°C 0.5 ml/min UV, RI	Block copolymers PS/PMMA.	Pasch H, Brinkmann C, Gallot Y [47], Pasch H, Augenstein M [48]
				Blends PS+PMMA, PVC+PMMA, PS in SEC mode.	Pasch H [49], Pasch H, Brinkmann C, Gallot Y [47]
PMMA CR: 2.5kD-100kD	MEK / cyclohexane 73 / 27 vol. S / N	Silica gel Nucleosil SI 100°A, 5 µm 20×0.4 cm	25°C 0.5 ml/min 20 µl UV, RI	Block copolymers DMMA – PM-MA.	Pasch H, Augenstein M [48]
PMMA CR: 30kD -280kD	MEK / cyclohexane 72 / 28 vol. S / N	Lichrospher 300°A+1000°A, 10 µm, 2×(20×0.4) cm	RI, VIS	Blends PMMA+ PtBMA, PM-MA+(P(nBMA-co-MMA)	Pasch H, Rode K [50]
PMMA CR: 0.9kD-100kD	MEK / cyclohexane 73 / 27 vol. S / N	Nucleosil Si-100°A, 5 µm, 20×0.4 cm	25°C 0.5 ml/min UV, RI	Block-copolymers PS-PMMA.	Pasch H, Brinkmann C, Gallot Y [47]
PMMA	THF / cyclohexane 63 / 37 vol. S / N	Nucleosil CN, 300°A+500°A, 7 µm, 25×0.4cm,	40°C 1 ml/min UV, ELSD	Graft copolymer EPDM-PMMA separated from PMMA.	Siewing A, Schierholz J, Braun D, Hellmann G, Pasch H [51]

Table 1. (continued)

Polymer Molar masses Critical range Used samples	Mobile phase Solvent / Nonsolvent Adsorli / Desorli	Stationary Phase Particle diameter Pore size Column dimension	Conditions Temperature Pressure Flow rate Inj. volume Detector	Analytical application	Investigators Reference
PMMA	THF / hexane 77.75 / 22.25 vol. S / N	Nucleosil Si 100,	RI	Block copolymers PDMA-PMMA.	Pasch H, Much H, Schulz G [52]
PMMA CR: 16kD-294kD	THF / n-hexane 81.8 / 18.2 wt.	Bare silica gel, Biospher, 12nm, 10 μm, 25×0.8 cm	30°C 20μl ELSD	Blends PMMA, PTHF and homo PTHF.	Janco M, Berek D, Önen A, Fischer CH, Yagci Y, Schnabel W [53]
PMMA, syndiotactic CR: 2.7kD-153kD	THF / n-hexane 83.2 / 16.8 wt.	Nucleosil, 100°A, 5μm, 25×0.4cm,	30°C 20μl ELSD	It-PMMA eluted in SEC mode. Separation for mixtures of st-PMMA and it-PMMA of lower molar mass.	Berek D, Janco M, Hatada K, Kitayama T, Fujimoto N [54]
PMMA US: 1.8kD-305kD CR: 0.7kD-12kD	THF /n-hexane 82 / 18 wt. S / N	Nucleosil, 120°A, 5 μm, 15×0.4cm, Nucleosil 300°A, 5 μm, 25×0.4cm	0.5 ml/min 25 μl UV, ELSD	Block copolymers PMMA-PtBMA, two-dimensional separation.	Falkenhagen J, Much H, Stauf W, Müller AHE [55]
PMMA US: 1.8kD-305kD CR: 1.8kD-51kD	THF / ACN 8.5 / 91.5 wt. S / S, D / A	Nucleosil C18, 100°A, 5 μm, 25×0.4 cm, Nucleosil C18, 1000°A, 7 μm, 25×0.4 cm		Block copolymers PMMA-PtB-MA. PtBMA eluted in SEC mode.	

Polymer / CR	Mobile phase	Column	Conditions	Notes	Reference
PMMA CR: 2.5kD–100kD	THF / toluene 36 / 64 wt. S / S, D / A	Silica gel Biospher SI-120, 12nm, 10 μm, 25×0.4 cm	20 μl RI, ELSD	Decreased recovery for PMMA above 100kD, no recovery above 500kD. The higher molar mass, the larger peak broadening. S-shaped curvature of the calibration curve in the vicinity of the critical conditions.	Berek D, Janco M, Meira GR [56]
PMMA CR: 1.5kD–620kD	THF / toluene 38 / 62 wt. S / S, D / A	Silica gel Biospher SI-120, 12nm, 10 μm or silica gel SGX-1000, 100nm, 10 μm, 25×0.4 cm	35°C 50 μl RI, ELSD	Limited sample recovery for high molar mass. No recovery problems when injected in pure THF or toluene.	Berek D [57]
PMMA, isotactic CR: 12kD–157kD	THF / chloroform 12.3 / 87.7 wt. S / S, D / A	Silica gels Nucleosil 12+30+100nm, 5 μm, 3×(25×0.4) cm	35°C 1 ml/min 20 μl ELSD	St-PMMA eluted in SEC mode. Mixture of st-PMMA and it-PMMA separated.	Berek D, Janco M, Hatada K, Kitayama T, Fujimoto N [54]
PMMA, isotactic CR: 17kD–258kD	THF / chloroform 12.4 / 87.6 wt. S / S, D / A	Silica gels 12+30+100nm	35°C	SEC mode at 12.4 wt. % THF and 30°C. Increase of temperature increases retention of PMMA.	Skvortsov AM, Gorbunov AA, Berek D, Trathnigg B [58]
PMMA, isotactic CR: 12kD–157kD	Chloroform / ethanol near crit. 97.2 / 2.8 wt. S / N, A / D	Silica gels Nucleosil 12+30+100nm, 5 μm, 3×(25×0.4) cm	30°C 1 ml/min 20 μl ELSD	Mixture of st-PMMA and it-PMMA separated for polymers up to 100kD.	Berek D, Janco M, Hatada K, Kitayama T, Fujimoto N [54]
PDMA CR: 10kD–220kD	THF / ACN 78.5 / 21.5 vol.	Nucleosil RP-18, 300+1000°A 5 μm, 2×(25×0.4) cm.	RI, VIS	Blends of PMMA, PDMA, PnBMA, P(MMA-co-nBMA).	Pasch H, Rode K [59]

Table 1. (continued)

Polymer Molar masses Critical range Used samples	Mobile phase Solvent / Nonsolvent Adsorli / Desorli	Stationary Phase Particle diameter Pore size Column dimension	Conditions Temperature Pressure Flow rate Inj. volume Detector	Analytical application (LC, TLC or SFC separation and characterization) and/or notices	Investigators Reference
PDMA CR: 11kD–100kD	THF / ACN 77 / 23 vol.	Nucleosil RP-18, 300°A+1000°A	RI	Polymer blends. Limiting factor of these separations is adverse effect of the solvent peak.	Pasch H, Rode K, Chaumien N [60]
PDMA CR: 10kD–110kD	Chloroform / ethanol 61 / 39 vol.	Nucleosil RP-18, 300°A+1000°A 5 µm, 25×0.4cm	25°C 0.5 ml/min 20 µl RI, UV	Block copolymers PDMA-PMMA.	Pasch H, Augenstein M, Trathnigg B [61]
PDMA CR: 11kD–100kD	MEK / cyclohexane 3.34 / 96.66 vol.	Lichrospher 300°A+1000°A	RI	Polymer blends.	Pasch H, Rode K, Chaumien N [60]
PtBMA CR: 23kD–200kD	ACN / DCM 9.3 / 90.7 vol. S / S, D / A	Lichrospher Si 300, 5 µm, 300°A, 35×0.05 cm	25°C	PtBMA is fully adsorbed from DCM on silica gel.	Zimina TM, Kever JJ, Melenevskaya EY, Fell AF [45]
			25°C 5 µl/min 0.5 µl UV	Block copolymers PS-PtBMA.	Zimina TM, Kever JJ, Melenevskaya EU, Zgonnik VN, Belenkii BG [62]
PtBMA US: 23kD–200kD	THF / DCM 4.5 / 95.5 vol.	Silica gel SGX, 5 µm, 300°A, 25×0.2 cm	28°C	Block copolymers PS-PtBMA. Peak splitting under near critical conditions. PtBMA is fully adsorbed from DCM on silica gel.	Zimina TM, Kever JJ, Melenevskaya EY, Fell AF [45]

Polymer	Solvent	Column	Detection	Notes	Reference
PtBMA CR: 48kD-213kD	THF / ACN 49.6 / 50.4 vol.	Nucleosil RP-18, 300°A+1000°A 5 µm, 2×(25×0.4) cm	RI, VIS		Pasch H, Rode K [59]
PtBMA CR: 2.9kD-145kD	THF / ACN 49.5 / 50.5 wt.	Nucleosil C18, 100°A, 5 µm, 25×0.4cm, Nucleosil C18, 1000°A, 7 µm, 25×0.4cm	0.5ml/min, 25 µl UV, ELSD	Block copolymers PMMA-PtBMA, two-dimensional separation. PMMA eluted in SEC mode.	Falkenhagen J, Much H, Stauf W, Müller AHE [55]
PtBMA CR: 2.9kD-69.7kD	THF / n-hexane 43 / 57 wt.	Nucleosil, 100°A, 5 µm, 15×0.4cm, Nucleosil 300°A, 5 µm, 25×0.4cm			
PtBMA CR: 48kD-185kD	MEK / cyclohexane 18.8 / 81.2 vol.	Lichrospher Si-300+ Si-1000, 10 µm, 2×(20×0.4) cm	RI, VIS	Blends of PtBMA, PDMA, PnBMA.	Pasch H, Rode K [59], Pasch H, Rode K, Chaumien N [60]
PtBMA CR: 20kD-200kD	Cyclohexane / toluene / MEK 9 / 1 / 2.3 vol.	Silica gel KSKG 120°A, 5-20 µm, TLC	TLC	Block copolymers PS-PtBMA.	Belenkii BG, Gankina ES, Zgonnik VN, Malchova II, Melenevskaya EU [63], Gankina ES, Belenkii BG, Malakhova I, Melenevskaya EU, Zgonnik VN [64]
PnBMA CR: 8kD-225kD	DMFA	Nucleosil C18, 30 nm, 5 µm, 25×0.6cm	154C 50 µl 1 ml/min ELSD		Macko T, Hunkeler D, Berek D [6]
PnBMA CR: 57kD-210kD	THF / ACN 53.1 / 46.9 vol.	Nucleosil RP-18, 300+1000°A 5 µm, 2×(25×0.4) cm	RI, VIS	Blends of PtBMA, PS, PnBMA	Pasch H, Rode K [59]
PnBMA US: 21kD-200kD	MEK / cyclohexane 14.3 / 85.7 vol. S / S	Lichrospher 300°A+1000°A	RI	Polymer blends. Limiting factor of these separations is adverse effect of the solvent peak	Pasch H, Rode K, Chaumien N [60]

Table 1. (continued)

Polymer Molar masses Critical range Used samples	Mobile phase Solvent / Nonsolvent Adsorli / Desorli	Stationary Phase Particle diameter Pore size Column dimension	Conditions Temperature Pressure Flow rate Inj. volume Detector	Analytical application (LC, TLC or SFC separation and characterization) and/or notices	Investigators Reference
PEMA	diethylmalonate	Silica gel C18		TGIC of PEMA at 25-40°C.	Chang T, Lee HC, Lee W, Park S, Ko C [38]
PEMA highly syndiotactic CR: 6kD-45kD	Acetone / acetone-d60 / cyclohexane 34 / 5 / 61 wt. S / S / N	Develosil NH$_2$, 120°A, 5μm, 25×0.8cm	35°C ELSD, NMR	Analysis of tacticity distribution. PEMA eluted in order: isotactic, heterotactic, syndiotactic.	Kitayama T, Janco M, Ute K, Niimi R, Hatada K, Berek D [65]
PEMA highly syndiotactic CR: 6.8kD-56kD	THF / cyclohexane 64 / 36 wt. S / N	Develosil NH$_2$, 120°A, 5μm, 25×0.8cm	35°C 20 μl ELSD	Analysis of tacticity distribution. Two-dimensional separation. Sample recovery dropped with increasing of molar mass, no elution for samples with M_w >56kD.	Janco M, Hirano T, Kitayama T, Hatada K, Berek D [66]
PEMA highly isotactic CR: 4.9kD-52kD	THF / cyclohexane 56 / 44 wt. S / N	Develosil NH$_2$, 120°A, 5μm, 25×0.8cm	35°C 20 μl ELSD	Analysis of tacticity distribution. For samples with M_w >52kD adsorption mode.	
PEMA CR: 1kD-1000kD	THF / n-hexane 43.4 / 56.6 vol. S / N	Silica gel Lichrospher Si-300°A + Si-1000°A, 10μm, 20×0.4cm	UV, RI	Di and tri-block copolymers of styrene/butadiene. PB eluted in SEC mode.	Braun D, Esser E, Pasch H [28]
PS CR: 15kD-1200kD	DMFA	Nucleosil C18, 30nm, 5μm, 25×0.6cm	95°C 50μl 1 ml/min ELSD	Separation of PS and PMMA. PMMA eluted in SEC mode.	Macko T, Hunkeler D, Berek D [6]

Sample	Eluent	Column	Conditions	Comments	Reference
PS CR: 3.5kD-135kD US: 3.5kD-3220kD	THF / n-hexane 50 / 50 wt. 49 / 51 wt. 48 / 52 wt. 47 / 53 wt. 46 / 54 wt. 45 / 55 wt. 39.7 / 60.3 vol. S / N	Nucleosil Silica 100°A, 5 μm, 25×0.46cm	3°C 13°C 23°C 36°C 54°C 75°C 25°C 1 ml/min 20 μl UV		Baran K, Laugier S, Cramail H [67]
				Non-functional PS separated from ω-aldehyde PS and α, ω-aldehyde PS at 13°C and 75°C.	Baran K, Laugier S, Cramail H [68]
PS US: 3.5kD-3220kD	THF / n-hexane 50.4 / 49.6 vol.	Silica gel NH$_2$ 100°A, 5 μm, 25×0.46cm	25°C 1 ml/min 20 μl UV		Baran K, Laugier S, Cramail H [67]
PS US: 1kD-135kD	THF / n-hexane 56.4 / 43.5 wt.	Nucleosil NH$_2$ Silica 100°A, 5 μm, 25×0.46cm	25°C 1 ml/min 20 μl UV	ω-hydroxyl PS separated from linear PS.	Baran K, Laugier S, Cramail H [68]
PS CR: 1kD-135kD	THF / n-hexane 44.2 / 55.8 wt.	Nucleosil Silica 100°A, 5 μm, 25×0.46cm			
PS CR: 1.8kD-35kD	THF / n-hexane 34.6 / 65.4 vol.	PS/DVB gel Eurogel P-RP-100, 100°A, 5 μm, 25×0.4 cm	0.5 ml/min UV, RI		Pasch H, Deffieux A, Henze I, Schappacher M, Rique-Lurbet L [69]

Table 1. (continued)

Polymer Molar masses Critical range Used samples	Mobile phase Solvent / Nonsolvent Adsorli / Desorli	Stationary Phase Particle diameter Pore size Column dimension	Conditions Temperature Pressure Flow rate Inj. volume Detector	Analytical application (LC, TLC or SFC separation and characterization) and/or notices	Investigators Reference
PS linear CR: 1.8kD–24kD	THF / n-hexane 45 / 55 vol.	Silica gel YMC, 100°A, 5μm, 25×0.4 cm	0.5ml/min 20μl UV, RI	Separation according to architecture (cyclic versus linear). Linear PS eluted at critical conditions faster, as cyclic PS in the same column and mobile phase.	Pasch H, Deffieux A, Henze I, Schappacher M, Rique-Lurbet L [69]
PS cyclic CR: 15kD–23kD	THF / n-hexane 45 / 55 vol.				
PS CR: 3.7kD–100kD US: 3.7kD–680kD	THF / n-hexane 47 / 53 wt. S/N	Nucleosil Silica 100°A, 5μm, 25×0.46 cm	25°C 1ml/min 20μl UV	PS with molar mass above 100kD eluted in SEC mode.	Baran K, Laugier S, Cramail H [70]
PS CR: 3.7kD–1200kD US: 3.7kD–3220kD	THF / n-hexane 50 / 50 wt.				
PS Linear CR: 5kD–14kD	THF / n-hexane 47.9 / 52.1 wt.			Separation cyclic versus linear PS. Cyclic PS eluted after linear PS.	Lepoittevin B, Dourges MA, Masure M, Hemery P, Baran K, Cramail H [71]
PS Linear CR: 1kD–9kD				Separation cyclic versus linear PS. Cyclic PS eluted earlier than linear due different functionality.	Lepoittevin B, Perrot X, Masure M, Hemery P [72]

Polymer	Mobile phase	Stationary phase	Conditions	Comments	Reference
PS Linear CR: 1kD–14kD	THF / n-hexane 50.5 / 49.5 wt.		RI, LS		Lepoittevin B, Hemery P [73]
PS CR: 1kD–900kD	THF / n-hexane 43.4 / 56.6 vol.	Kromasil Silica 100°A, 5μm, 25×2.0 cm	25°C 15ml/min 20μl UV	Analysis of multicyclic and grafted PS. Linear PS separated from grafted PS and from poly(chloromethylstyrene).	Lepoittevin B, Dourges MA, Masure M, Hemery P, Baran K, Cramail H [71], Lepoittevin B, Perrot X, Masure M, Hemery P [72]
PS CR: 1kD–1000kD		Silica gel Lichrospher, 10 μm, 300°A + 1000°A, 2×(20×0.4) cmFTIR	25°C RI, UV, cmFTIR	Preparative fractionation cyclic versus linear PS. Cyclic PS in adsorption mode.	Braun D, Esser E, Pasch H [28]
PS CR: 1.6kD–500kD	THF / n-hexane near crit. 48.2 / 51.8 vol.	Silica gel Nucleosil, 300°A + 1000°A, 2×(25×0.4) cm		Block copolymers styrene-butadiene. PB eluted in SEC mode.	Pasch H, Esser E, Kloniger C, Iatrou H, Hadjichristidis N [30]
PS CR: 4.4kD–122kD	THF / n-hexane near crit. 51.3 / 48.7 vol.	Silica gel Nucleosil, 1000°A, 25×0.46cm	50°C UV, ELSD	Analysis of mictoarm star polymers.	Gerber J [74]
PS CR: 20kD–330kD	THF / cyclohexane 15 / 85 vol. S / S	Silica gel Nucleosil, 100°A, 25×0.46cm	25°C		
PS CR: 2.2kD–68kD		Silica gel LiChrospher 500°A	31°C	Macrocycles of PS eluted in adsorption mode. Cyclohexane is theta solvent for PS at 34°C.	Blagodatskich IV, Gorshkov AV [75]
		Silica gel Resolve 90°A	23°C		

Table 1. (continued)

Polymer Molar masses Critical range Used samples	Mobile phase Solvent / Nonsolvent Adsorli / Desorli	Stationary Phase Particle diameter Pore size Column dimension	Conditions Temperature Pressure Flow rate Inj. volume Detector	Analytical application (LC, TLC or SFC separation and characterization) and/or notices	Investigators Reference
PS US: 15kD–1100kD	THF / n-heptane near crit. 30 / 70 vol. S/N	PS gel		Blends of PEMA, PS, PLMA and copolymer PS / nBMA.	Balke ST [16], Balke ST, Patel RD [76], Balke ST [77]
PS CR: 2.5kD–385kD US: 0.6kD–385kD	THF / isooctane 50 / 50 vol.	Nucleosil C18, 100°A+500°A +1000°A 3×(25×0.46) cm	7°C	Block copolymers PI-PS. Molar masses of PI determined are systematically smaller (deviation 0–33%) as molar mass of PI precursor.	Lee W, Cho DH, Chang T, Hanley KJ, Lodge TP [78]
PS US: 1.8kD–450kD	THF / ACN 47 / 53 vol. S/N	Novapak C18, 4µm, 60°A, 30×0.39 cm	35°C 0.5 ml/min UV		Philipsen HJA, Klumperman B, Van Herk AM, German AL [79]
PS CR: 19kD–426kD	THF / ACN 49.4 / 50.6 vol.	Nucleosil RP-18, 300°A+1000°A, 5 µm, 2×(25×0.4) cm	RI, VIS	Blends of polymers.	Pasch H, Rode K [59]
PS CR: 20kD–200kD	THF / ACN 49 / 51 vol.	Nucleosil C18, 300°A, 5 µm + 1000°A, 7 µm, 2×(25×0.4) cm	25°C 0.5 ml/min 20 µl RI, UV	Block copolymers PS-PMMA.	Pasch H, Gallot Y, Trathnigg B [80]
PS US: 1.7kD–2890kD	THF / ACN 49 / 51 vol.	Nucleosil C18, 100°A, 5µm, 25×0.45 cm	37°C 0.5 ml/min UV		Lee HC, Chang T [81]

Liquid Chromatography under Critical and Limiting Conditions: A Survey of Experimental...

Sample	Mobile phase	Column	Conditions	Purpose	Reference
PS CR: 1.6kD–200kD	THF / ACN 50 / 50 vol.	Silica gel Nucleosil C18, 300°A, 25×0.46 cm	37°C UV, ELSD		Radke W [82]
PS US: 1.8kD–450kD	DCM / ACN 58 / 42 vol. S/N	Novapak C18, 4 μm, 60°A, 30×0.39 cm	25°C		Philipsen HJA, Klumperman B, Van Herk AM, German AL [79]
	DCM / ACN 58 / 42 vol.	Nucleosil C18, 100°A, 7 μm, 25×0.4 cm	27°C		
	DCM / ACN 58 / 42 vol.	Nucleosil C18, 4000°A, 7 μm, 25×0.4 cm	31°C		Philipsen HJA, Klumperman B, Van Herk AM, German AL [79]
	DCM / ACN 50 / 50 vol.	Pl-gel, 10 μm, 100°A, 30×0.75 cm	25°C		
PS US: 56.7kD–1530kD	DCM / ACN 57 / 43 vol.	Nucleosil C18, 1100°A, 5 μm, 25×0.21 cm	30.5°C UV	TGIC of linear and star PS.	Lee HC, Chang T, Harville S, Mays JW [83]
PS CR: 5kD–200kD		Nucleosil C18, 100°A, 25×0.46 cm	43°C 0.5 ml/min	Cyclic PS separated from linear PS. Retention of cyclic PS increased with molar mass.	Lee HC, Lee H, Lee W, Chang T, Roovers J [84]
PS CR: 2kD–289kD		Nucleosil C18, 100°A, 5 μm, 25×0.4 cm	36°C 0.5 ml/min 50 μl UV, RI	TGIC of PS and PMMA.	Chang T, Lee HC, Lee W [85]
PS	DCM / ACN 58 / 42 vol.		33.6°C 0.5 ml/min UV, MALDI-TOF MS	Characterization of ring PS (5kD).	Cho DH, Park S, Kwon K, Chang T [86]

Table 1. (continued)

Polymer Molar masses Critical range Used samples	Mobile phase Solvent / Nonsolvent Adsorli / Desorli	Stationary Phase Particle diameter Pore size Column dimension	Conditions Temperature Pressure Flow rate Inj. volume Detector	Analytical application (LC, TLC or SFC separation and characterization) and/or notices	Investigators Reference
PS US: 0.58kD-390kD	DCM / ACN 57 / 43 vol.	Nucleosil C18, 10nm 30nm 50nm 100nm 25×0.46cm	37.9°C 43.1°C 41.0°C 41.5°C 0.5ml/min UV/VIS	Ring PS separated from linear PS. Ring PS in adsorption mode. S-shaped curvature of the calibration curve in the vicinity of the critical conditions.	Lee W, Lee H, Lee HC, Cho DH, Chang T, Gorbunov AA, Roovers J [87]
PS CR: 2kD-17.5kD	DCM / ACN 65 / 35 vol.	Kromasil 100 C18, 3.5 µm, 40×0.032cm	130°C 7µl/min 200nl UV		Bruheim I, Molander P, Theodorsen M, Ommundsen E, Lundanes E, Greibrokk T [88]
PS US: 1.8kD-450kD	DCM / hexane 61 / 39 vol. S / N	Resolve silica, 90°A, 5 µm, 15×0.39cm	45°C	Peak splitting due to difference in composition between the sample solvent and the eluent. Boiling point of DCM is 40.1C.	Philipsen HJA, Klumperman B, Van Herk AM, German AL [79]
PS US: 3.7kD-1700kD	DCM / n-hexane 69 / 31 wt.	Nucleosil Silica, 100°A, 5 µm, 25×0.46cm	25°C 1 ml/min 20µl UV	Peak splitting not observed.	Baran K, Laugier S, Cramail H [67]
PS	DCM / n-hexane 50.3 / 49.7 vol.	Silica gel NH2, 100°A, 5 µm, 25×0.46cm			

PS US: 1kD–135kD	DCM / n-hexane 74.3 / 25.7 wt.	Nucleosil Silica, 100°A, 5 µm, 25×0.46cm	25°C 1ml/min 20µl UV	Baran K, Laugier S, Cramail H [68]
PS US: 1kD–135kD	DCM / n-hexane 63.5 / 36.5 wt.	Nucleosil Silica NH$_2$, 100°A, 5 µm, 25×0.46cm		ω-acetal and α,ω-acetal PS separated from PS.
PS US: 1.8kD–450kD	DCM / n-hexane 48 / 52 vol.	µ-Bondapak NH$_2$, 125°A, 10µm, 30×0.39cm	25°C 0.5 ml/min 25 µl UV	Philipsen HJA, Klumperman B, Van Herk AM, German AL [79]
PS US: 1.8kD–450kD	THF / water 86 / 14 vol. S / N	Novapak C18, 4µm, 60°A, 15×0.39cm	58°C 0.5 ml/min UV	Philipsen HJA, Klumperman B, Van Herk AM, German AL [79]
PS CR: 0.5kD–9.1kD	THF / water 87 / 13 vol. S / N	Novapak C18, 60°A, 4 µm, 30×0.39cm;	30°C 0.5–1.0 ml/min 25µl UV	Cools PJCH, Herk AM, German AL, Staal WJ [37]
PS CR: 0.5kD–9.1kD	THF / water 89 / 11 vol.	u-Styragel HT, 10µm, 30×0.78×cm;		
	THF / water 89 / 11 vol.	Deltapak C18, 100°A, 5 µm, 15×0.39cm;		
	THF / water 88 / 12 vol.	Deltapak C18, 300°A, 5 µm, 30×0.39cm		

Table 1. (continued)

Polymer Molar masses Critical range Used samples	Mobile phase Solvent / Nonsolvent Adsorli / Desorli	Stationary Phase Particle diameter Pore size Column dimension	Conditions Temperature Pressure Flow rate Inj. volume Detector	Analytical application (LC, TLC or SFC separation and characterization) and/or notices	Investigators Reference
PS US: 1kD–135kD	THF / water 87.1 / 12.9 wt 87.4 / 12.6 wt. 87.0 / 13.0 wt. 86.4 / 13.6 wt. 87.7 / 12.3 wt.	Nucleosil Silica C18, 7 μm, 12.5×0.4 cm 100°A 300°A 500°A 1000°A Prontosil Silica C18, 5 μm, 12.5×0.4 cm 300°A	25°C 1 ml/min 20 μl UV	Block copolymers PS-PEO eluted in exclusion mode (using Prontosil) and in adsorption mode (using Nucleosil).	Baran K, Laugier S, Cramail H [68]
PS CR: 3.7kD–50kD US: 3.7kD–1700kD	MEK / n-hexane 37.1 / 62.9 wt. S / N	Nucleosil Silica 100°A, 5 μm, 25×0.46 cm	25°C 1 ml/min UV	PS with $M_w > 50$kD are not soluble in the eluent.	Baran K, Laugier S, Cramail H [67]
PS CR: 3.7kD–50kD US: 3.7kD–1700kD	Ethylacetate / n-hexane 37.7 / 62.3 wt. S / N	Nucleosil Silica 100°A, 5 μm, 25×0.46 cm	25°C 1 ml/min 25 μl UV	PS with $M_w > 50$kD are not soluble in the eluent.	
PS US: 1kD–135kD	Ethylacetate / n-hexane 38 / 62 wt.			Non-functional PS separated from ω-aldehyde PS and α,ω-aldehyde PS as well as from ω-Acid PS and α,ω-Acid PS.	Baran K, Laugier S, Cramail H [68]

Polymer	Eluent	Stationary phase	T, flow	Injection, detection	Reference	Notes
PS US: 1kD-135kD	Ethylacetate / n-hexane 50 / 50 wt.	Silica gel NH2, 100°A, 5μm, 25×0.46cm	25°C 1 ml/min	20μl UV	Radke W [82]	
PS CR: 7kD-100kD	Dimethylacetamid / ACN 81 / 19 vol.	Silica gel Nucleosil C18, 300°A, 25×0.46cm	50°C	UV, ELSD		
PS CR: M_w<100kD	Toluene / n-hexane 83.9 / 16.1 wt. S / N	Nucleosil Silica 100°A, 5μm, 25×0.46cm	25°C 1 ml/min	20μl RI	Baran K, Laugier S, Cramail H [67]	
PS	Toluene / n-hexane 77.3 / 22.7 vol.	Silica gel NH2, 100°A, 5μm, 25×0.46cm				
PS US: 1kD-135kD	Toluene / n-hexane 81.7 / 18.3 wt.	Nucleosil Silica NH2, 100°A, 5μm, 25×0.46cm	25°C 1 ml/min	20μl UV	Baran K, Laugier S, Cramail H [68]	
PS CR: 19.8kD-173kD	Cyclohexane / benzene / acetone 40 / 16 / 1.8 vol. S / S / N	Silica gel KSK TLC		TLC	Belenkii BG, Gankina ES, Tennikov MB, Vilenchik LZ [2]	SEC at lower content of acetone. Mixture cyclohexane + acetone dissolves PS.
PS CR: 10kD-498kD	Cyclohexane / toluene / MEK / pyridine 9/ 0.4 / 1.8 / 1 vol. S / S / S / S	Silica gel KSKG 120°A, 5–20μm, TLC		TLC	Belenkii BG, Gankina ES, Zgonnik VN, Malchova II, Melenevskaya EU [63]	Block copolymers PS-PtBMA. Cyclohexane is theta solvent for PS above 36°C.

Table 1. (continued)

Polymer Molar masses Critical range Used samples	Mobile phase Solvent / Nonsolvent Adsorli / Desorli	Stationary Phase Particle diameter Pore size Column dimension	Conditions Temperature Pressure Flow rate Inj. volume Detector	Analytical application (LC, TLC or SFC separation and characterization) and/or notices	Investigators Reference
PS CR: 10kD-200kD	Cyclohexane / toluene / MEK / pyridine 9 / 0.4 / 0.4 / 4 vol.	LiChrospher 300°A, 10μm, TLC	TLC	Block copolymers PS-PtBMA.	Belenkii BG, Gankina ES, Zgonnik VN, Malchova II, Melenevskaya EU [63], Gankina ES, Belenkii BG, Malakhova I, Melenevskaya EU, Zgonnik VN [64]
PS CR: 0.6kD-164kD	Chloroform / benzene / acetone, near crit. 40 / 16 / 2 vol. S / S / N	Silica gel S-80, 50nm or silica gel KSM, 12nm, TLC	TLC		Belenkii BG, Vilenchik LZ [89]
PS CR: polymerization degree 1-10000 US: 0.4kD-2145kD	Chloroform / tetrachloromethane 4.5 / 95.5 vol. S / S, D / A	Macroporous glass, pore radius 125°A, 25-32μm, 60x0.4cm	31°C 12.5atm 50ml/hour UV	Critical conditions are function of both temperature and pressure.	Nefedov PP, Zhmakina TP [7]
PS CR: 5kD-120kD	Chloroform / tetrachloromethane 5.35 / 94.65 vol. S / S, D / A	LiChrospher 300°A	30°C 0.5ml/min 10μl UV	Separation of a polymer blend.	Dudorina AV, Gorshkov AV, Filatova HH, Evreinov VV, Entelis SG [90]
PS CR: 0.6kD-59kD	THF / CO_2 46 / 54 mol. %	Silica, 200°A, 5μm, 1.8m×250μm	260atm 60nl UV, ELSD SFC	Mono and dicarboxy-terminated PS and PS-MA copolymers.	Yun H, Olesik SV, Marti EH [91]

Sample	Mobile phase	Stationary phase	Conditions	Notes	Reference
PS US: 12-114kD	THF / CO_2 46 / 54 mol. %	Silica, 200°A, 5μm, 25×0.2 cm	323 K – 252 atm; 333 K – 263 atm; 343 K – 280 atm; 353 K – 336 atm; 363 K – 349 atm 200 nl 0.36 ml/min UV/VIS, ELSD SFC	A slight increase of THF content (+0.3%) has large influence on value of pressure (+67 atm). Block copoly-styrene/poly (butylene-co-ethylene) analyzed.	Phillips S, Olesik SV [92]
PS CR: 4kD-160kD US: 4kD-1000kD	THF / CO_2 46 / 54 mol. %	DVB gel Jordi-Gel, 5μm, 500°A, 10×0.45cm	298 K 136 atm 200 nl UV SFC	Solvating effects of eluent are varied by changing its density (pressure).	Souvignet I, Olesik SV [17]
PS CR: 4kD-160kD US: 4kD-1000kD	THF / methanol 57 / 43 vol. S / N	DVB gel Jordi-Gel, 5μm, 500°A, 10×0.45cm	room temp. 136 atm UV SFC	Above 57 vol. % THF, PS with $M_w > 20kD$ are adsorbed.	Souvignet I, Olesik SV [17]
	THF / ACN 47.2 / 52.8 vol. S / N			Above 46 vol. % THF, PS are adsorbed.	
PS CR: 10-100kD	Xylene / CO_2 55 / 45 mol.%	Silica, 200°A, 5 μm, 25×0.2 cm	413 K 435 atm ELSD SFC	PE standards eluted in SEC mode. Block copoly-styrene/poly (butylene-co-ethylene) eluted in the size-exclusion mode.	Phillips S, Olesik SV [92]

Table 1. (continued)

Polymer	Mobile phase	Stationary Phase	Conditions	Analytical application	Investigators
Molar masses Critical range Used samples	Solvent / Nonsolvent Adsorli / Desorli	Particle diameter Pore size Column dimension	Temperature Pressure Flow rate Inj. volume Detector	(LC, TLC or SFC separation and characterization) and/or notices	Reference
ω-hydroxyl PS CR: 1kD-135kD	THF / n-hexane, near crit. 44.2 / 55.8 vol.	Nucleosil Silica 100 Å, 5 µm, 25×0.46cm	25°C 1ml/min 20µl UV	ω-hydroxyl PS separated from linear PS.	Baran K, Laugier S, Cramail H [68]
PVC polyvinylchloride	dimethylformamid	Silica gel C18		TGIC of PVC at 25–40°C.	Chang T, Lee HC, Lee W, Park S, Ko C [38]

Table 2. Critical Conditions for Special Polymers (Polymers Injected in the Mobile Phase.)

Polymer Molar masses Critical range Used samples	Mobile phase Solvent / Nonsolvent Adsorli / Desorli	Stationary phase Particle diameter Pore size Column dimension	Conditions Temperature Flow rate Inj. volume Detector	Analytical application and / or notices	Investigators Reference
Polyacetals Poly (1,3,6-trioxacyclo-octane) with hydroxyl end groups CR: 0.9kD-83kD	ACN / water 49.5 / 50.5 vol.	Silica gel Nucleosil C18, 10μm, 25×0.46cm	20 μl UV, RI	Separation according to functionality. Two-dimensional separation.	Krüger RP, Much H, Schulz G, Wehrstedt C [93], Krüger RP, Much H, Schulz G [94], Pasch H, Much H, Schulz G [52], Pasch H, Brinkmann C, Much H, Just U [95]
Poly(1,3,6-trioxacyclooctane) terminated with 9-anthrylmethyl groups CR: 1.6kD-70kD	ACN / water 60 / 40 vol.	Silica gel C18, 5μm, 90°A, 30×0.46cm	30°C 0.8ml/min DAD	Functionality distribution. Two dimensional separation. UV-active end groups with anthryl-methyl derivative introduced.	Gancheva VB, Vladimirov NG, Velichova RS [96], Gancheva VB, Vladimirov NG, Velichova RS [97]
Polyamide-6 Linear US: 2kD-15kD	Formic acid / 1-propanol near crit. 81.6 / 18.4 wt. S / N	Silica gel Nucleosil, 500°A, 5μm, 2×(20×0.4) cm	38°C 0.65ml/min 55 μl ELSD	Separation of the linear and cyclic (cyclic monomer – cyclic pentamer) structures. Cyclic samples eluted after linear macromolecules.	Mengerink Y, Peters R, deKoster CG, van der Wal SJ, Claessens HA, Cramers CA [98]
Chlorinated Polyethylene CR: 55kD-520kD	nexane / chloroform (stabilized by 1% ethanol) 59.45 / 40.5 vol.	Silica gel Silasorb 600, 5μm, 25×0.46cm	25°C 0.7 ml/min 50 μl	Elution did not depend on molar mass for polymers with the same amount of chlorine.	Brun Y [99]

Table 2. (continued)

Polymer / Molar masses / Critical range / Used samples	Mobile phase / Solvent / Nonsolvent Adsorli / Desorli	Stationary phase / Particle diameter Pore size Column dimension	Conditions / Temperature Flow rate Inj. volume Detector	Analytical application and / or notices	Investigators Reference
Polyesters PDEGA Poly(diethylene glycol adipate) CR: 560	Acetone / chloroform 35 / 65 vol.	Silica gel Si-100 25×0.4cm	24°C 1 ml/min 10 µl RI	Separation of zero, mono and bifunctional (hydroxyl groups) polymers.	Gorshkov AV, Prudskova TN, Filatova HH, Evreinov VV [100]
PDEGA	MEK / hexane 92 / 8 vol.	Silica gel Si-100	24°C 1 ml/min 10 µl RI	Separation according to functionality (end-OH-groups, zero, mono bifunctional) and topology (linear, cyclic).	Gorshkov AV, Evreinov VV, Entelis SG [101]
PDEGA CR: oligomers	Toluene / ethanol 90 / 10 vol.	Silica gel, TLC	TLC	Both mono and bifunctional oligomers elute at critical conditions but their retention is different.	Tikhonova TZ, Petrakova EA [102]
PDEGA Bifunctional hydroxy-containing CR: 0.3kD-2.4kD	MEK / hexane 84 / 16 vol.	Silica gel Si-100	24°C 1 ml/min 10 µl RI		Gorshkov AV, Evreinov VV, Entelis SG [103]
PDEGA CR: 560	MEK / chloroform 35 / 65 vol. MEK / hexane, 92 / 8 vol.	Silica gel Lichrospher 100°A, 10µm, 25×0.4cm	24°C 1 ml/min RI	Separation of zero, mono and bifunctional samples.	Filatova HH, Gorshkov AV, Evreinov VV, Entelis SG [104]

Sample	Eluent	Column	Conditions	Notes	References
	THF / ethylacetate 7 / 93 vol.				
	MEK / ethylacetate 27 / 73 vol.				
PBGA Poly(butylene glycol adipate) CR: 1kD-2kD	MEK / n-heptane 65 / 35 vol.	Lichrospher 100°A, 10 μm, 25×0.4cm	30°C 1 ml/min 10μl RI	Separation of zero, mono and bifunctional samples. Retention increases in order: R-R, R-OH, OH-OH.	Filatova HH, Minina EO, Gorshkov AV, Evreinov VV, Entelis SG [105]
	MEK / n-heptane 65 / 35 vol.	Zorbax-NH2, 80°A, 7μm, 25×0.4cm	30°C		
	MEK / n-heptane 70 / 30 vol.	Zorbax-Gold, 70°A, 3μm, 8×0.62cm	40°C		
Polyesters of adipic acid - 1,6- hexanediol CR: 1kD-3.9kD	Acetone / hexane 50.5 / 49.5 vol.	Silica gel Si-120, 7μm	RI	Separation by the end-groups functionality (cycles, ether, alkyl terminated chains). Two-dimensional characterization. Addition of hexane supports adsorption.	Kilz P, Krüger RP, Much H, Schulz G [106] Krüger RP, Much H, Schulz G [107].
Polyesters of adipic acid - 1,6- hexanediol CR: 1kD-6kD	Acetone / hexane 51 / 49 vol.	Nucleosil 50°A, 100°A, 300°A, 500°A (100°A optimal), 7μm, 25×0.4cm	303 K 20μl RI	Polyesters with different end-groups (OH, COOH, alkyl) and cycles separated. Two-dimensional separation.	Krüger RP, Much H, Schulz G [108]
Polyesters of adipic acid -	Acetone / hexane	Silica gel Separon SGX 120°A	35°C		Krüger RP, Much H, Schulz G [94]
1,4-butanediol	63 / 37 vol.				
1,3-butanediol	62 / 38 vol.				
1,2-propanediol	64 / 36 vol.				
1,3-propanediol	60 / 40 vol.				
1,2-ethanediol	70 / 30 vol.				
neopentyl glycol	53 / 47 vol.				
diethylene glycol	40 / 60 vol.				
dipropylene glycol	60 / 40 vol.				

Table 2. (continued)

Polymer Molar masses Critical range Used samples	Mobile phase Solvent / Nonsolvent Adsorli / Desorli	Stationary phase Particle diameter Pore size Column dimension	Conditions Temperature Flow rate Inj. volume Detector	Analytical application and / or notices	Investigators Reference
Polyesters of phthalic acid - ethylene glycol	Acetone / hexane 66 / 34 vol.	Spherisorb SAX 120°A	35°C		Krüger RP, Much H, Schulz G [94]
diethylene glycol	72.5 / 27.5 vol.	Spherisorb SAX 120°A			
triethylene glycol	100 / 0 vol.	Separon SGX 120°A			
PPOA Poly(olypropylene adipate) CR: 1100-3340	Benzene / ethanol near crit. 3 / 1 vol. Benzene / THF near crit. 1 / 1 vol.	Silica gel KSK, TLC	TLC	Separation of linear and branched polyesters.	Belenkii BG, Valchikhina MD, Vakhtina IA, Gankina ES, Tarakanov OG [3]
PPGA Poly(propylene glycol adipate)	Chloroform / MEK 90 / 10 vol.	Silica gel Lichrospher 100°A, 10μm, 25×0.4cm	24°C 1 ml/min 10 ul RI		Filatova HH, Gorshkov AV, Evreinov VV, Entelis SG [104]
PDEGA CR: 1.8kD	Chloroform / MEK 67 / 33 vol.			Separation of zero, mono and bifunctional polymers.	
Polyesters of terephthalic acid and phenolphtalein CR: 1kD-2kD	THF / hexane 70 / 30 vol.	Silica gel Silasorb 600, 6.2×0.2cm		Strong retention of bifunctional oligoesters.	Tennikova TB, Blagodatskikh IV, Svec F, Tennikov MB [109]

Polymer	Eluent	Column	Conditions	Remarks	References
	DCM / n-hexane 80 / 20 vol.	Cross-linked copolymer glycidyl methacrylate-ethylene dimethacrylate 6.2×0.2 cm		Mono and bifunctional (terminal hydroxy groups) oligomers in adsorption mode.	
	THF / n-hexane 70 / 30 vol.			Zero and bifunctional oligomers elute at different elution volumes independently on molar mass.	
	DCM / 2-propanol 99.5 / 0.5 vol.	Silica gel Silasorb 600, 6.2×0.2 cm		Mono and bifunctional oligomers in adsorption mode.	
PTHF Polytetrahydrofuran CR: 1kD	Ethylacetate / MEK 50 / 50 vol.	Lichrospher 100°A. 10 µm, 25×0.4 cm	24°C 1 ml/min RI	Separation of zero, mono and bifunctional (OH groups) polymers.	Filatova HH, Gorshkov AV, Evreinov VV, Entelis SG [104]
PTHF CR: 1kD – 283kD	THF / ACN 50 / 50 wt.	Nucleosil C18, 30+100 nm, 5+7 µm, 2×25×0.4 cm	30°C 20 µl ELSD	Separation of blends PMMA, PTHF.	Janco M, Berek D, Önen A., Fischer C.H, Yagci Y, Schnabel W. [53]
PTHF	Acetone / hexane 95 / 5 vol.	Nucleosil C18	RI	Analysis of trioxane-THF triblock copolymers.	Pasch H, Much H, Schulz G [52], Pasch H, Much H, Schulz G, Gorshkov AV [110]
PEO US: 0.2kD-20kD	ACN / water 58 / 42 vol.	Silica gel RP-C18, 25×0.46 cm	298K 1 ml/min 20 µl RI	PEO separated according to terminal groups -OH, -C_4H_9, -CH=CH_2. Separation of triblock copolymers PEO-PPO-PEO and copolymers PPO-PEO. Strong interaction of PEO with hydroxyl groups on RP column packing (peaks with elongated tails). Random copolymer PPO-PEO elute earlier than the block copolymer of the corresponding composition.	Gorshkov AV, Much H, Becker H, Pasch H, Evreinov VV, Entelis SG [111]

Table 2. (continued)

Polymer Molar masses Critical range Used samples	Mobile phase Solvent / Nonsolvent Adsorli / Desorli	Stationary phase Particle diameter Pore size Column dimension	Conditions Temperature Flow rate Inj. volume Detector	Analytical application and / or notices	Investigators Reference
PEO	ACN / water 70 / 30 vol.	Nucleosil 100 RP 18, 12.5×0.4cm	25°C 0.5 ml/min RI	Separation of fatty alcohol ethoxylate with respect to chain length as well as the end-groups.	Pasch H, Rode K [112]
PEO	ACN / water 42 / 58 vol.	Nucleosil 100 RP 18, 25×0.4cm	25°C 1 ml/min RI, MALDI-TOF-MS	Separation of PEO-PPO-PEO triblock copolymer according to degree polymerization of PPO block.	Pasch H, Rode K [112]
PEO CR: 0.6kD-20kD	ACM / water 46 / 54 vol.	Nucleosil C18, 25×0.4cm	30°C ELSD, MALDI-TOF-MS	Separation of polyethers according to end-groups (-H, -OH, -OCH$_3$, ..).	Krüger RP, Much H, Schulz G [107]
PEO CR: 0.2kD-10kD	ACN / water 43 /57 vol.	Nucleosil C18, 5µm, 25×0.4cm	0.5 ml/min 20µl RI	Separation of PEO-PPO-PEO triblock copolymers according to molar mass and block length. Two-dimensional separation.	Pasch H, Brinkmann C, Much H, Just U [95]
PEO CR: 0.3kD-7.6kD	ACN / water 46 / 54 vol.	Nucleosil C18, 6×0.4cm	25°C RI, UV	Analysis of alkoxy and aryloxy terminated PEO. Elution order with respect to terminal groups: OH(PEG)<C$_{10}$H$_{21}$<C$_{12}$H$_{25}$<C$_{13}$H$_{27}$<C$_{15}$H$_{31}$	Pasch H, Zammert I [113], Adrian J, Braun D, Pasch H [114]

Sample	Mobile phase	Column	Conditions	Comments	Reference
PEO US: 0.5kD-17.5kD	ACN / water 36.7 / 63.3 wt. D / A	Nucleosil C18, 300°A, 7µm, 12.5×0.4cm	25°C 1 ml/min 20 µl RI	α-hydroxyl PEO, ω-diphenyl PEO and α,ω-hydroxyl PEO eluted at different elution volume independently on their molar mass.	Baran K, Laugier S, Cramail H [68]
PEO	ACN / water 28.1 / 71.9 wt.	Nucleosil CN, 300°A, 7µm, 12.5×0.4cm		α-hydroxyl PEO, ω-diphenyl PEO and α,ω-hydroxyl PEO eluted at different elution volume independently on their molar mass.	
PEO US: 0.5kD-17.5kD	ACN / water 31.8 / 68.2 wt.	Nucleosil NH2, 300°A, 7µm, 12.5×0.4cm			
PEO	Methanol / water 86 / 14 vol.	Silica gel C18		Separation of PEO-PPO-PEO triblock copolymers. Two-dimensional separation.	Adrian J, Braun D, Pasch H [114]
PEO US: 0.5kD-17.5kD	Methanol / water 86 / 14 wt. A / D	Nucleosil NH2, 300°A, 7µm, 12.5×0.4cm	25°C 1 ml/min 20 µl RI		Baran K, Laugier S, Cramail H [68]
PEO	Methanol / water 85 / 15 wt. D / A	Nucleosil C18, 300°A, 7µm, 12.5×0.4cm			
PEO	Methanol / water 85 / 15 wt. A / D	Nucleosil CN, 300°A, 7µm, 12.5×0.4cm			

Table 2. (continued)

Polymer / Molar masses / Critical range / Used samples	Mobile phase / Solvent / Nonsolvent / Adsorli / Desorli	Stationary phase / Particle diameter / Pore size / Column dimension	Conditions / Temperature / Flow rate / Inj. volume / Detector	Analytical application and / or notices	Investigators Reference
PEO oligomers	Methanol / water 78.5 / 21.5 wt. S / S	Nucleosil C18, 300°A, 5µm, 25×0.4cm	45°C 0.5ml/min ELSD, RI, MALDI-TOF-MS	PEO-co-polymethylene separated according to groups $C_8 - C_{18}$.	Falkenhagen J, Friedrich JF, Schulz G, Krueger RP, Much H, Wiedner S [115]
PEO US: 0.5kD-17.5kD	1,2-dimethoxyethane / water 21.5 / 78.5 wt. D / A	Nucleosil CN, 300°A, 7µm, 12.5×0.4cm	25°C 1ml/min 20µl RI	α-hydroxyl PEO, ω-diphenyl PEO and α,ω-hydroxyl PEO separated from PEO.	Baran K, Laugier S, Cramail H [68]
PEO	1,2-dimethoxyethane / water 18.5 / 81.5 wt. D / A	Nucleosil C18, 300°A, 7µm, 12.5×0.4cm			
PEO CR: 1kD-4.5kD Macromonomers	THF / hexane 71 / 29 wt.	PS/DVB gel 30×0.78cm	30°C 1ml/min 20µl ELSD	PS-graft-PEO copolymers.	Murgasova R [116], Murgasova R, Capek I, Lathova E, Berek D, Florian D [117]
PEO CR: 300-600	Chloroform / pyridine 5 / 7 vol. (41.66 / 58.33 vol.)	Silica gel KSK for TLC	TLC		Belenkii BG, Valchikhina MD, Vakhtina IA, Gankina ES, Tarakanov OG [3]

Polymer	Eluent	Column	Conditions	Description	Reference
PEG CR: polymerization degree 2-75	Water / methanol 20 / 80 wt.	Silica gel Spherisorb ODS 2, 5µm, 25×0.46cm	0.5 ml/min 5µl RI, density detector	Characterization of ethoxylated fatty alcohols. Two-dimensional separation.	Trathnigg B, Thamer D, Yan X, Maier B, Holzbauer HR, Much H [118]
PEG	Water / methanol 10 / 90 wt.	Spherisorb ODS 2, 5µm, 80°A, 25×1 cm	2 ml/min RI, density detector	Characterization of ethoxylated fatty alcohols. Two-dimensional separation.	Trathnigg B, Kollroser M, Parth M, Röblreiter S [119], Trathnigg B, Kollroser M, Rappel C [120]
PEG CR: 0.2kD-4kD	Water / methanol 10 / 90 wt.	Spherisorb ODS 2, 5µm, 25×0.46cm	25°C 0.5 ml/min 50µl RI, density detector	PEG eluted from the second column Spherisorb ODS 2, 5um, 250×4.6mm in the eluent 10/90 wt. and 20/80 wt. in adsorption mode.	Trathnigg B, Thamer D, Yan X, Maier B, Holzbauer HR, Much H [118]
PEG CR: 0.2kD-4kD	Water / methanol 10 / 90 wt. near crit. 20 / 80 wt.	Spherisorb ODS 2, 3µm, 10×0.46cm	25°C 0.5 ml/min 50µl RI, density detector		
PEG CR: 0.2kD-4kD	Water / methanol near crit. 20 / 80 wt.	LiChrospher 100 C18, 5µm, 10×0.46cm	25°C 0.5 ml/min 50µl RI, density detector		
PEG US: 26kD-95kD	Water / ACN near crit. 58 / 42 vol.	Silica gel C18		TGIC	Lochmüller CH, Jiang C, Liu Q, Antonucci V, Elomaa M [121]
PEG	Acetonitrile / water 60 / 40 vol.	Silica gel Lichrosorb RP18, 10µm,	1 ml/min 10µl RI	Monomethyl-ethers of PEG (2–5kD) separated from PEG	Kazanskii KS, Lapienis G, Kuznetsova VI, Pakhomova LK, Evreinov VV, Penczek S [122]

Table 2. (continued)

Polymer Molar masses Critical range Used samples	Mobile phase Solvent / Nonsolvent Adsorli / Desorli	Stationary phase Particle diameter Pore size Column dimension	Conditions Temperature Flow rate Inj. volume Detector	Analytical application and / or notices	Investigators Reference
PEG CR: 1.5kD–8kD	Acetonitrile / water 60 / 40 vol.	Silica gel C18, 100°A, 25×0.46cm	68°C RI, UV, MAL- DI-TOF-MS	For diblock copolymers poly (L-lactide)-block-poly(ethylene oxide) retention depends on the total number of L-lactide units in the diblock copolymer.	Lee H, Lee W, Chang T, Choi S, Lee D, Ji H, Nonidez WK, Mays JW [123]
				Ln k of diblock copolymers depends linearly on poly (L-lactide) block length.	Lee H, Chang T, Lee D, Shim MS, Ji H, Nonidez WK, Mays JW [124]
	Acetonitrile / water 47 / 53 vol.		37°C	Poly(L-lactide)-block-poly (ethylene oxide)-block-poly (L-lactide) eluted according to number of L-lactide units in the triblock copolymer.	
	Acetonitrile / water 54 / 46 vol.		54°C	The retention of triblock copolymers depends on the number of L-lactide units. Peaks splitting due to isomerism of the poly (L-lactide) blocks at both ends of the PEO block.	
PEG CR: 0.5kD–40kD	Phosphate water buffer 0.05M, pH=5.8	Butyl-Toyopearl 650M, 45–90µm 12×0.4cm	43°C 0.2ml/min 10–30µl RI	Elution behavior of PEG is function of pH value of aqueous buffer.	Gorbunov AA, Solovyova LYa, Skvorcov AM [125]

Sample	Mobile phase	Stationary phase	Conditions	Comments	Reference
PEG CR: 0.3kD-40kD	Phosphate buffer 0.05M, pH=4.6, near crit.	Butyl-Toyopearl 650M, 45–90 μm 12×0.4 cm	40°C 0.2 ml/min 10–30 μl RI		
PEG CR: 0.06kD-3kD	Phosphate buffer 0.1M, pH=5.0, near crit.	SOLOZA K-33, 200 μm, 12×0.4 cm	1.5°C 0.2 ml/min 10–30 μl RI		
PEG CR: 0.2kD-2kD	Phosphate buffer 0.1M, pH=4.5 near crit.	SOLOZA K-0, 200 μm 12×0.4 cm	18.5°C 0.2 ml/min 10–30 μl RI		
PEG-monomethylether CR: polymerization degree 2-76	Water / methanol 20 / 80 vol.	Silica gel Spherisorb ODS 2, 5 μm, 25×0.46 cm	0.5 ml/min 50 μl RI, density detector	PEG, PEG-mono- and PEG-dimethylethers eluted at a little different elution volumes from the same column and eluent.	Trathnigg B, Thamer D, Yan X, Kinugasa S [126]
PEG-dimethylether CR: polymerization degree 2-76	Water / methanol 20 / 80 vol.				
PPG Polypropyleneglycol C 10kD-24kD	Phosphate buffer 0.1M, pH=5.0	SOLOZA KG-8, 200 μm, 12×0.4 cm	5°C 0.2 ml/min 10–30 μl RI	Elution behavior of PPG depends on temperature and pH value.	Gorbunov AA, Solovyova LYa, Skvorcov AM [125]
PPG US: 0.4kD-10kD	Hexane / isopropanol 92 / 8 vol.	Silica gel Nucleosil NH$_2$, 5 μm, 20×0.4 cm	ELSD	Analysis of the functional endgroups for PPO	Keil C, Esser E, Pasch H [127]

Table 2. (continued)

Polymer / Molar masses / Critical range / Used samples	Mobile phase / Solvent / Nonsolvent / Adsorli / Desorli	Stationary phase / Particle diameter / Pore size / Column dimension	Conditions / Temperature / Flow rate / Inj. volume / Detector	Analytical application and / or notices	Investigators Reference
PPG with 2 terminal hydroxy-groups CR: 0.4kD–4kD	MEK / ethylacetate 7 / 93 vol.	Silica gel LiChrosorb Si 100, 10µm, 25×0.46cm	30°C 20µl	Fractionation according to functionality. PPG eluted in pure ethylacetate in adsorption mode.	Gorshkov AV, Evreinov VV, Lausecker B, Pasch H, Becker H, Wagner G [128]
PPG with 2 terminal hydroxy-groups CR: 0.4kD–4kD	MEK / ethylacetate 5 / 95 vol.	Silica gel S-100	1 ml/min 10µl	Separation of linear and branched hydroxylated PPG according to functionality.	Gorshkov AV, Evreinov VV, Entelis SG [129]
PPG with 2 terminal hydroxy-groups CR: 0.4kD–4kD	MEK / ethylacetate 10 / 90 vol.	Silica gel Lichrospher 100°A, 10µm, 25×0.4cm	30°C 1 ml/min RI	Separation of linear and branched PPG according to functionality.	Filatova HH, Gorshkov AV, Evreinov VV, Entelis SG [104]
POPG Poly(oxypropylene glycol) CR: 0.3kD–4.5kD	Heptane / THF 55 / 45 vol. for zero functionality; 67 / 33 vol. for bi and tri-functional oligomers	Silica gel, TLC	TLC	Determination of functionality. Bi and tri-functional oligomers elute at critical conditions, but their retention is different.	Tikhonova TZ, Petrakova EA [102]
Polyoxyphemylene CR: 30kD	Chloroform / tetrachloromethane 71 / 29 vol.	Silica gel Lichrospher 300	30°C 0.5 ml/min 10µl UV	Separation of blend with PS.	Dudorina AV, Gorshkov AV, Filatova HH, Evreinov VV, Entelis SG [90]

Sample	Mobile phase	Stationary phase	Conditions	Purpose	Reference
PBTF Polybutyleneteftalate	Heptane / THF 65 / 35 vol.	Lichrospher 60°A: 10 μm, 25×0.4cm	24°C 1ml/min 10μ UV	Separation of zero, mono and bifunctional (OH groups) polymers. Heptane supports adsorption.	Gorshkov AV, Prudskova TN, Filatova HH, Evreinov VV [100], Gorshkov AV, Prudskova TN, Guryanova VV, Evreinov VV [130]
PBTF	Heptane / chloroform 10 / 90 vol.	Lichrospher 60°A, 10 μm, 25×0.4cm	24°C 1ml/min 10μ UV	Separation of zero, mono and bifunctional (OH groups) polymer. In heptane / chloroform bifunctional macromolecules are fully retained.	Gorshkov AV, Prudskova TN, Filatova HH, Evreinov VV [100]
PBTF CR: Oligomers	THF / n-heptane 65 / 36 vol.	Silica gel Si-60	1ml/min 10μl UV	Separation of zero, mono-, bi-functional homologues.	Guryanova VV, Pavlov AV [131]
Polycarbonate linear CR: oligomers	Chloroform / tetrachloromethane 17 / 83 vol.	Silasorb 600, D<100°A, 25×0.4cm,	1ml/min 10μl UV	Separation according to therminal chloroanhydride (COCl) and hydroxyl groups.	Gorshkov AV, Prudskova TN, Guryanova VV, Evreinov VV [132]
Polycarbonate	Chloroform / tetrachloromethane 3/70 vol.	Lichrospher Si 500, 500°A			
Polycarbonate CR: oligomers	Chloroform / tetrachloromethane 30 / 70 vol.	Silica gel Si-60	1ml/min 10μl UV	Separation of oligocarbonates with –H, –OH and –COCl end-groups.	Guryanova VV, Pavlov AV [131]
Polycarbonate CR: oligomers	Toluene / ethylacetate 60 / 40 vol.	Silica gel, TLC	TLC	Separation according to functionality (terminal hydroxygroups).	Tikhonova TZ, Petrakova EA [102]
Polycarbonate, linear, aliphatic CR: 0.5kD	Toluene / ethylacetate 80 / 20 vol.	Silica gel, TLC	TLC	Determination of monofunctional impurities.	Petrakova EA, Okuneva AG, Balkova EV [133]
	Toluene / MEK 7 / 3 vol.				

Table 2. (continued)

Polymer Molar masses Critical range Used samples	Mobile phase Solvent / Nonsolvent Adsorli / Desorli	Stationary phase Particle diameter Pore size Column dimension	Conditions Temperature Flow rate Inj. volume Detector	Analytical application and / or notices	Investigators Reference
Epoxy resins linear macromolecules CR: 0.36-1.91kD	Chloroform / more polar solvent (not specified in the both refs.)	Silica gel Zorbax, 8.2cm	32°C 0.5ml/min UV	Separation according to end-groups (14 functionality types resolved).	Gorshkov AV, Verenich SS, Evreinov VV, Entelis SG [134], Gorshkov AV, Verenich SS, Markevic MA, Petinov VI, Evreinov VV, Entelis SG [135]
bisglycidyl bisphenol A CR: 0.8kD-5kD	THF / n-hexane 74 / 26 vol.	Silica gel Nucleosil 50-5, 50°A, 5µm, 20×0.4cm	UV, ELSD	Separation of linear and branched oligomers (in order doubly branched, singly branched, linear oligomers) and . fully epoxydized, monodiol and bisdiol oligomers.	Adrian J, Braun D, Rode K, Pasch H [136], Adrian J, Braun D, Pasch H [137], Pasch H, Adrian J, Braun D [138]
Oligo(epichlorohydrin) CR: 0.9kD-1.8kD	THF / n-heptane 45 / 55 vol. acetone / n-heptane 42 / 58 vol.	Silica gel Silasorb SPH 600, 60°A, 7µ 3×(25×0.4cm)	303K RI	Separation of mono(-OH) and difunctional oligomers and cyclic products. Analyses with acetone-heptane were less informative.	Solomko SI, Kuzaev AI [139] Kuzaev AI, Olkhova OM, Solomko SI, Baturin SM [140]
Oligo(epichloro-hydrin) US: 0.5kD-3.5kD	DMFA / benzene 50 / 50 vol. MEK / n-heptane 90 / 10 vol.	Silica gel Separon SGX, 100°A, 7µm, 2×(15×0.33cm)	295K 0.2ml/ min RI	Macromolecules of seven types functionalities were separated.	Bektashi NR, Alieva DN, Dzhalilov RA, Ragimov AV [141]

Polymer	Solvent	Column	Conditions	Description	Reference
Calixarene without substituents	THF / water 80 / 20 wt.	Silica gel YMC C18, 120°A, 5 µm, 25×0.46cm	0.5 ml/min 10 µl ELSD, MALDI-TOF-MS	Calixarenes with substituents eluted in adsorption mode. Addition of 0.1 % trifluoroacetic acid to the mobile phase improves shape of peaks.	Falkenhagen J, Krüger RP, Schulz G, Gloede J [142]
p-methylcalixarene	THF / water 87 / 13 wt.				
p-tert-butyl-calixarene	THF / water 88 / 12 wt.			Calixarenes without substituents and p-methylcalixarenes eluted in SEC mode, p-octylcalixarenes in adsorption mode.	
p-octylcalixarene	THF / water 91.5 / 8.5 wt.			Calixarenes with substituents eluted in SEC mode.	
Polysulphone CR: oligomers	Coroform / tetrachloromethane 53 / 47 vol.	Silica gel Si-60	1 ml/min 10 µl UV	Separation of zero-, mono-, bi-functional homologues with –OH and –Cl end-groups.	Guryanova VV, Pavlov AV [131]
PVP Polyvinylpyrrolidone CR: 10kD-25kD	Phosphate buffer 0.1M, pH=6.0	SOLOZA K-33, 2 µm, 12×0.4 cm	22°C 0.2 ml/min 10–30 µl RI		Gorbunov AA, Solovyova LYa, Skvorcov AM [125]
Poly-2-vinylpyridine CR: 6kD-70kD	THF Near crit.	µ-Bondagel (silica with layer of polyether bonded onto surface)	1 ml/min 50 µl	PS elute from the column in SEC mode.	Mencer HJ, Grubisic-Gallot Z [143]
PDMS Polydimethyl-siloxane CR: 1kD	Toluene / isooctane 27 / 73 vol.	Nucleosil 100°A, 5 µm, 20×0.4 cm	0.5 ml/min RI	Linear, cyclic, OH-terminated chains separated.	Krüger RP, Much H, Schulz G, Rikowski E [144]

Table 2. (continued)

Polymer Molar masses Critical range Used samples	Mobile phase Solvent / Nonsolvent Adsorli / Desorli	Stationary phase Particle diameter Pore size Column dimension	Conditions Temperature Flow rate Inj. volume Detector	Analytical application and / or notices	Investigators Reference
Kaprolakton-diol CR: 0.5kD-2kD	Hexane / acetone 50 / 50 vol.	Silica gel LiChrospher Si-100, 100°A	1 ml/min, 5 µl RI	Fractionation according to number and type of terminal hydroxyl groups.	Gorshkov AV, Overim T, van Aalten Ch, Evreinov VV [145]
Kaprolakton-diol CR: 0.5kD-2kD	Diethylether / acetone 78 / 22 vol.	Zorbax NH$_2$, 70°A	1 ml/min, 20 µl RI	Fractionation according to terminal hydroxy-groups. Two-functional PFS are irreversibly adsorbed.	
PFS Polyphenylensulfone CR: polymerization degree 1-15; M<5kD	Chloroform / heptane 65 / 35 vol. S / N	Silica gel Silasorb 600, 60°A, 7 µm	1 ml/min 10 µl UV	Separation according to end-OH-groups (zero and mono functional). Two-functional PFS eluted from the column additionally with the eluent chloroform / methanol, 98 / 2 vol. Decreased polymer recovery for M>10kD [146].	Prudskova TN, Guryanova VV, Gorshkov AV, Evreinov VV, Pavlov AV [147]
PFS CR: polymerization degree 1-15	Chloroform / tetrachloromethane 53 / 47 vol. S / N			Separation according to end-OH-groups (zero and mono functional).	

PFS US: 15kD-60kD	Chloroform / tetra-chloromethane 50 / 50 vol. S / N	Silica gel Silasorb 600, 10 μm, 50-80°A, 25×0.46cm	10 μl UV	Functionality-type distribution of end groups in macromolecules.	Guryanova VV, Prudskova TN [146]
Polylactides US: 0.2kD-100kD	n-hexane / 1,4-dioxane 40 / 60 vol. S / S	Silica gel Nucleosil 100, 25×0.4cm	30°C DAD MALDI-TOF-MS	Analysis of original and thermally threated samples. Hexane-poor solvent;dioxane-good solvent.	Krüger RP, Much H, Schulz G [94], Wachsen O, Reichert KH, Krüger RP, Much H, Schulz G [148]

Table 3. Liquid Chromatography Systems under Limiting Conditions (Polymers Dissolved and Injected Only in One Component of the Mixed Mobile Phase)

Polymer	Mobile phase	Stationary phase Particle diameter Pore size Column dimensions	Conditions Temperature Flow rate Inj. volume Inj. conc. Detector	Analytical application and/or comments	Investigators Reference
PS US: 1.4kD–427kD Inj. in 2,2,4-trichlorobenzene	Solvent / Nonsolvent Adsorli / Desorli dimethylsulfoxide	PS/DVB gel Pl-Mixed A, 20μm, 25×0.8 cm	140°C 305 μl 1 ml/min ELSD	LC LCS	Macko T, Aust N, Lederer K [149]
PS CR: 0.37kD–500kD US: 0.37kD–400kD Inj. in THF	THF / n-hexane 50 / 50 vol. S / N	Silica gel SGX 1000, 100nm, 5μm, 30×0.8 cm	22°C 1 ml/min 20 μl 1 mg/ml UV	LC LCA	Bartkowiak A, Murgasova R, Janco M, Berek D, Spychaj T, Hunkeler D [150]
PS CR: 0.37kD–4100kD Inj. in THF	THF / n-hexane 36 / 64 vol. S / N	Silica gel SGX 500, 50nm, 5μm, 30×0.8 cm		LC LCA Molar mass independent elution found for PS 900kD, PS 4000 kD above exclusion limit of the column.	
PS CR: 0.37kD–10kD US: 0.37kD–1400kD Inj. in THF	THF / n-hexane 27 / 73 vol. S / N	PS/DVB, Shodex, 10μm, 30×0.8 cm		Hybrid LC LCA-LCS Separation of random copolymers PS-PMMA. At above 73 vol. % n-hexane is the eluent a nonsolvent for PS with $M_w > 150$ kD.	

Sample	Mobile phase	Column	Conditions	Mode / Notes	Reference
PS CR: 10kD–800kD Inj. in toluene	Toluene / methanol 68 / 32 vol. S / N	Silica gel, 80 nm, 10 μm, 25×0.6 cm	1 ml/min 10 μl 1 mg/ml RI		Hunkeler D, Macko T, Berek D [26]
PS Inj. in THF	THF / water near. crit. 64 / 36 wt. S / N	Silica gel, 100 nm, 10 μm, 25×0.6 cm	10 μl 1 mg/ml or 0.5 mg/ml RI, ELSD	LC LCS	Hunkeler D, Janco M, Gury-anova VV, Berek D [27]
PS CR: 0.9kD–12kD US: 0.9kD–1000kD Inj. in toluene	Toluene / methanol near crit. 50 / 50 wt. S / N			LC LCS PS with $M_w > 100$kD do not leave silica gel, when eluent contains above 30 wt. % of methanol.	Macko T [151]
PS CR: 1kD–1000kD Inj. in DCM	DCM / cyclohexane 49 / 51 vol. S / N	Silica gel SGX 500, 10 μm, 25×0.8 cm	30°C 1 ml/min 20 μl 1 mg/ml UV	LC LCS Cyclohexane dissolves PS at temperature above 35°C.	
PMMA CR: 15kD–800kD Inj. in toluene	Toluene / methanol near crit. 27 / 73 vol. S / N, A / D	Silica gel, 80 nm, 10 μm, 25×0.6 cm	10 μl 1 mg/ml RI		Hunkeler D, Macko T, Berek D [26]
PMMA CR: 15kD–300kD US: 15kD–700kD Inj. in toluene	Toluene / methanol near crit. 20 / 80 wt.	Silica gel, 100 nm, 10 μm, 25×0.6 cm	10 μl 0.5 or 1 mg/ml RI, ELSD	LC LCS For PMMA with $M_w > 200$kD recovery problem.	Hunkeler D, Janco M, Gury-anova VV, Berek D [27]
PMMA low stereoregular CR: 17kD–613kD US: 17kD–613kD Inj. in THF	THF / n-hexane 81 / 19 wt. S / N	Silica gel SGX 1000, 100 nm, 10 μm, 25×0.6 cm	20 μl 0.5 mg/ml ELSD	LC LCS Characterization tacticity of PMMA. Less soluble syndiotactic PMMA move with its initial solvents while better soluble isotactic PMMA elute in SEC mode.	Berek D, Janco M, Hatada K, Kitayama T, Fujimoto N [54]

Table 3. (continued)

Polymer Molar masses Critical range Used samples Injected in	Mobile phase Solvent / Nonsolvent Adsorli / Desorli	Stationary phase Particle diameter Pore size Column dimensions	Conditions Temperature Flow rate Inj. volume Inj. conc. Detector	Analytical application and/or comments	Investigators Reference
PMMA syndiotactic CR: 12.7kD-153 kD US: 12.7kD-153kD Inj. in THF	THF / n-hexane near crit. 82 / 18 wt.	Silica gel SGX 1000, 100nm, 10μm, 25×0.6cm			
PMMA CR: 6kD-350 kD Inj. in THF	THF / n-hexane 74 / 26 vol. S / N	Silica gel SGX 500, 10μm, 25×0.6cm	25°C 1.5 ml/min 1 mg/ml 20 μl UV	LC LCA Separation of random copolymers PS-PMMA.	Bartkowiak A, Hunkeler D [152]; Sauzedde F, Hunkeler D [153]
PMMA Atactic CR: 6kD-1000kD Inj. in THF	THF / n-hexane 58 / 42 vol.	PS/DVB, Shodex, 10μm, 30×0.8cm	25°C 1.5 ml/min 1 mg/ml 20 μl UV	LC LCS Separation of random copolymers PS-PMMA. Polymer is not soluble in the eluent.	Bartkowiak A, Hunkeler D [154], Bartkowiak A, Hunkeler D [152]
PMMA CR: 15kD-620kD Inj. in toluene	THF / toluene 50 / 50 wt. THF stabilized with 0.1wt. % butylated p-cresol S / S, D / A	Silica gel Biospher Si-120, 12nm, 10 μm, Separon SGX 1000, 100nm, 10μm, 25×0.4cm	30°C 1 ml/min 50 μl RI, ELSD	LC LCD PMMA with higher molar mass have limited recovery, if the mixed eluent is used as the sample solvent.	Berek D [57]

Liquid Chromatography under Critical and Limiting Conditions: A Survey of Experimental...

Sample	Eluent	Column	Conditions	Mode / Comments	Reference
PMMA CR: 15kD-620kD; Inj. in THF	THF / toluene 35 / 65 wt. THF stabilized with 0.1wt. % butylated p-cresol S / S, D / A			LC LCA	
PMMA CR: 16kD-613kD Inj. in THF	THF / toluene 35-31 / 65-69 wt. S / S, D / A	Silica gel, 10 nm 30×0.78cm	30°C 1 ml/min 50 µl RI, ELSD	LC LCA PMMA is fully retained by silica gel in pure toluene. Above 66 wt. % toluene limited recovery of the polymer.	Berek D, Hunkeler D [155]
PMMA CR: 16kD-613kD Inj. in THF	THF / Toluene near crit. 0 / 100 vol. S / S, D / A	PS/DVB 25×0.8cm	30°C 50 µl 1 ml/min RI, ELSD	LC LCA This result is possible obtain only with very interactive column packing.	Berek D [156]
PMMA CR: 16kD-613kD Inj. In THF	THF / chloroform 93-97 / 7-3 wt. S / S, D / A	Silica gel, 10nm 30×0.78cm	30°C 1 ml/min 50 µl RI, ELSD	LC LCA Separation of PS from PMMA. Peak splitting at compositions over 95 wt. % of chloroform. PMMA is fully retained by silica gel in chloroform.	Berek D, Hunkeler D [155]
PMMA US: 16kD-613kD Inj. In THF	THF / water Near crit. 64 / 36 wt.	Silica gel, 10um, 100 nm	10 µl 0.5 or 1 mg/ml RI, ELSD		Hunkeler D, Janco M, Guryanova VV, Berek D [27]
PMMA US: 16kD-613kD Inj. In chloroform	Toluene / chloroform (1%ethanol as stabilizer) 65 / 35 wt. S / S, A / D	PS/DVB 30×0.8cm	30°C 1 ml/min 50 µl RI, ELSD	LC LCA Separation of PS from PMMA. Peak splitting without plausible explanation. PMMA is fully retained within the column when eluent is toluene.	Berek D, Hunkeler D [155]

Table 3. (continued)

Polymer	Mobile phase	Stationary phase Particle diameter Pore size Column dimensions	Conditions Temperature Flow rate Inj. volume Inj. conc. Detector	Analytical application and/or comments	Investigators Reference
PMMA US: 16kD–613kD Inj. In chloroform	Toluene / chloroform 90 / 10 wt. S / S	PS/DVB 25×0.8 cm	30°C 50 µl 1 ml/min RI, ELSD		Berek D [156]
PAM Polyacrylamide CR: 7.9kD–197kD US: 7.9kD–725kD Inj. in water with Na_2SO_4	Deionized water with 0.02M Na_2SO_4 / methanol 42 / 58 vol.	Polyhydroxy-methacrylate sorbent Shodex OHpak 300×0.8 cm		Hybrid LC LCA-LCS PAM with $M_w > 20kD$ are insoluble in the eluent and have limited recovery from the column.	Bartkowiak A, Hunkeler D, Berek D, Spychaj T [157]
Polyethylene US: 53kD–500kD Inj. in 2,2,4-tri-chlorobenzene	2,2,4-trichloro-benzene / dimethylsulfoxide 50 / 50 vol. S / N	PS/DVB gel PL-Mixed A, 20µm, 25×0.8 cm	140°C 305 µl 1 ml/min ELSD	LC LCS	Macko T, Aust N, Lederer K [149]
Polypropylene US: 150kD–800kD Inj. in 2,2,4-tri-chlorobenzene				LC LCS	
PI CR: 2kD–900kD Inj. in DCM	DCM / ACN 60 / 40 vol.	Silica gel C18, 5 µm, 25×0.4 cm	30°C 0.5 ml/min 1 mg/ml 20 µl UV	LC LCS PS injected in DCM eluted in SEC mode, PB injected in DCM eluted in adsorption mode.	Macko T [151]

Sample	Mobile phase	Column	Conditions	Notes	Reference
PEG CR: 4.4kD-23kD US: 0.1kD-23kD; Inj. in water	Water / methanol 0 / 100 vol.	Polyhydroxy-metha-crylate sorbent Shodex OH pak, 300×0.8 cm		Between 4.4kD-23kD molar mass independed retention, under 4.4kD SEC mode.	Bartkowiak A, Hunkeler D, Berek D, Spychaj T [157]
PEO CR: 5kD-100kD US: 5kD-825kD Inj. in water	Water / methanol 3 / 97 vol. S / S	Polyhydroxy-metha-crylate sorbent Shodex OH pak, 300×0.8 cm		LC LCA PEO soluble in both methanol (at higher temperature) and water. PEO with M_w>12kD have limited recovery.	Bartkowiak A, Hunkeler D, Berek D, Spychaj T [157]
PVAc Polyvinylacetate Inj. in THF	THF / water 64 / 36 wt. S / N	Silica gel, 100 r.m, 10μm			Hunkeler D, Janco M, Berek D [158]
PVAc Inj. in THF	THF / hexane 80 / 20 wt. S / N	Silica gel, 100 nm, 10μm			Hunkeler D, Janco M, Berek D [158]
Polyester of adipic acid and dipropoxylated bisphenol with two alcohol end groups CR: oligomers Inj. in DCM	THF / n-heptane, bi-49 /51 vol.	Silica gel Nucleosil, 100A, 5μm, 200×0.46 cm	35°C 2.5 ml/min 10 μl UV	Accordings to authors [159], the solvent peak plays no role in this system and the system corresponds to critical conditions.	Philipsen HJA, Claessens HA, Jandera P, Bosman M, Klumperman B [159]
	THF / n-heptane, near crit. 70 / 30 vol.	Jordi gel DVB Polyamine, 500°A, 5 μm, 25×0.46 cm			
Pollulan CR: 5.8kD-21kD US: 5.8kD-1600kD Inj. in deionized water 30 / 70 vol. with Na_2SO_4	Deionized water with 0.02M Na_2SO_4 / methanol	Polyhydroxy-metha-crylate sorbent Shodex OHpak 300×0.8 cm		Hybrid mechanism For M_w>25kD limited recovery of the polymer.	Bartkowiak A, Hunkeler D, Berek D, Spychaj T [157]

3
Correlations

In the general mobile phase-polymer-sorbent system, mutual interactions between all components are possible. Theoretical calculations for model pores with attractive or repulsive interactions between model chains and pore walls describe, qualitatively, the LC separation of macromolecules, to a reasonable extent [12,14,159–162]. However, the quantitative description, allowing the prediction of CEEC in a real LC system, is not yet available. Selected relationships between the aforementioned components are described with experimental parameters such as a solubility parameter, solvent strength, polarity index, radius of gyration or Mark-Houwink constants [163–165].

3.1
Polymer – Sorbent versus Solubility Parameter of Mobile Phase

The solubility parameter (σ) reflects interactions between the solvent and the polymer in question. Usually, the polymer is dissolved more easily when the solubility parameter of a polymer and solvent are similar ($\sigma(\text{solvent}) - \sigma(\text{polymer}) \leq 2$). Solubility parameters are based on experimental data and are tabulated for many of solvents and polymers [163,164]. Philipsen et al. [79] and Baran et al.[70] have observed, after studying the critical conditions with few different solvent mixtures, that the solubility parameters were close, indicating that the critical chromatography mode could occur at a critical solubility parameter of the eluent. In Fig. 2A,B,C the dependence between solubility parameters of eluents at critical conditions on various type of sorbents are shown.

As is revealed in Fig. 2, data for both a sorbent and a polymer are concentrated in some regions of σ values. It is hypothesized that sorbents of the same type (for example silica gels) have different adsorptive properties and/or they were used at different temperatures. This causes variations in the critical compositions of the mobile phases and, thus, a dispersion of the σ values corresponding to one sorbent type in Fig. 2. Taking this into account, it may be concluded that values of solubility parameters corresponding to a pair polymer-sorbent are, in the majority of cas-

Fig. 2. A Correlation between the solubility parameters of the mobile phase and the type of sorbent for polystyrene under critical conditions. The solubility parameter of PS is also shown. Data inside the figure refer to the number of LC systems (i.e. various mobile phases for the same sorbent and the polymer) plotted. Symbols: PS (squares), silica gel (circles), silica gel C18 (triangles), silica gel NH2 (inverted triangles), PS/DVB gel (diamonds). **B** Correlation between the solubility parameters of the mobile phase and the type of sorbent for some commodity polymers under critical conditions. Symbols: Polymer (squares), silica gel (circles), silica gel C18 (triangles), silica gel NH2 (inverted triangles). **C** Correlation between the solubility parameter of the mobile phase and the type of sorbent for some specialty polymers under critical conditions. Symbols: polymer (squares), silica gel (circles), silica gel C18 (triangles) ▶

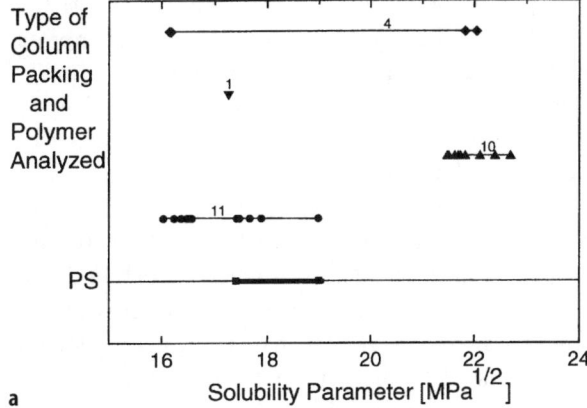

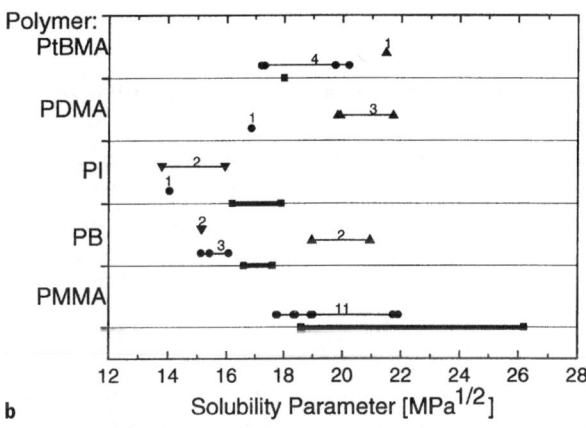

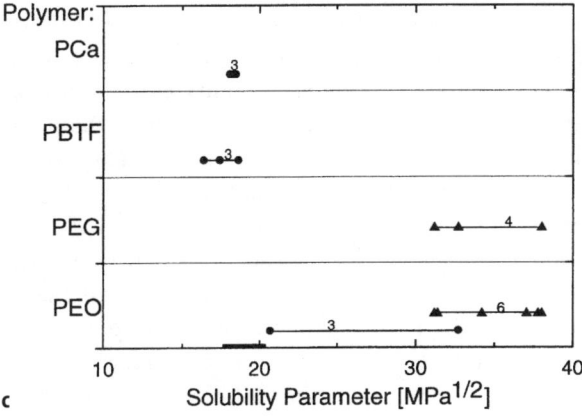

es, concentrated in a narrow range of the σ values (average value of σ±3 MPa$^{1/2}$ for one pair polymer-sorbent in Tab 2A,B; σ±6 MPa$^{1/2}$ in Tab 2C). Furthermore, the σ values corresponding to silica gels and silica gels NH2 are always smaller compared to the σ value corresponding to silica gels C18. For PB, and partially for PS, PMMA, the σ values for silica gel are even smaller then the σ value of corresponding polymer. The σ value of mobile phases corresponding to systems with silica gel C18 are much larger as the σ value for PS, PB. It may be concluded, therefore, that PS and PB require poorer mobile phases for reaching CEEC with silica gel C18 then with both silica gel and silica gel NH2. This reflects situation that in majority cases, mixtures good solvent/precipitant are used for obtaining CEEC.

Interestingly, for PI, i.e. a nonpolar polymer, CEECs have never been identified using silica gel C18, i.e. nonpolar sorbent. Similarly, for polar polymer PEG, CEEC with polar sorbent such as silica gel were never published. The authors hypothesize that the adsorption of these polymers on a sorbent with similar polarity is too strong.

3.2
Polymer-Sorbent contra Elution Strength and the Polarity Index of Mobile Phase

The extent of interactions between silica gel and solvent is traditionally quantified with elution strength (ε) [165], while the polarity index (P)' of an mixed eluent represents its ability to interact by dispersive forces, dipole-dipole interactions and hydrogen bonds with different molecules. A high value of P' signifies a solvent which strongly interacts with others molecules.

The dependence of both elution strength and polarity index (both parameters were calculated according to the procedure given in [166,167] on solubility parameter is shown in Fig. 3a,b for LC systems corresponding to PS, PMMA. The ε values are less dissipated as P' values, although no clear correlation is observed. An explanation for dispersion of points in Fig. 3 could be that values of ε were derived for a standard silica surface, while adsorptive properties of silica gels may differ. Small differences in adsorptive properties may have larger impact on retention under CEEC as under normal conditions [56]. Possibly, other parameters, such as number of silanol groups on the corresponding silica gels could yield better correlation. In the case of silica gels C18, additionally, number of alkyl groups should be measured. Unfortunately, the number of silanol groups or carbon coverage is still not a part of information delivered by manufacturers of sorbents.

In Fig. 4 the dependence between ε of pure component of binary eluent and volume ratio of the second component is shown. The data are dissipated around the line, although Baran et al. [67] has found perfect linear dependence using the same column packing (3 points plotted in Fig. 4). Again, difference in adsorptive properties of silica gels (and/or measurements at different temperatures) may be reason for dissipation of points. Such linear dependence can be useful at forecasting of

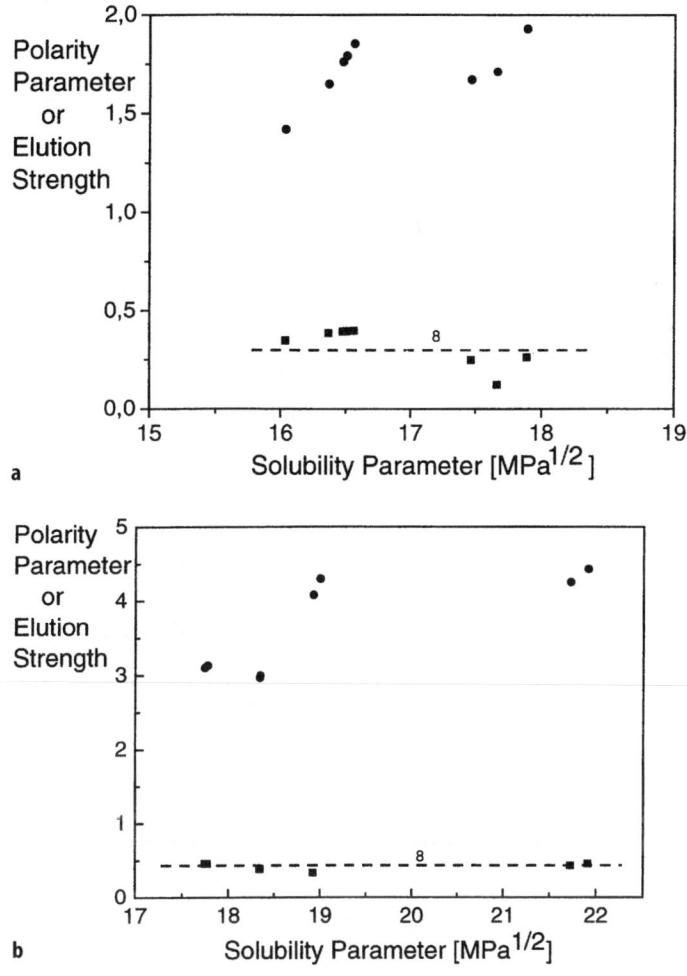

Fig. 3. a The dependence of both elution strength and polarity index of mobile phase on solubility parameter for PS under critical conditions. Stationary phases: Silica gels. Symbols: Elution strength (squares), polarity parameter (circles). **b** The dependence of both elution strength and polarity index on the solubility parameter for PMMA under critical conditions. Stationary phases: Silica gels. Symbols: Elution strength (squares), polarity parameter (circles)

composition of binary eluents for CEEC. For ensuring the linearity of the plot, the same silica gel column and temperature should be used.

3.3
Polymer-Sorbent versus Thermodynamic Quality of Mobile Phase

The thermodynamic quality of solvents for a polymer is represented by values of the Mark-Houwink constants. Constants a and K, included in Table 4, were deter-

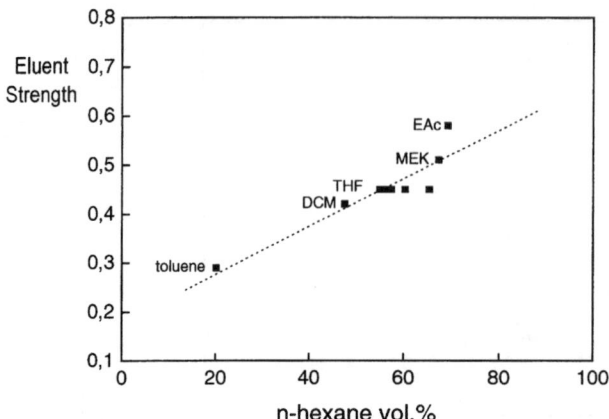

Fig. 4. Dependence elution strength of the first component of binary eluent on the volume ratio of the second component (n-hexane) of the mixed eluent. Stationary phases: Silica gel

Table 4. Mark-Houwink Constants under Critical Conditions

Mark-Houwink constants a	log K	Polymer	Mobile phase	Sorbent Temperature	Ref.
0.71		PS	Toluene / n-hexane 83.9 / 16.1 wt.	Silica gel 25°C	[67]
0.77			DCM / n-hexane, 69.0 / 31.0 wt.		
0.53			THF / n-hexane 47.0 / 53.0 wt.		
a<0.5			Ethylacetate / n-hexane 37.7 / 62.3 wt.		
a<0.5			MEK / n-hexane 37.1 / 62.9 wt.		
0.51			THF / n-hexane 50.0 / 50.0 wt.	Silica gel 3°C	
0.50			THF / n-hexane 49.0 / 51.0 wt.	13°C	
0.51			THF / n-hexane 47.0 / 53.0 wt.	36°C	
0.567	−3.388	PDMA	THF / ACN 78.5 / 21.5 vol.	Silica gel C18	[59]
0.690	−3.879	PnBMA	THF / ACN 53.1 / 46.9 vol.		
0.725	−4.136	PtBMA	THF / ACN 49.6 / 50.4 vol.		
0.503	−2.972	PS	THF / ACN 49.4 / 50.6 vol.		
0.663	−1.827	P2VP	THF (near critical conditions)	μ-Bondagel	[143]

mined by off-line measurement [67], or by employing viscosimetric detector [59,143]. As shown in Table 4, the critical conditions phenomenon is not related to a thermodynamic quality of the mobile phase. The Mark-Houwink constants take on different values depending on the nature of the mobile phase [67].

Similarly, the hydrodynamic size of polymer coils, expressed as the radius of gyration under the critical conditions, also depends on the eluent nature [67]. Thus, the critical conditions can not be related to a specific molecular conformation of PS and the hydrodynamic dimensions of macromolecules with the same molar mass are different in various mixed mobile phases corresponding to CEEC.

3.4
Upper Molar Mass Limit versus Thermodynamic Quality of Mobile Phase

The difference between the applied molar mass range and critical molar mass range (symbols US and CR in Table 1–3) reflects separation difficulties. For example, it has often been reported that the polymer was partially, or fully, retained within the column. Several authors also indicate a lack of solubility in the eluent or a molar mass independent retention observed only over a limited range of molar masses. In several published works, it is also stated that separation range at critical conditions is limited to 100 kD. However, as shown data in Table 5, polymer samples with much higher molar mass are eluted under critical and limiting conditions.

These results indicate that critical conditions have potential to separate homopolymers at least up to 2000 kD. The highest molar mass limit was reached in two thermodynamically good solvents of different polarity. One of the solvents

Table 5. The highest Reported Molar Masses of Polymers separated under Critical Conditions

Polymer and Highest Molar Mass	Mobile Phase	Thermodynamic Quality of Solvents	Column Packing	Ref.
PS up to 2145 kD	Chloroform / carbon Tetrachloride	polar solvent / non polar solvent	Porous glass	[7]
PS up to 1200 kD	THF / n-hexane	polar solvent / non polar nonsolvent	Silica gel	[70]
PS up to 1000 kD	THF / n-hexane			[28]
Chlorinated PE up to 520 kD	Chloroform / n-hexane			[99]
PMMA up to 1000 kD inj. in THF	THF / n-hexane		PS/DVB gel	[154]
PMMA up to 400 kD	Chloroform / methanol	polar solvent / polar nonsolvent	Silica gel	[45]

supports adsorption of the analyzed polymer. Such pairs of solvents have been, to date, rarely selected by investigators, although here seems to be an opportunity for the reduction of the problems associated with the limited recovery and increase of the molar mass separation range under critical conditions. The second option for an increase of the separation range may represent use of a thermodynamically good solvent for injection of the analyzed polymer. Interestingly, under limiting conditions of solubility sometimes high upper molar mass limit may be obtained, although the mobile phase is nonsolvent for the polymer. For example, Bartkowiak et al. [154] has found elution PMMA 1000kD injected in THF into mobile phase THF/n-hexane, 58/42 vol. (Table 5), while PMMA was soluble in the mobile phase only up to 30 vol.% of n-hexane.

Water soluble polymers were studied using classic form of LCCC only up to molar masses of 40kD [125]. Molar mass independent elution was observed for polyacrylamide up to 173kD using limiting conditions approach [157].

4
Mixed Mobile Phases for Critical Conditions

Mixtures of a solvent and a non-solvent (S/N) were used in majority of the cases for "tuning" of the polymer retention. As shown in Tables 1-3, non solvents are added in amounts from 0.6 to 70%. A decrease in the thermodynamic quality of the mobile phase by adding of the precipitant (non-solvent) is occasionally connected with problems of the poor solubility of higher molar mass samples or even with the inability to reach the critical condition due precipitation of polymer samples. The precipitation may be supported, not only by the high concentration of the non solvent in the mobile phase, but also by high concentration of non solvent adsorbed on the sorbent surface. In mixtures of solvent and precipitant, the extent of the preferential solvation of the polymer coils may be the reason why both solubility and movement of the polymer in solvent/precipitant mixtures are possible. An evaluation of coefficient of preferential solvation requires separate measurements, which have been not performed in connection with CEEC.

The use of a pair of thermodynamically good solvents (adsorli/desorli) results in the perfect solubility of a selected polymer. One solvent should support adsorption of the polymer onto the column packing, the second one desorption. Such mixtures can enable critical behavior up to very high molar masses, as shown in Table 5. At present only a the small number of solvent/solvent systems is known (Table 1: PMMA (6 systems), PtBMA (at least 1 system), PS (1 system)). A screening of solvents, for example, using silica and polymer of interest, could enable to identify suitable solvents, where the polymer is adsorbed. Specifically, it is known that PS is adsorbed on porous glass from carbon tetrachloride [7], PMMA is adsorbed on silica gel from DCM [45] and from chloroform [155] or toluene [57] and PtBMA from DCM [45]. PS, PBMA, PMMA, PTHF are adsorbed on silica from carbon tetrachloride [168], as well as PVC, PEMA, and PVAc [169]. PEO adsorbs on silica from pure water [168]. PMMA is adsorbed from toluene on some PS/DVB columns [155].

The use of two good solvents for a polymer as components of a binary mobile phase, however, can not guarantee the full recovery of the polymer. For example, employing THF/toluene for PMMA on silica gel caused limited or no recovery of PMMA above ca 200 kDa, although a molar mass independent elution was confirmed [57]. It seems that the adsorption of PMMA on silica gel was too strong. The problem has been in this case overcome by using pure THF or toluene as sample solvents for PS injection. On the other hand, the use of two polar good solvents THF/chloroform for PMMA injected in the mobile phase onto silica gel did not result in recovery problems [58].

In aqueous mobile phases, a new parameter, pH comes into play, which leads to ionic effects and thus an ionic exclusion, expulsion and attraction. The variation of pH enables transition from exclusion to adsorption mode through critical region [125].

5
Single Mobile Phases

There exist a limited amount of experimental data which indicate that critical conditions in the single eluent are feasible. Inspection of Tables 1–3 reveals that near critical conditions were observed in single mobile phases by Spychaj [44] (for PMMA on silica gel in ethylacetate), Mencer and Grubisic-Gallot [143] (for P2VP in THF on µ-Bondagel), Krüger et al. [94] (for polyester in acetone on silica gel) and Berek [156] (for PMMA injected in THF into toluene, PS/DVB). Recently, Chang and coworkers proposed temperature gradient interaction chromatography (TGIC) method for the selective separation of macromolecules. TGIC utilizes a temperature gradient to control the retention of macromolecules near the system critical conditions [38]. In addition to mixed mobile phases single component eluents were also used in TGIC [38,170]]. It has been very recently experimentally confirmed that some from single solvents used for TGIC show critical behavior for polymers at the constant temperature [6,39], for example, as seen in Fig. 1. For this reason we have incorporated the single eluents used for TGIC into Table 1.

The use of single eluents for the critical conditions has the potential to improve CEEC separations since the elimination of both the solvent peaks and the effects of both preferential solvation and sorption of mobile phase components on LC separation can be eliminated. Moreover, many detectors are easily applicable with such eluents. A new generation of column packings, i.e. sorbents grafted with temperature sensitive or pH sensitive polymers which change their retentive properties in a large extent after relatively small change of temperature or pH in single eluents [171], could also be suitable for critical conditions in single mobile phases. The use

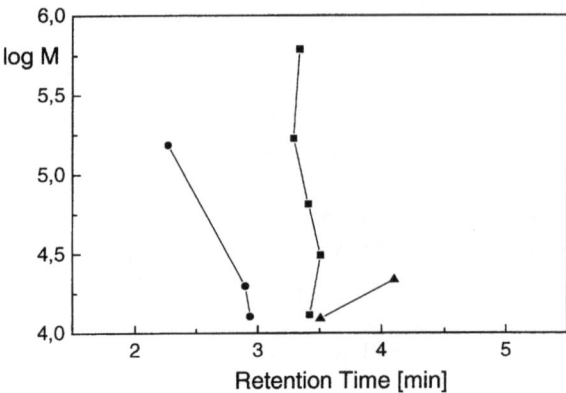

Fig. 5. Influence of the stereoregularity of PMMA samples on their elution behavior under critical conditions [6]. Mobile phase: Acetonitrile. Stationary phase: Nucleosil C18. Temperature: 68 °C. Symbols: (squares) atactic PMMA, (circles) syndiotactic PMMA, (triangles) isotactic PMMA

of single eluents for the critical conditions posses an attractive potential which has not yet been examined. For example, pure ACN on silica gel C18 enables different elution of izotactic, syndiotactic and atactic PMMA at higher temperature (Fig. 5) [6].

6
Polymer Peaks and Solvent Peaks under Critical Conditions

As a rule, in a mixed mobile phase a solvent peak appears near the void volume of the column. The appearance of the solvent peak may due to one of several effects, the first of which is the preferential solvation of polymers [172]. After dissolution in a mixed solvent, the polymer binds into its solvation shell one part of mixture to a larger extent. After the separation of the solvated polymer from rest of injected solvent, the solvent peak appears on chromatogram as was demonstrated by SEC [172] and under suitable condition [173] its area, or height, may be correlated with coefficient of preferential solvation [172]. An evaporation of one component from the sample bottle or displacement effects may also lead to appearance of a solvent peak [173]. The solvent peak represents a local change of composition of the mobile phase. Under critical conditions small changes of the mobile phase composition (for example, 0.1% wt.) have a large influence on polymer retention, thus the solvent peak could influence the elution of the macromolecules. If so, this could imply that a tabulated critical composition is not precisely that, which really correspond to the critical conditions. The real, acting critical composition of eluent may be, and likely is, the composition somewhere, in the middle, of the solvent peak. The presence of the solvent peak influences especially pronouncedly the elu-

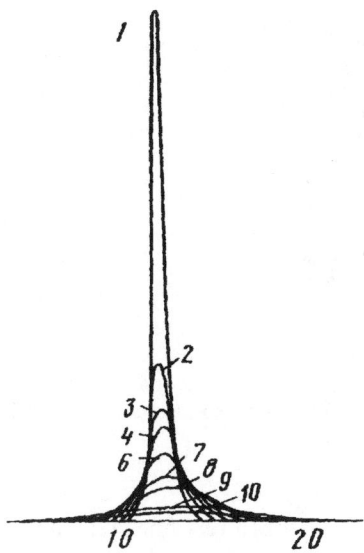

Fig. 6. Shapes of peaks of PS standards under near critical conditions. Column packing: Porous glass. Mobile phase: tetrachloromethane/chloroform, 95.5/4.5 vol. Temperature: 30.4°C. Detection: UV. Symbols: Mw=0.4kDa (1), 2kDa (2), 5kDa (3), 10.3kDa (4), 19.85kDa (5), 51kDa (6), 98.2kDa (7), 173kDa (8), 411kDa (9), 867kDa (10). (From [7] with permission)

tion of polymer under limiting conditions. In this case is the solvent peak very large and interactions polymer – sample solvent, polymer – mobile phase have a contradictory character (for example, polymer - thermodynamically good sample solvent versus polymer – precipitant).

The disruption of the polymer separation by the solvent peak may be the reason why broad and narrow polymer standards did not elute under the critical conditions as a very narrow peak, as has been observed by the first critical conditions investigators and, subsequently, by others [7,21,56,62,79]. Namely, CEEC should yield a very narrow peaks for polymeric samples of different polydispersity, with the width of the polymeric peak determined only by the chromatographic efficiency of LC column. Indeed, the width of the polymer peaks, more or less, increases with the molar mass under critical conditions, as is illustrated in Fig. 6.

An additional reason for peak broadening under the critical conditions may be the temperature gradient inside of column generated by viscous heat dissipation [174] and, to a much less extent pressure gradient [175]. As is known, critical conditions are quite sensitive to temperature [7,38,39,79,125] and, in some systems, to pressure [7,17,92]. Furthermore, another reason for broadening could be differences in polymer architecture, because isomers may manifest itself under critical conditions (Table 1). In any case, the question as to why polymer peaks are not significantly narrower under critical conditions, remains unanswered and requires

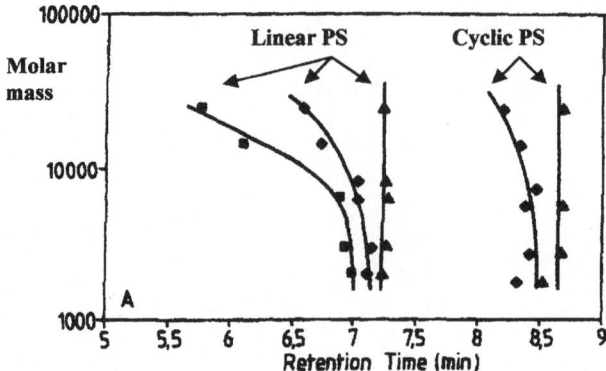

Fig. 7. Different retention of linear and cyclic PS. Eluent: THF/n-hexane. Column: Silica gel. (From [69] with permission)

further study. As is the case in microscopy, LCCC requires a very fine preparation of the sample and adjustment of parameters for obtaining of the very specific separations (i.e. very narrow peaks).

In the majority of the systems, polymers at the critical conditions elute with the solvent peak. On the other hand, Pasch et al. [69] have observed that linear PS eluted and cyclic PS eluted independently of molar mass – Fig. 6 – from the same column and at the same eluent composition. The same effect has been found for zero and bifunctional polyesters [109]; at TLC of bi- and tri-functional oligomers of polyoxypropylene glycols [102]; for PEG and PEG mono- or di-methylethers [126] as well as for various functional PEOs [68]. Similarly, Berek et al. [54] have observed that st-PMMA eluted at higher retention volume than the total volume of liquid in the column. This interesting phenomenon has been not explained. If understood, it could be utilized for elimination of co-elution of the solvent peak with the polymer peak.

7
Limiting Conditions

The molar mass independent elution of polymers may also be obtained in case when polymer is injected **in only one component of mixed mobile phase** [26] (Fig. 8).

In such cases, the corresponding critical mobile phase may be even precipitant for the polymer or may cause its full adsorption, if the polymer would be injected using critical mobile phase as a sample solvent [154]. It has been considered that in such cases the polymer can not exclude from the solvent peak, since eluent acts as a barrier, which is unsurpassed for the polymer. The term "limiting conditions" was coined for this phenomenon since the polymer elutes in a micro-gradient of

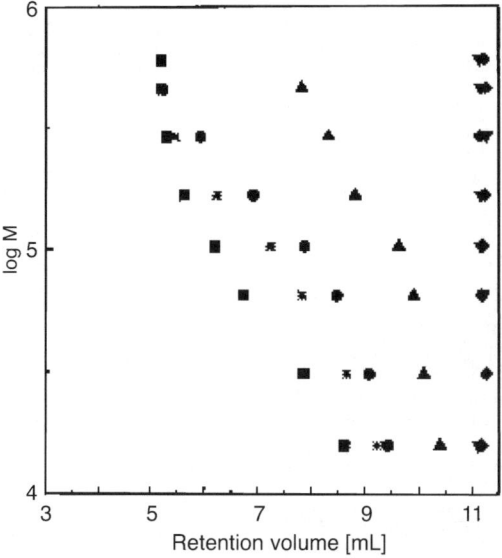

Fig. 8. LC LCA calibration curves for PMMA of low stereoregularity with bare silica gel column packing. Pure THF (squares) and THF/toluene with various amount of toluene. Composition in wt. %: 60 (stars), 61.5 (circles), 63 (triangles), 67 (inverted triangles), 69 (diamonds). Samples were injected in THF. Limiting conditions of adsorption were maintained between 65 and 69% of toluene. (From [155] with permission)

composition at the limit of its solubility [154]. According to this explanation, **limiting conditions of solubility** (mobile phase is precipitant, sample solvent is good solvent for polymer [152,154], **limiting conditions of adsorption** (mobile phase supports adsorption of polymer, sample solvent is desorli) [155,157] and **limiting conditions of desorption** (mobile phase is desorli for polymer, sample solvent is adsorli) [57] were differentiated [24]. The family of LC methods utilizing CEEC is schematically presented in Figs. 9 and 10. The elution behavior under limiting conditions could be eventually explained considering the initial solvation of polymer coils, when the solvation shell of polymer is quit stabile. In all cases, mutual interactions between the polymer – the sample solvent and the polymer – eluent have much more antagonistic character than that is in classical form of LCCC. When limiting conditions are applied on characterization of copolymers, peaks of copolymers move outside of the solvent peak.

Limiting conditions have the advantage that polymers are more readily soluble. Additionally, the molar mass independent elution is observed over a broader range of compositions of mixed eluents [155]. On the other hand, decreased polymer recovery was, at times, identified and mixed mechanism, when adsorption and precipitation have taken place in the system [157]. LC systems with limiting conditions observed to date are collected in Table 3.

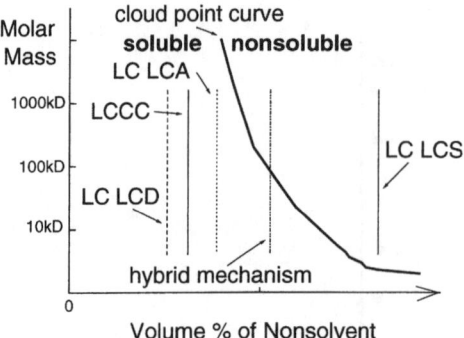

Fig. 9. Schematic representation of LC methods according to solubility of the analyzed polymer in the mobile phase. The choice of a thermodynamically good solvent for dissolution and injection of the polymer enables its elution even in a eluent which does dissolve the polymer

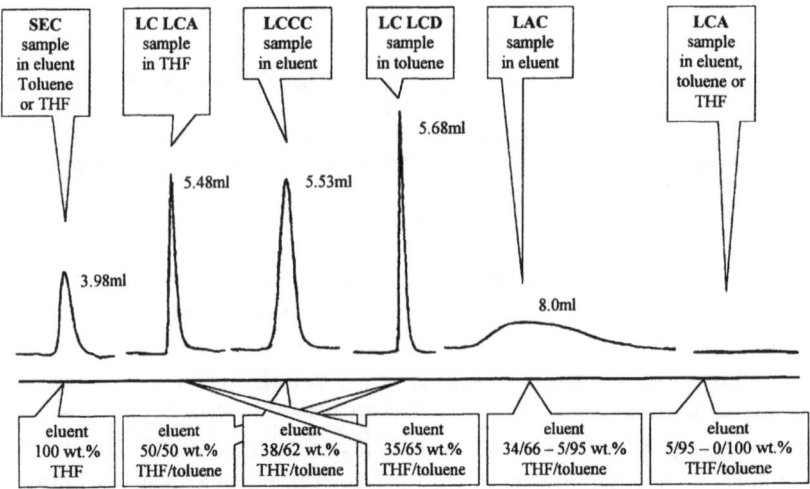

Fig. 10. LC methods based on CEEC represented on chromatograms of PMMA eluted in different modes from silica gel column. Detector: ELSD. The figure is constructed according to data published by Berek [57]. THF and toluene are thermodynamically good solvents for PMMA, herefore LC LCS is not possible in this system

8
Comparison of Molar Masses Calculated from Separation under Critical Conditions and other Methods

Under critical conditions for one homopolymer, which constitutes a part of a binary copolymer, a second polymer may be characterized [61–64,80,176]. The molar mass of a part of copolymer, which is not under critical conditions, however, is eluting in SEC mode, may be determined in the column at critical conditions or after reinjection, in the additional SEC column [22]. Clearly, corresponding homopolymer standards for SEC calibration should be at the hand. Experimental reports for the CEEC analysis of block copolymers, which support the feasibility of the individual block characterization, were published. For example, Zimina et al. [45] described that block copolymers PS-PMMA eluted under critical conditions for PMMA according to size of PS blocks. The authors have determined the molar mass of PS in the block copolymers and compared the chemical composition based on NMR with obtained from LCCC. The agreement was very good.

Pasch et al. have compared of copolymer molar masses from SEC and from LCCC for block copolymers PDMA-PMMA [61] and PS-PMMA [80]. From the molar mass of the individual blocks, which were measured under LCCC, the total molar masses of the block copolymers were calculated. Moreover, the expected nominal chemical composition was compared with the chemical composition determined by LCCC. While the molar mass of copolymers from SEC was used for comparison in some studies [61,80], molar mass of PS precursor used for preparation of corresponding copolymers has also been selected as reference value in [47]. Very good agreement of values was found, as documented by data in Table 6A,B extracted from [47,80].

The average molar masses of components of block copolymers PMMA-PtBMA estimated by LC under critical conditions (for polymer eluted in SEC mode) agree very well with values SEC obtained by subtracting the molar mass of the PtBMA precursor from that of the block (deviation 2–4%) [55]. Agreement up of individual component ratio of 1:5 was found. On the other hand, Lee et al. [39,78] have found systematic difference 0–33% between by critical conditions measured the molar mass of precursor and expected molar mass of precursor. Differences between measured, and expected, molar masses increased with the concentration of the "invisible" part of copolymers (i.e. part of copolymer under critical conditions). Additional studies are needed to generalize this phenomenon.

Table 6A. Comparison of Molar Masses of Block Copolymers determined by various Methods

Sample Block copolymer PS-PMMA	Expected M_w PMMA [kD]	LCCC[a] M_w PMMA [kD]	Precursor M_w PS [kD]	LCCC[b] M_w PS [kD]	SEC[c] M_w Copolymer [kD]	SEC[d] M_w Copolymer [kD]	LCCC[e] M_w Copolymer [kD]
B1	55 kD	49 kD	114 kD	119 kD	165kD	152kD	168kD
B2	89 kD	97 kD	81 kD	91 kD	182kD	141kD	188kD
B3	133 kD	143 kD	48 kD	61 kD	188kD	140kD	204kD

[a] Determined under critical condition for PS using PMMA standards[80]
[b] Determined under critical conditions for PMMA using PS standards[47]
[c] SEC with light scattering detection
[d] SEC, Styragel, THF, PS calibration

Table 6B. Comparison of Compositions of Block Copolymers determined by various Methods

Sample Block copolymer PS-PMMA	Composition expected	Determined LCCC[e]	Determined SEC (dual detection-Density-RI)
B1	67 / 33	71 / 29	69 / 31
B2	51 / 49	48 / 52	53 / 47
B3	29 / 71	30 / 70	30 / 70

[e] Determined from the molar mass of the PS and the PMMA block

We may conclude that the molar masses obtained under critical conditions reflect, correctly, homo- and co- polymer characteristics, however deviations may be observed. These deviations may be caused by the method itself, due to parameters, who influence is not sufficiently well know, such as preferential solvation of polymer chains, unexpected differences in the structure or composition of the analyzed samples, or differences in adsorptive properties of column packings. Nevertheless, the data obtained are valuable due to very limited accessibility to such data by other analytical methods.

9
Detection

As was mentioned earlier, polymer, solvents or additives are under critical conditions, eluted at the same or near the elution volume. Universal detectors (RI, UV) monitor all components what may lead to incorrect conclusions since a solvent or an impurity peak may be considered as a polymer peak. Furthermore, a polymer peak may be deformed or split as consequence of overlapping of responses for both the polymer peak and the solvent peak (compare Figs. 11A,C with Figs. 11B,D). Fig. 11B shows the increase of broadening of polymer peaks under LCCC, while Fig. 11D illustrates the shape of polymer peaks under limiting conditions.

ELSD does not detect solvents and yields, for macromolecules, a strong signal. Therefore ELSD is recommended as a detector at looking for critical conditions. Solvent peak may also be "invisible" to viscosity [50] or LALLS [59] detectors, however, their application for mixed eluents is more problematic than ELSD. The connection of LC under critical conditions, on-line or off-line, with the specific detectors such as DAD, FTIR, MALDI-TOF-MS and NMR helps greatly to identify the chemical structure of eluted substances. The applicability of these detectors is widespread with a large amount of useful information's obtained in such way (for a review see [25]).

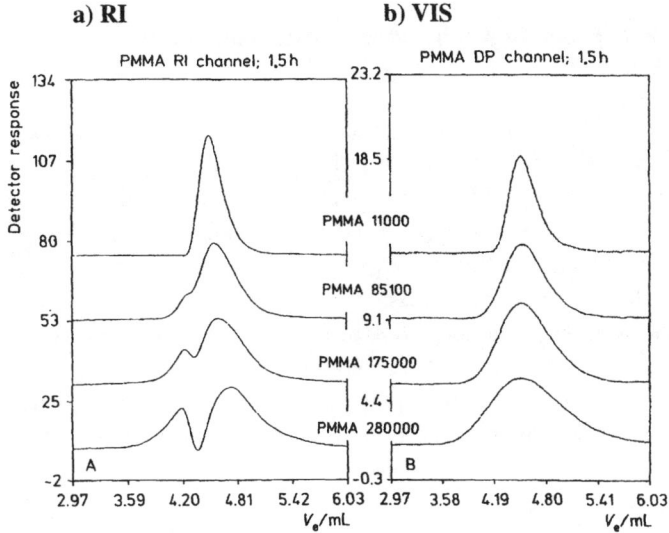

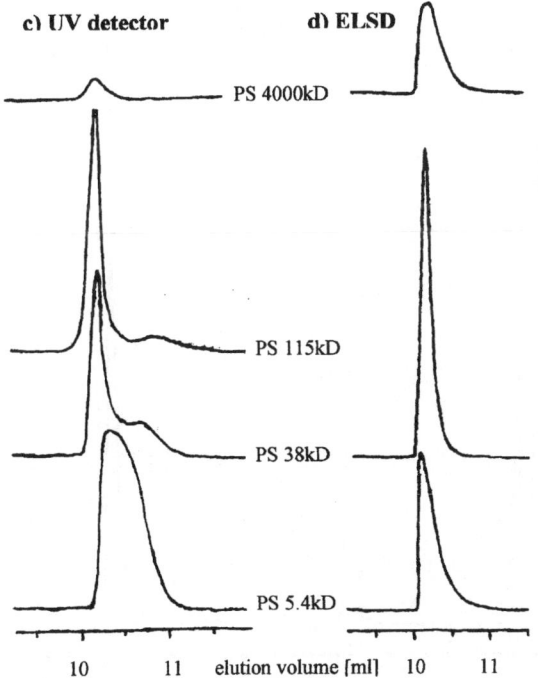

Fig. 11A–D. Chromatogram of PMMA 280kDa injected in eluent the detected by **A** RI detector, **B** viscosimeter. Mobile phase: MEK/cyclohexane. Column packing: Silica gel. From [50] with permission. A chromatogram of polystyrene sample injected in dichloromethane detected by **C** UV, **D** ELSD. Mobile phase: Dichloromethane/n-hexane. Column packing: Silica gel [151]

10
Applications of Liquid Chromatography under Critical Conditions to Polymer Separation

Applications summarized in Table 1, 2 and 3, document that CEEC may be found for many individual polymers as well as structural segment of a polymer sample. The unique sensitivity of LCCC to different characteristics of macromolecules, has been, to date primarily utilized for analysis of block copolymers (AB or ABA) as well as analysis of functionality of homopolymers (Fig. 12). Blends of polymers and structurally different macromolecules, i.e., cyclic versus linear versus branched and even with different tacticity, have been separated. Graft copolymers has been studied [43,117] as well as random copolymers [111,153,154] (Fig. 13).

The majority of the applications handle polymers soluble in organic solvents, while CEEC for water soluble polymers has been less extensively elaborated. Polyethylene and polypropylene, worldwide the most widely produced synthetic polymers, have not been characterized by LC system at CEEC, likely due to the need to use higher temperatures. The first LC system under limiting conditions for these polyolefins was just recently identified (Table 3) [149].

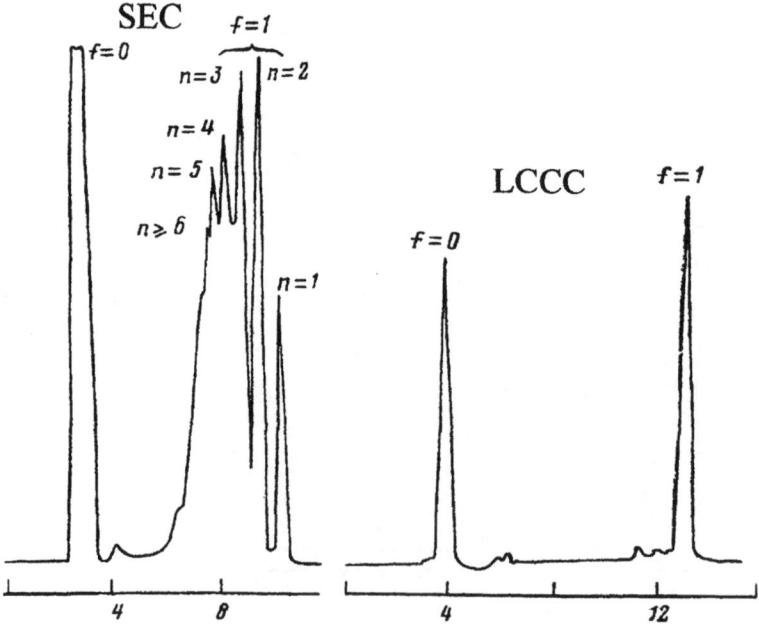

Fig. 12. Samples of polyfenylenesulfones (polymerization degree n = 1–6) separated under critical conditions. Symbols: without hydroxyl groups (f=0), one end hydroxyl group (f=1). A comparison with the SEC chromatogram illustrates that LCCC separates the samples independently of the size of macromolecules (From [130] with permission)

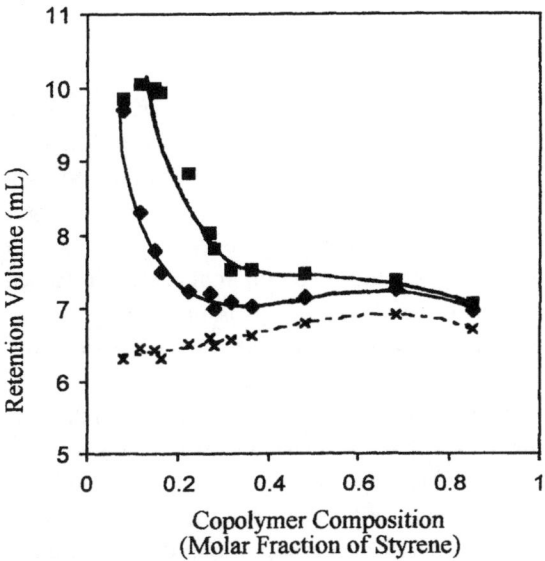

Fig. 13. Retention volume obtained by liquid chromatography under limiting conditions of adsorption (squares, diamonds) and by SEC (crosses) for random homopolymers of various composition. (Adapted from [153] with permission)

The SEC analysis of components separated according to the chemical composition under critical conditions may be carried out after transferring of effluent from column under critical conditions into a SEC column. This procedure requires additional sample valve with loading loops plus the adjustment of flow velocity and column dimensions [106]. The procedure is known as two-dimensional separation (for references see [22,25]). A fully automated chromatographic system for two-dimensional separation was developed by Kilz et al. [106,177] and is commercially available [178]. On the other hand, a reinjection of analyzed fraction from column working at critical conditions into the additional SEC column is not necessary, when two different systems under critical conditions are found, i.e. one for each components of copolymer (using two different columns or two different eluents). Corresponding complementary LCCC systems may be selected from Table 1–3. In this case, parts of copolymer, which are not under CEEC, should be eluted in SEC modus. The distribution of molar mass of both components of copolymers may be then measured directly under critical conditions, as have been demonstrated in [47,67,78,80]. If necessary, LCCC may be scaled up to preparative quantities [71].

11
Conclusions

All known LC systems under critical and limiting conditions reported over the last 25 years have been summarized. This collection, of over 180 systems, indicates for approximately 37 homopolymers, the most appropriate systems from the point of view separation up to high molar mass and with full recovery. It also enables to reduce the time which would be necessary for identifying the composition of the mobile phase. Suitable alternatives for the separation of different parts of a studied copolymers under conditions of entropy-enthalpy compensation can be identified. Furthermore, the data have been tested from point of view of correlation's between some experimental characteristics of the critical systems.

The solubility parameters of critical mobile phases are concentrated in relatively narrow windows of values for majority of pairs sorbent–polymer. Taking into account that sorbents of the same type differ in their adsorptive properties, and that sigma values are not corrected for influence of temperature, the correlation with the solubility parameters is very good. Therefore, it is hypothesized that all potentially suitable critical mobile phases for a polymer – sorbent pair should posses solubility parameter near to value known from previous LCCC experiments. Generally, the solubility parameter of eluent is not correlated with the solubility parameter of the polymer analyzed. Similarly, according to the Mark-Houwink constants, the critical eluents may correspond to thermodynamically poor or very good solvents. Under limiting conditions the mobile phase may be even precipitant for the analyzed polymer.

Values of the elution strength and the solubility parameter of critical eluents corresponding to a polymer are dissipated, however, values of elution strength are scattered to a smaller extent. A good correlation has been confirmed between elution strength of a pure eluent component and volume ratio of the corresponding mixed eluent. It seems that this dependence may be used for prediction of new potentially critical mobile phases when silica gels with identical properties are used. Namely, silica gels of different origin may have different adsorptive properties which is reflected, very sensitively, in different CEEC. Currently available sorbents need to be characterized in respect to their adsorptive properties. Values, such as the number of silanol groups or the carbon loading could be helpful for a comparison. A deliberate variation of adsorbent activity, either using a column from a series of columns with defined, standardized, but in small steps different adsorption properties, would make an adjustment of CEEC in a selected mobile phase much easier. Furthermore, interesting possibilities connected with the new generation of column packings, which alter substantially their retentive properties with temperature or pH, also exist.

The properties of mixed eluents which permit CEEC up to high molar masses have been identified. Mixed mobile phases composed from two thermodynamic

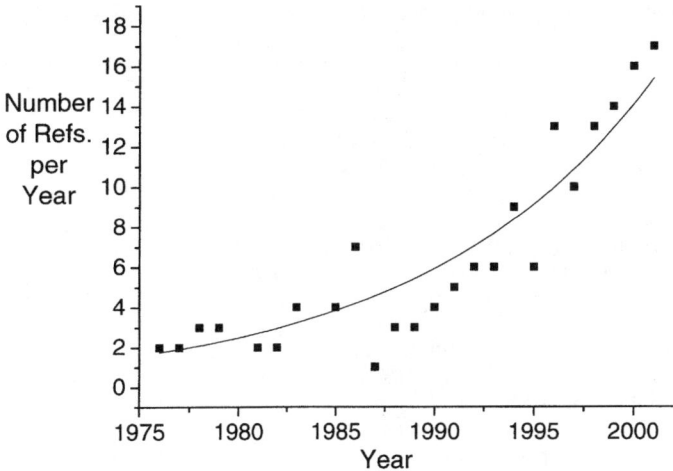

Fig. 14. Increase in the number of literature references about critical and limiting conditions

good solvents, in combination with a not strongly adsorbing sorbent, could eliminate problems with non solubility and limited recovery of polymer samples. Combinations of solvent plus precipitant, but both either polar or non polar liquids, as shown, also enable the elution of high molar mass samples. Therefore, the authors have, herein, presented some proposals and guidelines based on review and prospective needs. For example, single mobile phases may eliminate problems with both presence and influence of solvent peaks. Moreover, an use of absolute detectors is in single eluents less complicated. Eventually, the role of the solvent peak and its impact on LC separation could be deliberately increased by using various sample solvents.

The compilation of data confirms that critical conditions are a generally valid phenomenon in LC of polymers. Critical conditions are highly sensitive to very small differences in chemical and structural composition of macromolecules. For this reason LC under conditions of entropy-enthalpy compensation will likely play the increasingly important role in the future of polymer separations including for polyolefins (Fig. 14).

Acknowledgments. One of the authors (T.M.) acknowledges to Dr. M.D. Palamareva and Dr. H.E. Palamarev (University of Sofia, Bulgaria) for providing the LSChrom software for the calculation of eluent strength and polarity parameters.

References

1. Belenkii BG, Gankina ES, Tennikov MB, Vilenchik LZ (1976) Dokl.Akad.Nauk SSSR (Moscow) 231:1147
2. Belenkii BG, Gankina ES, Tennikov MB, Vilenchik LZ (1978) J.Chromatogr. 147:99

3. Belenkii BG, Valchikhina MD, Vakhtina IA, Gankina ES, Tarakanov OG (1976) J.Chromatogr. 129:115
4. Tennikov MB, Nefedov PP, Lazareva MA, Frenkel SYa (1977) Vysokomol.Soed.A 19:657
5. Skvortsov AM, Belenkii BG, Gankina ES, Tennikov MB (1978) Vysokomol.Soed.A 20:678
6. Macko T, Hunkeler D, Berek D (2002) Macromolecules 35:1797
7. Nefedov PP, Zhmakina TP (1981) Vysokomol.Soed.A 23:276
8. Skvortsov AM, Gorbunov AA (1979) Vysokomol.Soed.A 21:339
9. Gorbunov AA, Skvorcov AM (1986) Vysokomol.Soed.A 28:2453
10. Gorbunov AA, Zhulina EB, Skvorcov AM (1982) Polymer 23:1133
11. Skvortsov AM, Gorbunov AA (1986) J.Chromatogr. 358:77
12. Gorbunov AA, Skvorcov AM (1995) Adv.Colloid Interface Sci. 62:31
13. Entelis SG, Evreinov VV, Kuzaev AI (1989) Reactive Oligomers, VSP, Utrecht, The Netherlands
14. Entelis SG, Evreinov VV, Gorshkov AV (1986) Adv.Polym.Sci. 76:129
15. Klein J, Treichel K (1977) Chromatographia 10:604
16. Balke ST (1982) Sep.Purif.Methods 11:1
17. Souvignet I, Olesik SV (1997) Anal.Chem. 69:66
18. Belenkii BG (1979) Pure & Appl.Chem. 51:1519
19. Nefedov PP, Lavrenko PN (1979) Transport Methods in Analytical Chemistry of Polymers (in Russian), Khimiya, Leningrad
20. Belenkii BG, Gankina ES, Kasalainen GE, Tennikov MB (1996) Nonstandard methods based on SEC principles. In: Potschka M, Dubin PL (Eds.) Strategies in Size Exclusion Chromatography. ACS, Washington, DC, p 274
21. Berek D (1996) Macromol.Symp. 110:33
22. Pasch H, Trathnigg B (1997) HPLC of Polymers, Springer Verlag, Berlin
23. Pasch H (1997) Adv.Polym.Sci. 128:1
24. Berek D (2000) Prog.Polym.Sci. 25:873
25. Kilz P, Pasch H (2000) Coupled Liquid Chromatographic Techniques in Molecular Characterization. In: Meyers RA (Ed.) Encyclopedia of Analytical Chemistry. Wiley, New York, p 7495
26. Hunkeler D, Macko T, Berek D (1991) Polym.Mat.Sci.Eng. 65:581
27. Hunkeler D, Janco M, Guryanova VV, Berek D (1995) Limiting conditions in the LC of polymers. In: Provder T, Barth HG, Urban MW (Eds.) Chromatographic Characterization of Polymers, Hyphenated and Multidimensional Techniques. ACS, Washington, DC, p 13
28. Braun D, Esser E, Pasch H (1998) Int.J.Polym.Anal.Charact. 4:501
29. Esser KE, Braun D, Pasch H (1999) Angew.Makromol.Chem. 271:61
30. Pasch H, Esser E, Kloniger C, Iatrou H, Hadjichristidis N (2001) Macromol.Chem.& Phys. 202:1424
31. Gorshkov AV, Evreinov VV, Entelis SG (1985) Russian J.Phys.Chem. 59:1985
32. Gorshkov AV, Evreinov VV, Entelis SG (1985) Zh.Fiz.Khim. 59:1475
33. Gorshkov AV, Evreinov VV, Entelis SG (1985) Russian J.Phys.Chem. 59:869
34. Estrin YaI, Kasumova LT (1994) Russian J.Phys.Chem. 68:1620
35. Pulda J, Reisova A (1999) Chemicke Listy 93:445
36. Pulda J (1998) Polym.Mat.Sci.Eng. 78:61
37. Cools PJCH, Herk AM, German AL, Staal WJ (1994) J.Liq.Chromatogr. 17:3133
38. Chang T, Lee HC, Lee W, Park S, Ko C (1999) Macromol.Chem.& Phys. 200:2188
39. Lee W, Park S, Chang T (2001) Anal.Chem. 73:3884
40. Czichocki G, Heger R, Goedel W, Much H (1997) J.Chromatogr.A 791:350
41. Pokorny S, Janca J, Mrkvickova L, Tureckova O, Trekoval J (1981) J.Liq.Chromatogr. 4:1
42. Lee W, Cho DH, Chun BO, Chang T, Ree M (2001) J.Chromatogr.A 910:51
43. Adrian J, Esser E, Hellmann G, Pasch H (2000) Polymer 41:2439
44. Spychaj T (1978) *"Secondary processes in gel permeation chromatography with inorganic gels"*, PhD Thesis, Polymer Institute SAS, Bratislava, Slovakia

45. Zimina TM, Kever JJ, Melenevskaya EY, Fell AF (1992) J.Chromatogr. 593:233
46. Zimina TM, Fell AF, Castledine JB (1992) Polymer 33:4129
47. Pasch H, Brinkmann C, Gallot Y (1993) Polymer 34:4100
48. Pasch H, Augenstein M (1993) Macromol.Chem. 194:
49. Pasch H (1993) Polymer 34:4095
50. Pasch H, Rode K (1996) Macromol.Chem.& Phys. 197:2691
51. Siewing A, Schierholz J, Braun D, Hellmann G, Pasch H (2001) Macromol.Chem.& Phys. 202:2890
52. Pasch H, Much H, Schulz G (1993) J.Appl.Polym.Sci.: Appl., Polymer.Symp. 52:79
53. Janco M, Berek D, Önen A, Fischer CH, Yagci Y, Schnabel W (1997) Polym.Bull. 38:6810
54. Berek D, Janco M, Hatada K, Kitayama T, Fujimoto N (1997) Polym.J.(Tokyo) 29:1029
55. Falkenhagen J, Much H, Stauf W, Müller AHE (2000) Macromolecules 33:3687
56. Berek D, Janco M, Meira GR (1998) J.Polym.Sci.A, Polym.Chem. 36:1363
57. Berek D (1998) Macromolecules 31:8517
58. Skvortsov AM, Gorbunov AA, Berek D, Trathnigg B (1998) Polymer 39:423
59. Pasch H, Rode K (1998) Polymer 39:6377
60. Pasch H, Rode K, Chaumien N (1996) Polymer 37:4079
61. Pasch H, Augenstein M, Trathnigg B (1994) Macromol.Chem. 195:743
62. Zimina TM, Kever JJ, Melenevskaya EU, Zgonnik VN, Belenkii BG (1991) Vysokomol.Soed.A 33:1349
63. Belenkii BG, Gankina ES, Zgonnik VN, Malchova II, Melenevskaya EU (1992) J.Chromatogr. 609:355
64. Gankina ES, Belenkii BG, Malakhova I, Melenevskaya EU, Zgonnik VN (1991) J.Planar Chromatogr. 4:199
65. Kitayama T, Janco M, Ute K, Niimi R, Hatada K, Berek D (2000) Anal.Chem. 72:1518
66. Janco M, Hirano T, Kitayama T, Hatada K, Berek D (2000) Macromolecules 33:1710
67. Baran K, Laugier S, Cramail H (2000) Int.J.Polym.Anal.Charact. 6:123
68. Baran K, Laugier S, Cramail H (2001) J.Chromatogr.B 753:139
69. Pasch H, Deffieux A, Henze I, Schappacher M, Rique-Lurbet L (1996) Macromolecules 29:8776
70. Baran K, Laugier S, Cramail H (1999) Macromol.Chem.& Phys. 200:2074
71. Lepoittevin B, Dourges MA, Masure M, Hemery P, Baran K, Cramail H (2000) Macromolecules 33:8218
72. Lepoittevin B, Perrot X, Masure M, Hemery P (2001) Macromolecules 34:425
73. Lepoittevin B, Hemery P (2001) J.Polym.Sci.A, Polym.Chem. 39:2723
74. Gerber J (2002) Personal communication. German Institute for Polymeric Materials, Darmstadt, Germany
75. Blagodatskich IV, Gorshkov AV (1997) Vysokomol.Soed.A 39:1681
76. Balke ST, Patel RD (1983) In: Craver C.D. (Ed.) Polymer Characterization. ACS, Washington, DC, p 281
77. Balke ST (1987) Orthogonal Chromatography and Related Advances in Liquid Chromatography. In: Provder T (Ed.) Detection and Data Analysis in SEC. ACS, Washington, DC, p 59
78. Lee W, Cho DH, Chang T, Hanley KJ, Lodge TP (2001) Macromolecules 34:2353
79. Philipsen HJA, Klumperman B, Van Herk AM, German AL (1996) J.Chromatogr.A 725:13
80. Pasch H, Gallot Y, Trathnigg B (1993) Polymer 34:4986
81. Lee HC, Chang T (1996) Macromolecules 29:7294
82. Radke W (2002) Personal communication. German Institute for Polymeric Materials, Darmstadt, Germany
83. Lee HC, Chang T, Harville S, Mays JW (1998) Macromolecules 31:690
84. Lee HC, Lee H, Lee W, Chang T, Roovers J (2000) Macromolecules 33:8119
85. Chang T, Lee HC, Lee W (1997) Macromol.Symp. 118:261
86. Cho DH, Park S, Kwon K, Chang T (2001) Macromolecules 34:7570

87. Lee W, Lee H, Lee HC, Cho DH, Chang T, Gorbunov AA, Roovers J (2002) Macromolecules 35:529
88. Bruheim I, Molander P, Theodorsen M, Ommundsen E, Lundanes E, Greibrokk T (2001) Chromatographia 53:S-266
89. Belenkii BG, Vilenchik LZ (1983) Modern Liquid Chromatography of Macromolecules, Elsevier, Amsterdam
90. Dudorina AV, Gorshkov AV, Filatova HH, Evreinov VV, Entelis SG (1995) Vysokomol.Soed.B 37:1957
91. Yun H, Olesik SV, Marti EH (1998) Anal.Chem. 70:3298
92. Phillips S, Olesik SV (2002) Anal.Chem. 74:799
93. Krüger RP, Much H, Schulz G, Wehrstedt C (1990) Macromol.Chem. 191:920
94. Krüger RP, Much H, Schulz G (1996) Macromol.Symp. 110:155
95. Pasch H, Brinkmann C, Much H, Just U (1992) J.Chromatogr. 623:315
96. Gancheva VB, Vladimirov NG, Velichova RS (1996) Macromol.Chem.& Phys. 197:1757
97. Gancheva VB, Vladimirov NG, Velichova RS (1996) Macromol.Chem.& Phys. 197:1771
98. Mengerink Y, Peters R, deKoster CG, van der Wal Sj, Claessens HA, Cramers CA (2001) J.Chromatogr.A 914:131
99. Brun Y (1999) J.Liquid.Chromatogr.& Related.Technol. 22:3067
100. Gorshkov AV, Prudskova TN, Filatova HH, Evreinov VV (1987) Zh.Fiz.Khim. 61:3390
101. Gorshkov AV, Evreinov VV, Entelis SG (1983) Vysokomol.Soed.A 0:632
102. Tikhonova TZ, Petrakova EA (1991) Russian J.Phys.Chem. 65:1463
103. Gorshkov AV, Evreinov VV, Entelis SG (1985) Russian J.Phys.Chem. 59:552
104. Filatova HH, Gorshkov AV, Evreinov VV, Entelis SG (1988) Vysokomol.Soed.A 30:953
105. Filatova HH, Minina EO, Gorshkov AV, Evreinov VV, Entelis SG (1995) Vysokomol.Soed.B 37:358
106. Kilz P, Krüger RP, Much H, Schulz G (1995) Two-Dimensional Chromatography for the Deformulation of Complex Copolymers. In: Provder T., Barth H.G., Urban M.W. (Eds.) Chromatographic Characterization of Polymers, Hyphenated and Multidimensional Techniques. ACS, Washington, DC, p 223
107. Krüger RP, Much H, Schulz G (1996) Int.J.Polym.Anal.Charact. 2:221
108. Krüger RP, Much H, Schulz G (1994) J.Liquid.Chromatogr. 17:3069
109. Tennikova TB, Blagodatskikh IV, Svec F, Tennikov MB (1990) J.Chromatogr. 0:0
110. Pasch H, Much H, Schulz G, Gorshkov AV (1992) LC-GC Int. 5:38
111. Gorshkov AV, Much H, Becker H, Pasch H, Evreinov VV, Entelis SG (1990) J.Chromatogr. 523:91
112. Pasch H, Rode K (1995) J.Chromatogr.A 699:21
113. Pasch H, Zammert I (1994) J.Liq.Chromatogr. 17:3091
114. Adrian J, Braun D, Pasch H (1998) LC&GC Int. January:32
115. Falkenhagen J, Friedrich JF, Schulz G, Krueger RP, Much H, Wiedner S (2000) Int.J.Polym.Anal.Charact. 5:549
116. Murgasova R (1998) Copolymers of macromonomers and their molecular characterization in solution, Ph.D. work (In Slovak), Polymer Institute, SAS, Bratislava, Slovakia
117. Murgasova R, Capek I, Lathova E, Berek D, Florian D (1998) Eur.Polym.J. 34:659
118. Trathnigg B, Thamer D, Yan X, Maier B, Holzbauer HR, Much H (1994) J.Chromatogr.A 665:47
119. Trathnigg B, Kollroser M, Parth M, Röblreiter S (1999) Two-dimensional liquid chromatography of functional polyethers. In: Provder T (Ed.) Chromatography of polymers, hyphenated and multidimensional techniques. ACS, Washington, DC, p 190
120. Trathnigg B, Kollroser M, Rappel C (2001) J.Chromatogr.A 922:193
121. Lochmüller CH, Jiang C, Liu Q, Antonucci V, Elomaa M (1996) Crit.Reviews Anal.Chem. 26:29
122. Kazanskii KS, Lapienis G, Kuznetsova VI, Pakhomova LK, Evreinov VV, Penczek S (2000) Vysokomol.Soed.A 42:915

123. Lee H, Lee W, Chang T, Choi S, Lee D, Ji H, Nonidez WK, Mays JW (1999) Macromolecules 32:4143
124. Lee H, Chang T, Lee D, Shim MS, Ji H, Nonidez WK, Mays JW (2001) Anal.Chem. 73:1726
125. Gorbunov AA, Solovyova LYa, Skvorcov AM (1998) Polymer 39:697
126. Trathnigg B, Thamer D, Yan X, Kinugasa S (1993) J.Liq.Chromatogr. 16:2439
127. Keil C, Esser E, Pasch H (2001) Macromol.Mater.Eng. 286:161
128. Gorshkov AV, Evreinov VV, Lausecker B, Pasch H, Becker H, Wagner G (1986) Acta Polymerica 37:740
129. Gorshkov AV, Evreinov VV, Entelis SG (1988) Zh.Fiz.Khim. 62:490
130. Gorshkov AV, Prudskova TN, Guryanova VV, Evreinov VV (1983) Zh.Fiz.Khim. 0:182
131. Guryanova VV, Pavlov AV (1986) J.Chromatogr. 365:197
132. Gorshkov AV, Prudskova TN, Guryanova VV, Evreinov VV (1986) Polym.Bull. 15:465
133. Petrakova EA, Okuneva AG, Balkova EV (1991) Russian J.Phys.Chem. 65:1449
134. Gorshkov AV, Verenich SS, Evreinov VV, Entelis SG (1988) Chromatographia 26:338
135. Gorshkov AV, Verenich SS, Markevic MA, Petinov VI, Evreinov VV, Entelis SG (1989) Vysokomol.Soed.A 31:1878
136. Adrian J, Braun D, Rode K, Pasch H (1999) Angew.Makromol.Chem. 267:73
137. Adrian J, Braun D, Pasch H (1999) Angew.Makromol.Chemie 267:82
138. Pasch H, Adrian J, Braun D (2001) GIT Spezial - Separation 2:104
139. Solomko SI, Kuzaev AI (1994) Russian J.Phys.Chem. 68:1624
140. Kuzaev AI, Olkhova OM, Solomko SI, Baturin SM (1994) Russian J.Phys.Chem. 68:1629
141. Bektashi NR, Alieva DN, Dzhalilov RA, Ragimov AV (2000) Polym.Sci.Ser.B 42:276
142. Falkenhagen J, Krüger RP, Schulz G, Gloede J (2001) GIT Labor-Fachzeitschrift 4:380
143. Mencer HJ, Grubisic-Gallot Z (1982) J.Chromatogr. 241:213
144. Krüger RP, Much H, Schulz G, Rikowski E (1999) Monatshefte für Chemie 130:163
145. Gorshkov AV, Overim T, van Aalten Ch, Evreinov VV (1989) Vysokomol.Soed.A 31:818
146. Guryanova VV, Prudskova TN (1994) Vysokomol.Soed.B 36:1731
147. Prudskova TN, Guryanova VV, Gorshkov AV, Evreinov VV, Pavlov AV (1986) Vysokomol.Soed.B 28:757
148. Wachsen O, Reichert KH, Krüger RP, Much H, Schulz G (1997) Polym.Degrad.& Stabil. 55:225
149. Macko T, Aust N, Lederer K (2001) Submitted to Macromol.Symp.
150. Bartkowiak A, Murgasova R, Janco M, Berek D, Spychaj T, Hunkeler D (1999) Int.J.Polym.Anal.Charact. 5:137
151. Macko T (2000) Unpublished results.
152. Bartkowiak A, Hunkeler D (2000) Int.J.Polym.Anal.Charact. 5:475
153. Sauzedde F, Hunkeler D (2001) Int.J.Polym.Anal.Charact. 6:295
154. Bartkowiak A, Hunkeler D (1999) Liquid chromatography under limiting conditions: A tool for copolymer characterization. In: Provder T (Ed.) Chromatography of Polymers, hyphenated and multidimensional Techniques. ACS, Washington, DC , p 201
155. Berek D, Hunkeler D (1999) J.Liquid.Chromatogr.& Related.Technol. 22:2867
156. Berek D (1999) Interactive properties of polystyrene/divinylbenzene and divinylbenzene-based commercial SEC columns. In: Wu C (Ed.) Column Handbook for Size Exclusion Chromatography. Academic Press, New York, p 445
157. Bartkowiak A, Hunkeler D, Berek D, Spychaj T (1998) J.Appl.Polym.Sci. 69:2549
158. Hunkeler D, Janco M, Berek D (1996) Review of critical conditions of adsorption and limiting conditions of solubility in the liquid chromatography of macromolecules. In: Potschka M, Dubin PL (Eds.) Strategies in size exclusion chromatography. ACS, Washington, DC, p 250
159. Philipsen HJA, Claessens HA, Jandera P, Bosman M, Klumperman B (2000) Chromatographia 52:325
160. Cifra P, Bleha T (2000) Polymer 41:1003
161. Guttman CM, Di Marzio EA, Douglas JF (1996) Macromolecules 29:5723

162. Cifra P, Bleha T (2001) Macromolecules 34:605
163. Barton AFM (1985) Handbook of Solubility Parameters and other Cohesion Parameters, CRC press, Inc., Boca Raton, Florida
164. Brandrup J, Immergut EH, Grulke EA (1999) Polymer Handbook. 4th ed., Wiley, New York
165. Snyder LR (1968) Principles of Adsorption Chromatography, Marcel Dekker, New York
166. Palamareva MD, Palamarev HE (1989) J.Chromatogr.A 477:235
167. Meyer VR, Palamareva MD (1993) J.Chromatogr. 641:391
168. van der Beek GP, Cohen Stuart MA, Fleer GJ, Hofman JE (1991) Macromolecules 24:6600
169. Inagaki H (1977) In: Tung LH (Ed.) Fractionation of Synthetic Polymers. M. Dekker, New York, p 649
170. Lee W, Lee HC, Park T, Chang T, Chang JY (1999) Polymer 40:7227
171. Kanazawa H, Sunamoto T, Matsushima Y, Kikuchi A, Okano T (2000) Anal.Chem. 72:5961
172. Berek D, Bleha T, Pevna Z (1976) J.Polym.Sci., Polym.Lett. 14:323
173. Berek D, Bleha T, Pevna Z (1976) J.Chromatogr.Sci. 14:560
174. Poppe H, Kraak JC, Huber J, van der Berg JHM (1982) Chromatographia 14:515
175. Macko T, Berek D (2001) J.Liq.Chromatogr.& Relat.Technol. 24:1275
176. Skvortsov AM, Gorbunov AA (1990) J.Chromatogr. 507:487
177. Kilz P, Krüger RP, Much H, Schulz G (1993) Polym.Mat.Sci.Eng. 69:114
178. Prospect literature (2001), Polymer Standard Service, Mainz, Germany

Editor: H.H. Kausch
Received: April 2002

FTIR Microspectroscopy of Polymeric Systems

Rohit Bhargava[1], Shi-Qing Wang[2], Jack L. Koenig[3]

[1] Laboratory of Chemical Physics, NIDDK, National Institutes of Health, Bethesda, MD 20892, USA, E-mail: rohit@sunder.niddk.nih.gov
[2] Department of Polymer Science, University of Akron, Akron, OH, USA, E-mail: wang@polymer.uakron.edu
[3] Department of Macromolecular Science, Case Western Reserve University, Cleveland OH 44106, USA, E-mail: jlk6@po.cwru.edu

Abstract Fourier transform infrared (FTIR) spectroscopy is a mature analytical technique employed to examine polymeric materials. With the coupling of an infrared interferometer to a microscope equipped with specialized detectors, FTIR spectroscopy has been employed widely to examine microscopic areas in polymers for the last twenty years. Following the emergence of instrumentation techniques that employ focal plane array (FPA) detectors, FTIR microspectroscopy has experienced a recent renaissance in terms of the capability of instrumentation and visualization afforded for examining multicomponent polymer systems. We present an overview of the principles of instrumental configurations used to achieve spatially resolved spectral information, their relative advantages and limitations. Illustrative examples are presented that demonstrate the capabilities of FTIR microspectroscopy and the insight this technique provides into the composition, formation, and behavior of polymeric materials. Finally, some emerging techniques that may permit microspectroscopic analyses on a different spatial scale are reviewed.

1	Introduction	139
2	FTIR Microspectroscopy Using a Single Element Detector	140
2.1	Introduction	140
2.2	Instrumentation	141
2.2.1	Sampling Techniques	142
2.2.2	Data Processing	144
2.2.3	Errors in FTIR Mapping	145
2.3	Applications	147
2.3.1	Microsamples, Additives, Contaminants and Degradation	147
2.3.2	Single Polymer Fibers	150
2.3.3	Polymers in Biological Areas	150
2.3.4	Semi-crystalline Polymers	151
2.3.5	Phase Separated Polymer Blends	152
2.3.6	Multilayered Polymer Systems	154
2.3.7	Diffusion	155
2.3.8	Local Orientation	156
2.3.9	Polymer Surfaces and Interfaces	157

© Springer-Verlag Berlin Heidelberg 2003

2.3.10	Polymer-Liquid Crystal Systems	157
2.3.11	Filled Systems	159
2.3.11.1	Carbon Black, Silica and Inorganics Filled Systems	159
2.3.11.2	Fibers and Fillers	160
2.4	Summary	161

3	**FTIR Imaging**	**161**
3.1	Introduction	161
3.1.1	Mapping Versus Imaging	162
3.2	Instrumentation	163
3.2.1	Sampling Techniques	165
3.2.2	Sources of Error	165
3.2.2.1	Interferometer Noise	165
3.2.2.2	Detector Noise	166
3.2.2.3	Failing Bump Bonds	166
3.2.2.4	Intensity Distribution, Pixel Saturation and Dynamic Range	167
3.2.2.5	Sample	167
3.2.3	Post-Collection Operations	167
3.3	Applications	171
3.3.1	Contaminants/Defects in Polymers	171
3.3.2	Amorphous Polymer Blends	172
3.3.3	Semi-Crystalline Polymers and Their Blends	172
3.3.4	Polymer Laminate Films	172
3.3.5	Polymer-Liquid Systems	174
3.3.6	Embedded Polymer Systems	174
3.3.7	Polymer-Liquid Crystal Systems	175
3.3.8	Phase Separation in Polymers	177
3.4	Summary	178

4	**Other Infrared Microspectroscopy Approaches**	**179**
4.1	Hadamard Transform Infrared Microscopy	179
4.2	Synchrotron Infrared Microscopy	180
4.3	Solid State Focal Plane Array Imaging	180
4.4	Scanning Probe Microspectroscopy or Photothermal Imaging	182
4.5	Near Field Infrared Mapping	183
4.6	Infrared Diode-Based Imaging	184
4.7	Summary	185

5	**Concluding Remarks**	**186**
5.1	Instrumental Directions	186
5.2	Applications	186
5.3	Processing Techniques and Strategies	187

5.4   Summary 187

References 188

1
Introduction

Most utility polymeric articles available today contain multiphase polymeric systems comprised of semi-crystalline polymers, copolymers, polymers in solution with low molar mass compounds, physical laminates or blends. The primary aim of using multicomponent systems is to mould the properties available from a single polymer to another set of desirable material properties. The property development process is complex and depends not only on the properties of the polymer(s) and other components but also on the formation process of the system which determines the developed microstructure, and component interaction after formation. Moreover, the process of polymer composite formation and the stability of the composite is a function of environmental parameters, e.g., temperature, presence of other species etc. The chemical composition and some insight into the microscopic structure of constituents in a polymer composite can be directly obtained using Infrared (IR) spectroscopy. In addition, a variety of instrumental and sampling configurations for spectroscopic measurements combine to make infrared spectroscopy a versatile characterization technique for the analysis of the formation processes of polymeric systems, their local structure and/or dynamics to relate to property development under different environmental conditions. In particular, Fourier transform infrared (FTIR) spectroscopy is a well-established technique to characterize polymers [1, 2].

While dispersive IR spectroscopic instruments have been less popular for long, most advances in the last thirty years were confined to FT systems due to their ease of use, small experimental times, high throughput, and high reproducibility. Obtaining FTIR spectra from polymeric samples is usually relatively rapid and straightforward. Moreover, the technique has attained high precision and accuracy in measurement to be sufficiently reproducible for most industrial, research and development purposes. Several experimental and post-data collection techniques for specialized applications not amenable to routine IR spectroscopic analysis have been developed over the years and a vast database of knowledge exists. However, IR spectroscopy had primarily been a "bulk" examination technique as obtaining spectral information from microscopic areas was difficult. The last fifteen-twenty years have seen considerable activity and development in the capability to collect IR spectra from small, specific regions of a sample. These developments have resulted in the development of a sub-field of considerable research activity – IR microspectroscopy. The set of techniques enabling microspectroscopic examination have demonstrated ability or great potential in examining the chemical structure to relate it to physical characteristics of materials and their fabricated products [3].

In this review, we present the current state of technology for infrared microscopic characterizations, illustrative applications, and possible future directions for this field with respect to characterizations of polymers. Infrared microspectroscopy, or the infrared spectroscopy of microscopic areas, can be broadly divided into three approaches based on the instrumentation used. The first class, using a large single element detector, is also the oldest. Such instrumentation has been commercially available for almost 15 years. The next approach, using array detectors, has been employed for a little over 5 years with commercial instrument availability at less than 5 years. Other approaches to microscopic detection are still being developed and are available in very few researchers' laboratories. FTIR microspectroscopy combines the spatial specificity of microscopy with the powerful chemical specificity of spectroscopy. Hence, any developed instrumentation or experiment is designed to emphasize some aspect of the spatial or chemical aspects of analysis. A balance between the need for sensitive spectral information essential for molecular characterization and the need for high spatial resolution visualization essential to morphological analysis is often achieved by instrumental and temporal limitations. Specific applications usually determine which of the spectral or the spatial characteristics are to be emphasized and what tradeoffs are tolerated. Hence, in this review, the three broad classes of infrared microspectroscopy are examined in sequence with emphasis on instrumentation capabilities and experimental possibilities. In analyzing applications, we focus primarily on the information-rich and well-characterized mid-infrared region of the spectrum, particularly suited for polymer analyses, and refer the reader to reviews on near-infrared applications [4].

2
FTIR Microspectroscopy Using A Single Element Detector

2.1
Introduction

While the first reports of a micro-capability spectrometer appeared 50 years ago [5], the technique made no major advances in applications to polymers until the coupling an interferometer and microscope to a digital computer enabled Fourier transform spectroscopy, time averaging, and mapping. The light throughput is used more effectively in the FT process and the development of more stable, sensitive, fast-response cryogenic detectors allowing for reproducible measurements further spurred interest in microspectroscopy. Infrared interferometers coupled to modern infrared microscopes incorporating sensitive detectors and a digital computer became commercially available in the 1980s and are, to date, popular analysis tools. There are estimated to be ~5000 systems currently in use. Spectral information from a small area of the sample can be obtained by restricting the area illuminated by the infrared beam using opaque apertures of pre-defined size. This allows microspectroscopic examination of small samples and can routinely be used to examine specimens down to the microgram range. However, single point examinations are of limited use and statistical viability in examining multicomponent polymers. By moving the sample in a known, pre-determined manner relative to the aperture, a point-by-point examination of a contiguous area of any size can be carried out. This technique is termed FTIR Mapping as by plotting the absorbance magnitude of a specific vibrational mode over the area, a map of that chemical species' relative concentration can be obtained. Maps of chemical species' abundance are also termed chemical maps or functional group maps. While optical microscopes use differences in refractive index, selective phase staining or polarized light for contrast, the spectral signature of a material provides the contrast in FTIR mapping. The non-invasive, non-destructive characteristics of IR spectroscopy are maintained while adding microscopic examination capabilities.

2.2
Instrumentation

Three major additions to the spectrometer are required for microspectroscopic mapping to be accomplished. First, the radiation from a modulated source has to be diverted to a microscope to focus on the desired sample area. A rapid scan interferometer is generally used as a source of radiation for SNR considerations, though the technique is not modulator specific. A step-scan interferometer, filter, or dispersive system may also be employed. No modifications are required to a standard interferometer and an attached computer can trigger interferometer scanning at a particular sample spatial position. Second, radiation reaching the detector is limited to be only from the area of interest on the sample. This is accomplished by using opaque masks or apertures, usually equipped with adjustable slit dimensions. Third, a positioning stage is required to precisely and reproducibly position the sample many times. Each stage movement is sought to be accomplished in as short a time period as possible. A computer is used to control the stage, interferometer, equipped to acquire optical images, and collect and store data. The IR microscope (shown schematically in Fig. 1) is very similar to the optical microscope. As there is no means to visualize the infrared image ab-initio, the sample has to be first visualized using optical methods. To obtain IR spectra from microscopic regions, apertures are employed first to selectively limit the wide field of view of the optical signal. IR radiation can then only pass through the slits and spectra are acquired by measuring this throughput as a function of interferometer retardation. The dependence of the infrared sampling on the optical image requires that the optical and infrared paths be parfocal and collinear. The major difference between optical microscopes and the one required for IR use is the incorporation of all-reflecting

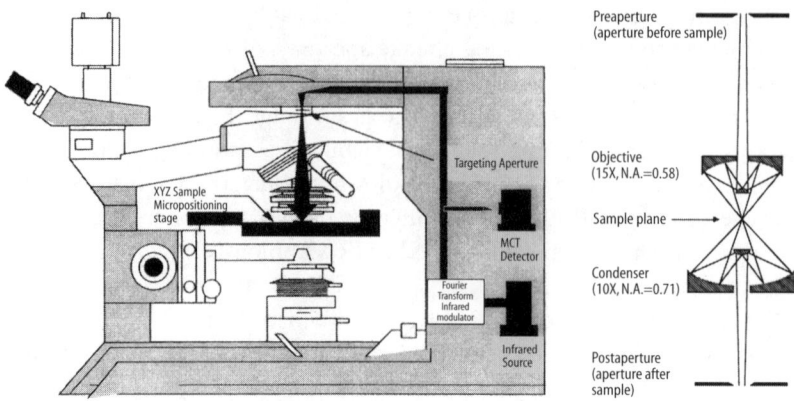

Fig. 1. Schematic diagram of an infrared microscope (*left*) reproduced from Koenig JL, Microspectroscopy of Polymers. The details of the focusing optics (*right*) are reproduced from Ref. [18]

optics to allow maximal IR transmission and the incorporation of aspherical reflecting surfaces in a Cassegrain-type configuration to minimize optical aberrations. Refractive elements, if required for some special needs, are made from IR transmitting materials that are resistant to moisture (e.g. CaF_2). Apertures used are usually coated with highly absorbing carbon black to eliminate stray radiation. Polarizers or any other filters can be inserted into the beam path and polarized infrared measurements can be conducted similar to polarized measurements using visible radiation.

2.2.1
Sampling Techniques

Radiation incident on a sample results in radiation interacting with the sample to be transmitted, reflected, refracted, absorbed or emitted. Almost any of these modes [6] (Fig. 2) may be used for examining the absorbance characteristics in a microscopic configuration. However, microscopy places unique demands and some techniques have been more popular than others. Transmittance and reflectance techniques remain the most popular due to the ease of sample preparation and conduct of experiments. Light transmission through the sample allows for easy sample positioning, good light throughput and yields spectra that require little processing effort. However, sample preparation is more involved as most materials are strongly absorbing at their characteristic frequencies. The samples have to be thin, usually requiring experience and expertise in sample preparation in addition to the time and effort required to prepare optimal thickness samples. Microtoming a sample is often required and an appropriate embedding medium [7] may

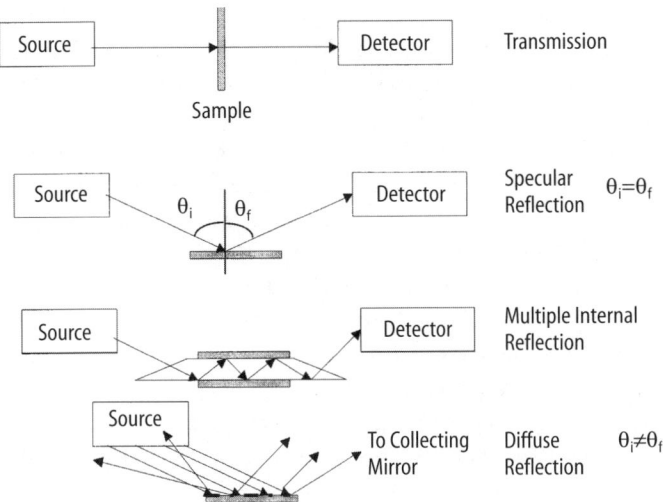

Fig. 2. Major Sampling Techniques for Infrared Spectra Collection in a Microscopic configuration

sometimes be used to handle a small sample before that step. While spectra may be obtained by flattening a sample [8] or using a diamond cell [9] to press the samples into thin layers, unwanted side effects of this approach may be manifest in terms of loss of orientation, interference fringes and distorted polarization. Once a good sample is obtained, quantification of data obtained from it is straightforward. The absorbance, A, of a specific vibrational mode is related to the incident, I_i, and transmitted, I_t, intensities of radiation sample parameters as,

$$A = abc = \log_{10}\left(\frac{I_i}{I_t}\right) \qquad (1)$$

Where, a is the absorption coefficient, b the path length of radiation through the sample and c the concentration of the absorbing species.

Reflectance spectroscopy is commonplace for samples that cannot be prepared for transmittance measurements. However, reflectance measurements must be carefully conducted as the reflected beam is not only indicative of the composition of the sample but is also affected by surface conditions at the sample plane. This makes the reflectance spectra, though indicative of material chemistry, difficult to interpret and generally less useful for quantitative analysis. Since the polarization of the beam is maintained for reflectance, especially specular reflectance methods, examination of orientation at polymer surfaces using reflection techniques is attractive [10]. Reflection-absorption modes involve the transmission of the infrared beam through the sample and subsequent reflection to pass through the sample again. Usually, sample preparation is difficult for such experiments and they are useful when samples are thin in their native state and/or supported on a reflecting surface. Specialized accessories have been developed to control the angle of incidence and condense the beam after reflection. Using appropriate equipment, spectra from films as thin as a few nanometers can be observed. Attenuated total reflection (ATR) has been combined with the microscope to carry out ATR microspectroscopy [11, 12]. An alternative approach has been developed that uses cartridges with hemispherical ATR elements and can be used in any microscope with reflectance capabilities [13]. Micro-Photoacoustic spectroscopy and time resolved microspectroscopy have also been reported. Adding to the versatility of the technique, temperature control may be achieved using a standard microscopy cell with IR transmitting windows [14]. There are no limitations to employing constant humidity and pressure (and vacuum) cells if the need arises. Special accessories [15, 16] can be readily incorporated by small changes in the optical configuration to examine different types of samples. FTIR microspectroscopy using polarized light provides information on the orientation of molecular species, which is especially useful for examining liquid-crystalline or semi-crystalline polymers and their composites. Infrared polarizers can be incorporated in microspectroscopic instrumentation to allow polarized radiation to impinge on the sample. Usually, two measurements, between which the polarizer is rotated 90° are carried out to yield dichroic ratios of specific absorption modes. These can then be related to the relative orientation of polymer repeat units. However, the calculation of accurate dichroic ratios is predicated on the absence of stray radiation, polarization accuracy through the microscope optical train and focused optics at the sample. These conditions are difficult to maintain in a routine manner and thus, polarization experiments have not been widely carried out.

2.2.2
Data Processing

Data processing of mapped areas is mostly spectra based. Baseline correction of individual spectra, thickness corrections of individual area elements and other such plotting steps are routinely applied. The size of the data sets depends on the spatial resolution and area mapped. Given the large mapping time and low spatial resolution in most cases, the mapping technique is ideally suited to examining samples that have large spatial features, the discrimination of which does not require high resolution spectroscopy. Most often, the data sets are of a size smaller than 1 MB. Data processing is fast and many of the techniques applied to single spectra can be applied to extract information. Computation speeds and power requirements are moderate as only a few hundred spectra (at most) are ever processed. More complex data processing may be employed to improve the SNR, spatial resolution and component extraction. Surface contaminants as small as 10 µm could be profiled by a data reduction program used to obtain projections about spatial locations. A method to co-add spectra from different positions was also suggested to improve spectral quality [17]. Since spectra collected are so few in number, it is usually more time effective to simply collect data for a longer time than apply any sophisticated techniques to improve the signal to noise ratio and extract information.

2.2.3
Errors in FTIR Mapping

The most glaring source of error comes from the component of the technique that facilitates microspectroscopy-apertures. Apertures are the largest source of diffraction in the IR microscope and can lead to the detector sampling light from outside the apertured region. The detector also samples the secondary lobes of the diffraction pattern. Thus, spectral information from the delineated area is spread over a larger area than the aperture. A case study revealed the effects of this stray light sampling as far as 40 µm away from the sample (Fig. 3) [18]. Consequences of this stray light on spectra were studied under conditions of different aperturing modes and sizes. This problem can be circumvented by using a second aperture to delineate the same region (redundant aperturing). The spectral purity is now increased at a cost to the amount of light allowed through the system. Hence, spectral quality (SNR) is degraded. This requires very large collection times (loss in temporal resolution) or larger apertures (loss in spatial resolution). Thus, the three interdependent errors (stray light, low SNR and low spatial resolution) limit the effectiveness of the technique as a mapping tool to examine the spatial distribution of chemical species. Stray light not only compromises spatial fidelity, but may also lead to errors in determination of absorbance [19].

With dual aperturing in the mid-IR range, the practical resolution limit is close to 15 µm. This lower limit may be further increased to ~10 µm, but at a substantial time cost. Typical collection time for a spectrum from a region is ~30 s. A further time of a few seconds is required to collect the data, process it, store it in the correct sequence, refresh data display and step to the next region before starting the next scan. Hence, to map a 500 µm × 500 µm region with a resolution of ~15 µm, it will take this mapping technique approximately 10 hours. Near field acquisition techniques have been suggested to improve spatial resolution and mask stray light. One such study [20] analyzed the acquisition of a physically masked sample (i.e. no mapping was possible). Another [21] proposed a near-field aperture turret to allow for mapping. Once again, the cost of improved resolution and accuracy is higher alignment and collection time.

Quantitative analyses require that spectra be measured from samples that are non-luminous, their absorbance be invariable with changes in concentration, the absorption coefficient of the species be independent of the intensity of incident radiation, the spectral response be uniform across the radiation cross-section, the instrument be devoid of stray light and respond in a linear manner to different lev-

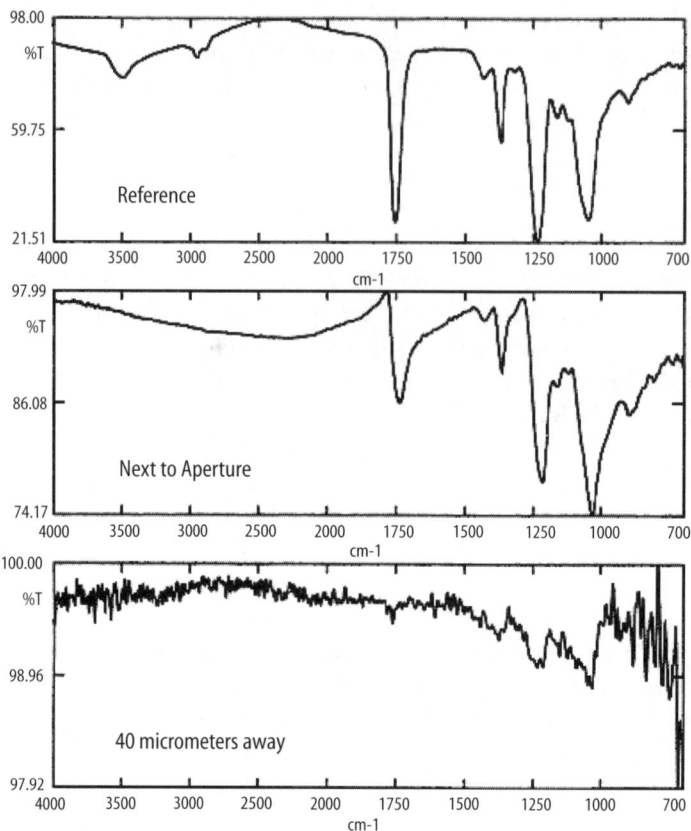

Fig. 3. The effects of aperturing can be seen on spectra collected with the aperture examining a polymer film (*top*) aligned with the film edge (*middle*) and 40 μm away from the edge. (reproduced from Ref. [18])

els of light intensity. Some or many of the factors above are violated by sample non-uniformity, instrumental factors and the behavior of chemical species (interactions) leading to incorrect measurements. The effects of deviations can be theorized, quantified and experimentally verified. While diffraction has been found to be a major source of frustration, Siedel aberration may also serve to limit mapping fidelity [22]. In particular, centrally obscured reflecting optics (Fig. 1) are prone to spherical aberration effectively requiring adjustments to the optics after mapping small areas on the sample and limit the field of view for wide field analysis if the sample is not disturbed. In addition, the central obscuration reduces light throughput, further starving the detector of already scarce radiation. Numerical apertures of reflecting systems are typically limited (typically to less than 0.7) and magnification of no more than 50×. This limitation is not important for routine FTIR microscopy in the mid-infrared and hence, reflective optics prove to be better suited – scientifically and economically – compared to refractive optics utilizing complex lens design. If the required wavelength range for infrared analysis allows for the use of materials (e.g. glass or CaF_2) that may prove to be economically feasible, refractive optics that provide higher numerical apertures and spatial resolutions may be readily employed. However, chromatic aberration would have to be corrected for by appropriate microscope design [23]. Errors may also arise during use due to improper location of the spot to be mapped, substrate optical effects that shift the beam focus and loss of accuracy in system alignment over time.

2.3
Applications

A compilation [24] and discussion [25] of the applications of FTIR microspectroscopy to polymers is available. A concise introduction to the diverse applications can also be found [26]. Some newer and representative examples of various applications are presented below to illustrate the possibilities of IR microscopy. We have included more studies that seek to map an area rather than obtain spectra at discrete points. We believe the true utility of microspectroscopic instrumentation lies in examining a large area to find localized spectral changes akin to microscopy.

2.3.1
Microsamples, Additives, Contaminants and Degradation

Many polymeric samples are often available in small quantities for analysis. For example, new synthetic polymers are usually available in small quantities and forensic specimens may be limited in availability. In many situations, the identification and/or quantification of such samples are critical. However, samples smaller than the diameter of the probing beam lead to spectral artifacts, which complicate identification and quantification. Hence, many of these "microsamples" cannot be appropriately analyzed by conventional wide beam spectroscopy in their native state and microspectroscopy presents perhaps the only route to obtain artifact free spectra and consequently, the best chance of material identification.

Specially designed sample handling accessories can be used to localize the beam and/or control sample dimensions for obtaining a spectrum of the microsample. Hence, the sample does not need to be mixed with/embedded in a matrix. While small sized samples and concentrated contaminants can be detected, FTIR microspectroscopy can also be used to examine minute quantities of samples. While it has been claimed that concentrations in the nanogram range [27] can be detected, we believe that concentration levels down to ~0.1% (w/w) of the volume of the beam at the sample can usually be quantified routinely. The sample preparation for such sensitive measurements is of crucial importance. Specifically, it is usually required that the material be isolated and pressed into a thin film over a spot size of ~15 µm to obtain the best spectra. Many specialized sample handling accessories are available and can be employed to accomplish this task or provide alternate sampling geometries if pressing the sample into a film is not desirable or viable. One such microsampling accessory, the Diamond Anvil Cell (DAC), allows for small sample handling and applies pressure to the sample, which can be employed to change the dimensions of the sample.

The sensitivity of FTIR spectroscopy to microscopic defects is greatly increased compared to a macroscopic measurement – if the defective area is examined. As opposed to single element, wide beam spectroscopy where the spectrum from the entire sample is obtained, the polymer does not overshadow the contaminant if a spectrum from a microscopic area of the sample is obtained. The contaminant can be later identified from a library of known compounds, providing clues to process control and helping to maintain quality control. This approach has been used to identify gel inclusions in poly(ethylene) [28, 29], contaminant on the surface of a semi-conductor device [30], acrylic fiber on a microcircuit die (Fig. 4) [31], contaminants in poly(vinyl chloride) (PVC) [32] and a mold release agent on the surface of a polyurethane [33]. An additive (erucamide) in linear low density polyethylene (LLDPE) that was found to migrate to the surface [34] thereby changing the surface properties could be readily identified. Various contaminants, whether in a

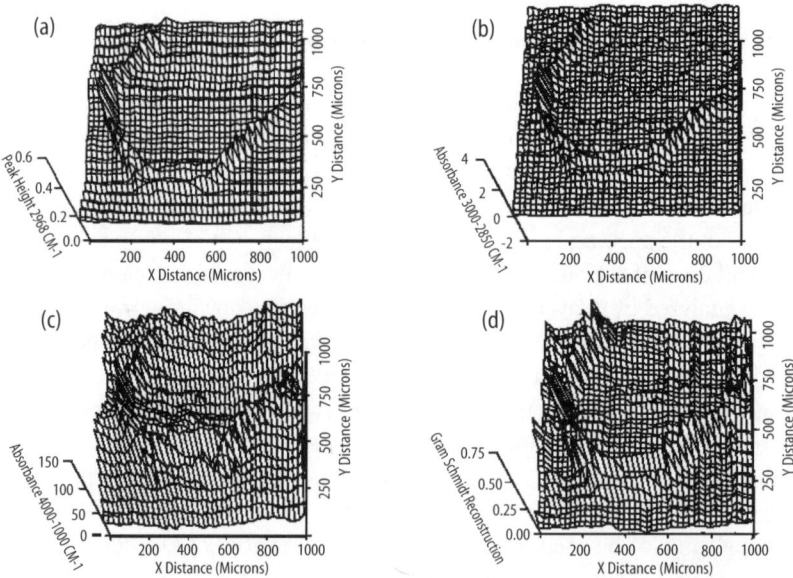

Fig. 4. The distribution of a contaminant is examined by reconstructing images of an acrylic fiber contaminant on a microcircuit die using various methods of visualization. Image generation methods include the use of **a** absorbance intensity at 2968 cm^{-1}, **b** integrated absorbance between 2850 and 3000 cm^{-1}, **c** integrated absorbance from 1000 to 4000 cm^{-1}, and **d** the Gram–Schmidt method. (reproduced from Ref. [31])

film on the surface, as inclusions in the bulk or simply attached to a polymer device can be identified using appropriate techniques to obtain a sample with the contaminant. In a few minutes of experimental time, the presence and identity of the contaminant can be readily determined.

Contaminants may be identified by two approaches: optical microscopy is first used to identify an anomaly followed by its isolation in the IR field of view by apertures. A spectrum is then acquired from the small, delineated area to identify the contaminant. Clearly, the identification of the small area containing the suspected contaminant using light microscopy is critical. Optical contrast needs to be sufficiently high to allow such delineation and the question as to the contaminant's identity can be resolved by microspectroscopy. Alternately, the whole sample area is mapped and a contaminant can be identified if an anomalous signal is observed from a small area in the map. Mapping is especially useful when the optical detection of contamination is difficult due to the heterogeneous nature of the sample or little difference in refractive indices between the sample and the contaminant. General mapping of sample areas has also been known to lead to serendipitous discoveries of contamination or to the discovery of more than one type of contaminant/degradation product. For example, the degradation of poly(propylene) (PP) was studied using FTIR mapping [35] indicating the presence of many different types of degradation products. Similarly, localized degradation photoxidation of polymers [36] can also be examined whether on the surface or in the bulk after microtoming. Other case studies reported have included lubricant in nylon [37], pyrolysis products from microsamples, inclusions in PVC, inclusions in a rubber sample and growth of foreign material in a brewery pipe [38]. Process contaminations can be readily examined, for example, anomalous material on a magnetic disk during manufacture was found to be cellulosic with some kaolin clay, probably from a coated paper used in the manufacturing process [29]. Another contaminant reported in the study was similar in visual appearance; but was found to be a polyamide that matched the Nylon resin used. Yet another contaminant on the disks, which could not be distinguished by optical microscopy alone, was found to be a polyglycol. Thus, different residues due to manufacturing defects can be readily identified and the process step where the contamination occurred can be monitored and/or modified.

The degradation of a polymeric sample often begins at the surface and proceeds into the bulk. Injection molded samples of polyamide 6,6 [39] were aged at elevated temperatures and examined to determine changes. Skin-core oxidation was observed and evaluated by infrared spectroscopy and by imaging chemiluminescence. The oxidation depth profiles determined by the two techniques using the carbonyl stretching frequency (FTIR) and the peroxide depth profiles (chemiluminescence) showed good agreement and the results indicated a diffusion-controlled oxidation process. Thus, not only can a material be monitored for contamination and/or degradation but also, in some cases, the mechanism and the source of the anomaly can be probed to allow characterization of the process which lead to the appearance of the anomaly or failure of the product.

2.3.2
Single Polymer Fibers

Single component, multicomponent fibers [40] or fibers with dyes [41] or pigments [42] that have diameters down to 10 µm can be detected and analyzed by FTIR microspectroscopy. Cotton, wool and acrylic fibers could be readily differentiated by examining their infrared spectra [43]. Fiber blends (cotton/terry vs. cotton/polyester) could also be differentiated. PET fibers [46, 44] and a polysulfone fiber in an epoxy matrix could be characterized. Groups of fibers classified according to size generally require different methods of sample preparation [44]. Fibers larger than ~30 µm in diameter did not require much sample preparation. For smaller fibers, better results were obtained by effectively increasing the sample width by flattening the sample using a DAC. ATR microspectroscopy can also be employed to provide information when transmission experiments are difficult or surface weighted information is required [45].

While the detection and identification of single fibers can be readily accomplished using microspectroscopy, the quantitative analysis of orientation is complex [46]. Polarized intensities obtained from a large film using microscopy and conventional wide beam illumination were found to agree well. However, a lensing effect similar to that of using variable path length cells [47] due to the curvature of fibers is observed when fibers are examined. This lensing effect complicates identification and severely affects the determination of orientation. To eliminate the lensing effect, flattening the fiber is suggested to make the surfaces parallel and the specimen larger. However, this process may result in loss of the original orientation in the fiber. The effects of polarization scrambling by the system optics, diffraction effects and stray light can be reduced by proper system design to yield better quantification of orientation. The use of physical masking at the sample plane was shown to result in high photometric reproducibility. By applying many of these experimental protocols, high quality spectra and structural information can be obtained reliably for many fibers [48].

2.3.3
Polymers in Biological Areas

Polymers are used for biological applications in a variety of situations from drug carriers to components of artificial limbs. In an instance of replacement of live tissue, PTFE grafts are used to treat lower-extremity ischaemia when autologus saphenous veins are unavailable. As a step toward understanding the ensuing complications and tissue build-up, a graft was simulated in a flow device [49]. The concentration of lipids within the grafts was monitored by FTIR microspectroscopy. Lipid uptake was found to be rapid initially and then slowed down over time. These model studies are important in developing an understanding of the transport in porous vascular prosthesis and FTIR spectroscopy and microscopy present a sensitive analysis tool to detect composition changes in biomaterials.

The oxidation and subsequent degradation of polymers may lead to a loss of mechanical properties and product failure for biological devices. One area where performance is critical and is affected strongly by polymer degradation is the use of implanted prosthetic devices. Ten new polyethylene prosthetic components (PEs), sterilized by two different routes were examined using FTIR-ATR and microscopy [50]. All the samples showed some degradation due to the formation process. However, by microscopic measurements and derivatization techniques, the ethylene oxide (EO)-sterilized prosthetic components were shown to be different from the gamma-radiation sterilized. EO sterilized devices had low levels of depth oxidation similar to degradation during preparation of samples. Gamma-sterilized devices showed higher oxidation levels, variable from sample to sample, on the surface and in the bulk. The oxidation is described both by carbonyl species distribution and by hydroperoxide concentration. Not only could differences in degradation but the scheme of degradation could also be inferred by IR analysis. Chain scission of gamma-radiated PE, estimated by IR analysis, results in reduction of molecular weight and lowers its abrasion resistance. Thus, the effects of the sterilization methods could be observed. The example illustrates the applicability of FTIR microspectroscopy in product development where FTIR microspectroscopy is especially attractive given the relatively low cost of instrumentation and the large amount of information provided in most instances. In a similar vein, the effects of near-colon activity on a cross-linked polymer were examined in-vitro [51]. A combination of conditions of use and rapid and accurate analysis is a cornerstone of the development process and FTIR microspectroscopy can often play a role.

2.3.4
Semi-Crystalline Polymers

Poly(vinylidene fluoride) is an interesting polymer that has a large number of crystal phases, which determine its performance properties. The crystal structure of PVDF has been analyzed to identify its various phases [52]. The different types of spherulites seen optically were assigned to different crystal modifications using FTIR identification. Blends of PVDF with PMMA were analyzed to determine differences in interaction based on PMMA tacticity [53]. On the basis of shifts in the carbonyl region, it was suggested that i-PMMA had stronger interactions with PVDF than s-PMMA. The crystalline and amorphous regions of another polymer with potentially wide application, Nafion [54], were examined by FTIR microscopy. The performance of the fuel cell is critically dependent on the water uptake in the polymer electrolyte. In the crystalline regions, sulfonic acid groups are present at the end of the side chains. Thus, the crystalline regions contain no water molecules. In the amorphous regions there is a complete proton transfer from the acid to the water molecules. As a result, sulfonate groups are obtained and water retention is facilitated. Such microstructure determinations, which affect performance properties and are critical for quality control, can be routinely carried out using FTIR microscopy.

External nucleating agents often facilitate polymer crystallization and a variety of analytical techniques can be employed to monitor the effects of these agents. The effect of seeding agents (saccharin, phthalimide and boron nitride (BN)) on the rate of crystallization of polyhydroxybutyrate (PHB), and its blends with poly (hydroxybutyrate-co-valerate) (CHB/HV) [55] was investigated using optical microscopy and DSC. However, most of these techniques are incapable of providing the distribution of chemical species across the various morphological features while the chemical constitution in separate local areas can be readily determined by IR microspectroscopy. It is well known that the interaction of the nucleating agent with the (crystallizing) polymer can determine the effectiveness of the crystallization method. Thus, in the study described above, the solubility of the nucleating agent in the molten polymer was determined. Saccharin and phthalimide were found to be poorer nucleating agents compared to BN. They were found to be soluble in molten PHB, but were rejected from the crystalline PHB. On the other hand, small crystallites of saccharin and phthalimide developed within the boundaries of the spherulite, but most material accumulated in the inter-spherulitic domains. Thus, not only the interaction but also morphological location of nucleating agents can be determined. Just as crystal structure heterogeneity can be detected, the homogeneity can also be determined. Thin Polyimide (PI) films formed by vapor co-deposition of the precursor molecules pyromellitic dianhydride(PMDA) and 4,4'-oxydianiline(ODA) were examined for spatial homogeneity [56]. No evidence for PDMA phase separation or crystallization variation was found. FTIR microspectroscopy, combined with polarized radiation from the interferometer, is a powerful means of simultaneously determining local composition and relative organization in crystalline polymers often providing a complementary tool to X-ray scattering [57], DSC [58] or microscopy studies [52, 59].

2.3.5
Phase Separated Polymer Blends

Phase separated blends of poly(vinyl alcohol) [PVA] and poly(vinyl acetate) [PVAc] were studied and evidence of intra-molecular and inter-molecular specific interactions was reported to depend on blend composition [59]. Blends with high PVA content were reported to have large intra-molecular specific interactions while blends with a high PVAc content showed inter-molecular interactions. The morphology was characterized by Fluorescence optical microscopy (FOM). PVA-rich domains were identified by green fluorescence of fluorescein while PVAc-rich domains were identified using the blue fluorescence of anthracene. The two polymers did not require any staining or tags to be distinguished by infrared spectroscopy. However, the FTIR mapping spatial resolution is often insufficient to determine morphological information like size and shape requiring additional microscopic approaches. A model polymer blend system [60], poly(methyl methacrylate) [PMMA] and poly(styrene-co-acrylonitrile) [PSAN], was studied in an attempt to characterize the phase diagram of the blend system. Poor correlation was found between the phase diagram determined in this manner and the phase diagram obtained by optical microscopy. The inconsistency probably arises from the large spot size of the infrared beam and the lack of spatial fidelity of apertured spectra. Olefinic blends are often used in utility polymeric articles. In a model blend that was studied, the refractive indices of the two components- isotactic PP and syndiotactic PP- were close enough for their phase separated blend to be deemed single phase by optical microscopy. However, chemical mapping of the sample [52] shows clear phase separation based on spectral differences (see Fig. 5).

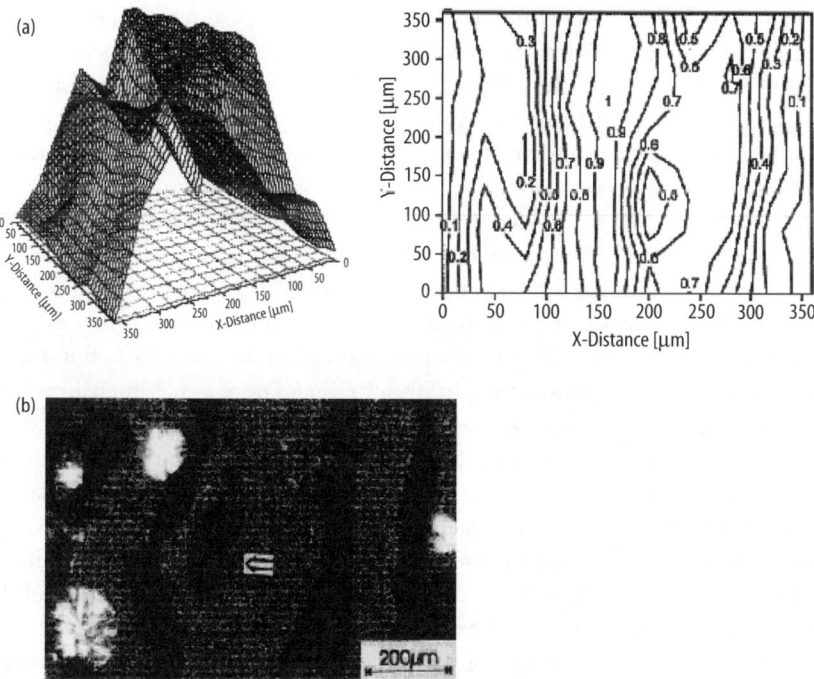

Fig. 5. The functional group image **a** of a region in a polymer blend of i-PP and s-PP. A contour plot is another visualization useful for the same. **b** The optical photograph reveals that the dark portion is s-PP surrounded by i-PP (Ref. [52])

These examples also illustrate the difference in spatial resolution and contrast mechanisms between optical and infrared microscopies. While optical microscopy is capable of higher spatial resolution, its discrimination is limited to a difference in the average of a property of materials, namely refractive index, unless specially labeled to detect a property of the label. Infrared microspectroscopy derives its contrast mechanism from the intrinsic composition of the material but suffers from a poorer spatial resolution. A judicious use of the two complementary techniques is often required to achieve good characterization. While the example above illustrated the detection of differences, FTIR microspectroscopy can also be used to determine homogeneity. For example, compositional differences in a PP-PE film could not be detected between the surface and up to 500 µm into the bulk of the sample [61].

2.3.6
Multilayered Polymer Systems

Multilayered polymer films are increasing used in packaging to take advantage of the properties of individual polymers. Often, the two (or more) polymers used are incompatible and a tie layer of another polymeric adhesive has to be used between them. This leads to complicated laminate systems with more layers than required for simple barrier properties. The analysis of such films has been reported many times in the literature [37, 62]. A laminate is usually microtomed along the thickness direction and mapped. A microtomed laminate has often been a popular one to compare microscopy techniques with other FTIR or microscopic techniques. In one such early comparison using a laminate [63], it was shown that the spectra obtained by microscopic analysis were much superior to the ones obtained by an ATR-FTIR technique. Additionally, microtoming allowed for examination of an additional dimension, where sections taken progressively in one direction could be sequentially compared. Poly(urethane) coated on an ethylene-acrylic acid (EAA) (15 % acrylic acid) was studied in this manner to find that the polyurethane had penetrated 60 µm into the EAA layer [64]. Dimensions in the tens of microns are directly accessible by FTIR microscopy. For dimensions smaller than tens of microns, spectral subtraction may yield qualitative information about the identity of a middle layer, the location of which has to be identified optically. Mathematical analysis methods have also been developed for detection of such layers. A factor analysis method can be used in cases where the thickness of the layer is smaller than the resolution of the microscope [65]. The resolution of a 2–3 µm thick inner layer, from a four-layer polymer laminate was achieved by self-modeling multivariate analysis [66]. Quality control of packaging materials, the performance of which may be critical to the packed materials, can be routinely carried out to verify local microscopic structure, detect defects and contaminants.

2.3.7
Diffusion

The approach of forming an interface by laminating films of two polymers, allowing them to diffuse and then microtoming a vertical section to obtain a sample for IR mapping is a popular one to examine diffusion. Spectral profiles along the diffusion direction can be obtained and the concentration of the diffusing species directly inferred. This procedure has been used to analyze the interdiffusion of poly(acrylic acid) (PAA) and poly(ethylene glycol) (PEG) [21]. The concentration profile was found to be consistent with Fickian diffusion. The interdiffusion of poly(ethylene-co-methacrylic acid) [EMAA] and PVME [67] could also be studied in a similar manner. The diffusion process also affects the adhesion process strongly. Effects of the diffusing species' (PEG) molecular weight and contact time on diffusion across the interface between two hydrogels were investigated [68]. Results indicated that diffusion enhanced the adhesion between the hydrogel layers. However, the study of diffusion is often hampered by the limited spatial resolution afforded by FTIR mapping systems and the large times required to map the diffusion gradients. Thus, examinations of diffusion have been limited to fast polymer or oligomer pair diffusions. Polymer dissolution by small molecular weight solvents has been too rapid to examine using mapping techniques and the diffusion profiles for very limited low molecular mass substances have been mapped. The diffusion of an anti-oxidant into PE was one of the first reported examples [69]. The diffusion Cryasorb UV531 in PP [70], and Bovine Serum Albumin (BSA) in amylopectin [71] has also been examined. Diffusion of olive oil into polypropylene was investigated and found to be Fickian with constant diffusivity [72].

The processes of reaction and diffusion occur at the same time in a variety of systems. These issues are particularly important in the formation of blend systems and are central issues in the performance property enhancement of such systems. A study of the competitive effects of the rates of the two processes can be easily carried out using FTIR microspectroscopy. The rate of diffusion can be monitored by the time evolution of the absorbance (concentration) profiles while the rate of reaction can be monitored as a time evolution of the reactant (or product) absorbance (concentration). Reaction of a random copolymer of styrene and maleic anhydride (SMA) with bis(amine)-terminated poly(tetrahydrofuran) (PTHF) is one such studied system [73]. Temperature was varied while studying the effects of two different PTHF molecular weights. The reaction rate constants were obtained from the initial slope of conversion-time plots. In addition, it was shown that the rate of diffusion was faster as diffusion of PTHF into the SMA phase occurred prior to the imide formation. The imide was formed in the SMA phase and quantitatively estimated. A corresponding decrease in the carbonyl stretching vibration of the maleic anhydride peak was seen.

2.3.8
Local Orientation

4,4-diaminodiphenyl methane (MDA) and 4,4-diphenyl methane diisocyanate (MDI) were polymerized using vapor deposition to form poly(urea). The dipole orientation of NH and CO dipolar groups caused by corona poling was analyzed. In diode corona poling, the dipoles are aligned greatest in the area directly below the needle electrode while in triode corona poling, a grid electrode is set up between the needle electrode and planar electrode. Hence, it is expected that orientation is accomplished uniformly over the entire surface. However, using infrared reflectance microspectroscopy, dipole orientation was found to follow the form of a net mesh structured like a grid electrode [74]. Orientation due to other formation processes, for example in extrusion molding, can be analyzed. Sheets of a thermotropic liquid crystalline polymer, consisting of 4-hydroxybenzoic acid, phenol and dicarboxylic acid units were examined [75]. The (relative) orientation of molecules and sub-molecular species is important for the properties of many novel optical and electrically responsive devices. Local structure of polymeric materials can usually be analyzed in the polymers native state or on the device itself. Given that devices are becoming smaller, the use of microspectroscopy to determine local structure can be expected to become more relevant.

The local orientation in polymers can also be examined in the context of polymer failure. Two types of polyisoprene samples were stretched until cracking was initiated and the crack tip was subsequently examined using FTIR mapping [76]. Unfilled polyisoprene samples revealed a uniform infrared spectrum for regions surrounding the crack tip at the resolution of the FTIR microscope. Silica filled samples, however, had demonstrated enhanced absorbance levels of the C=C stretching frequency ahead of the crack tip, the region corresponding to the highest stress. Preferential orientation was also observed ahead of the crack tip by employing dichroic ratio measurements. Hence, the molecular changes induced by stress close to the failure region could be observed by combining the sub-molecular sensitivity of FTIR spectroscopic measurements with the spatial discrimination afforded by the microscope accessory.

The local orientation distribution in polymeric samples due to self-organization (e.g., liquid crystalline polymers) or due to processing-induced orientation can be examined using polarized measurements. A relative measure of the distribution of orientation in a sample, such as comparison of surface orientation to orientation in the polymer bulk [77], can be accomplished by polarized microspectroscopy. An examination of extrusion-molded sheets of a liquid-crystalline copolyester [77] revealed higher surface orientation compared to the bulk. The orientation function can then be related to processing conditions, for example draw-down ratios [78]. Since the draw-down ratio can be related to physical properties, a measure of the orientation function provides a route to relate the microscopic chemical organization to macroscopic properties. By suitably changing the processing conditions, the spatial differences in orientation can be decreased and processing conditions for making polymers with uniform physical properties can be determined.

2.3.9
Polymer Surfaces and Interfaces

By their very nature, surfaces are usually heterogeneous. Sometimes, the structure or changes in structure across a surface or at the interface of two surfaces is important in determining the properties of the polymer composite. Usually, IR spectral surface analysis can be carried out using non-spatially resolved IR spectroscopy. Analyses of surfaces (with respect to depth below the surface) can be carried out by microtoming the sample. Similarly, microtoming is a useful technique for polymer interfaces and polymer-metal interfaces if the metal is present as a thin film. Reflection techniques can also be used to obtain spectra from samples without microtoming. A spatially resolved method was used to determine effects of different surface treatments on the same epoxy film [79] and chemical effects of water ingress [80]. $-SO_2OH$ clusters at the surface and embedded below the surface were observed as a function of environmental conditions for Nafion-H, silica-supported Nafion-H and two types of Nafion-H silica nanocomposites [81]. Specific interactions were observed, probably arising from the interface, for silica systems as evident by shifting and splitting of the OH peaks. Biodegradation of thin LDPE film in soil was studied and found to occur relatively fast [82]. This was attributed to the synergistic action of oxidative and/or photo-oxidative degradation on biological activity attributed to the increasing hydrophilicity of the film surface.

2.3.10
Polymer-Liquid Crystal Systems

At this point, it is instructive to examine the contact method (Fig. 6). When a solvent is brought into contact with a polymer film, it diffuses into the polymer. If the film is constrained between two IR transparent substrates and the liquid is forced into the gap by capillary action, the contact and subsequent diffusion of the liquid can be easily monitored by means of chemical imaging using light transmitted in the perpendicular direction. After contact, diffusion between the two materials starts. The absorbance (concentration) profile can be measured with accuracy to yield diffusion profiles [83]. However, for liquids, where capillary action is facilitated, the examination of concentration profiles in-situ is difficult. If the concentration profiles developed at higher temperatures are examined at a temperature that prevents further diffusion, the profiles are frozen-in. The diffusion of E7 into poly(butyl methacrylate) (PBMA) was studied [84] in this manner and compo-

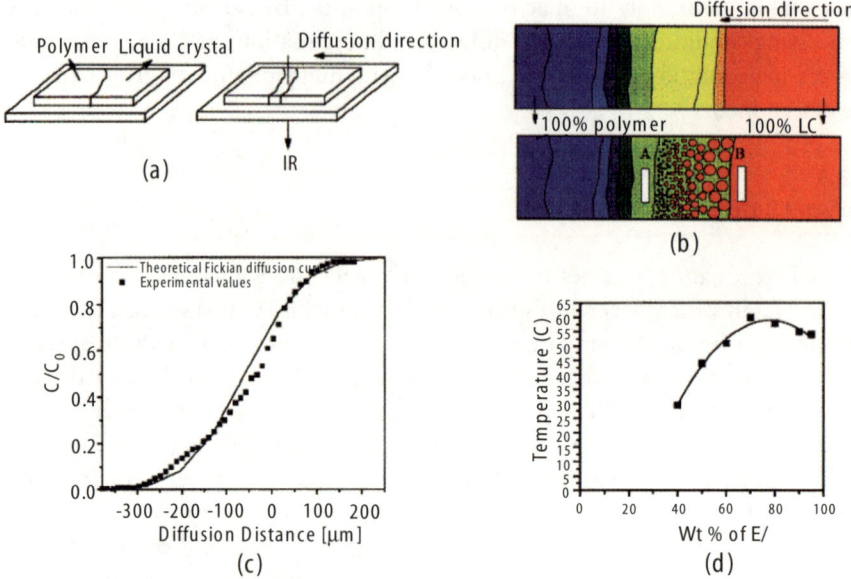

Fig. 6. The contact method (**a**) involves the introduction of two soon-to-be diffusing materials between substrates. Capillary action brings the liquid and solid in contact. This sets up an area of varying concentration due to diffusion (**b**) *top*. Upon Polymerization or cooling, this some part of this region may phase separate (**b**) *bottom*. The apertures seen in (**b**) can be sequentially moved to extract diffusion profiles (**c**) or phase diagrams (**d**) (from Ref. [84])

nents of E7 were shown to diffuse at different rates [85]. The component partitioning could be detected based on small spectral differences between the four components monitored as a function of time.

The contact method in conjunction with FTIR microspectroscopy has also been used to determine the phase diagram of an LC-cured pre-polymer pair. An LC mixture (E7) and pre-polymer (NOA65) were allowed to diffuse for some time to achieve a concentration gradient. The pre-polymer was subsequently polymerized by UV initiation. Under the optical microscope of the spectrometer, phase separation was observed in certain regions. The regions on either side of the phase-separated band are single phase and the highest concentrations in these regions, determined by microscopic FTIR measurements, yields the solubility limit. Repeating this procedure over different temperatures gave the phase diagram [86]. PDLCs were characterized to obtain spectra from the droplet and matrix regions as well as examine local LC orientation (see Fig. 7) [87]. The influence of the matrix material on local order and resulting droplet configuration could also be identified [88]. The time resolved response of microscopic regions in large and small PDLC droplets could also be obtained [89].

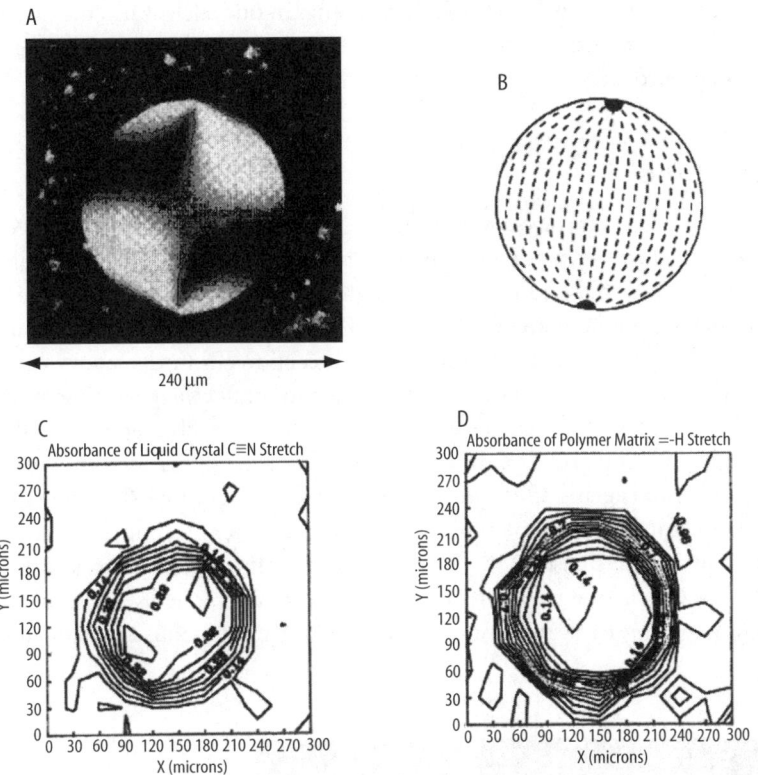

Fig. 7. Infrared Contour Maps of a liquid crystal specific peak (*left, bottom*) and a matrix (epoxy) specific peak (*right, bottom*) of a PDLC droplet, whose polarized optical image is shown (*top, left*) with the corresponding orientation visualized (*top, right*) (from Ref. [87])

2.3.11
Filled Systems

2.3.11.1
Carbon Black, Silica and Inorganics Filled Systems

Two major fillers used in the rubber industry are silica and carbon black. Carbon black is black because it absorbs/scatters all radiation, including infrared, impinging on it. Hence, simple transmission spectroscopy of carbon black filled specimens is not straightforward and is usually not possible unless the samples are very thin. Carbon black filled samples have not been readily examinable using microspectroscopic methods. Silica filled systems are more amenable to microscopic techniques [76] and can be examined to determine silica-polymer(rubber) interactions. The presence of inorganic materials, e.g., transition metal complexes in polymers, can be examined by FTIR mapping. In one such study, the distribution of the metal complex was found to be homogeneous [90], thus validating the formation procedure.

2.3.11.2
Fibers and Fillers

Fiber reinforced epoxies can be used to obtain materials with good mechanical properties. However, a point of failure is the poor adhesion between the fibers and epoxy. Hence, surface adhesion was sought to be improved using surface modification techniques and analyzed using FTIR mapping experiments. The fiber-matrix interphase was examined for chemical reactions in Kevlar-epoxy composites [91]. The amine groups in Kevlar were shown to accelerate the curing process (see Fig. 8). The interphase of as-received and modified glass fiber-epoxies could be obtained by least squares and difference analysis of thickness corrected maps [92]. Environmental ageing effects due to humidity on these composites were simulated and the composites mapped to find preferential water accumulation close to the fiber-matrix interphase [93]. Using a silane coupling agent decreased the effects [94]. The effects of environmental conditions on fibers themselves can be readily examined using FTIR microspectroscopy [95]. Simultaneous reaction and orien-

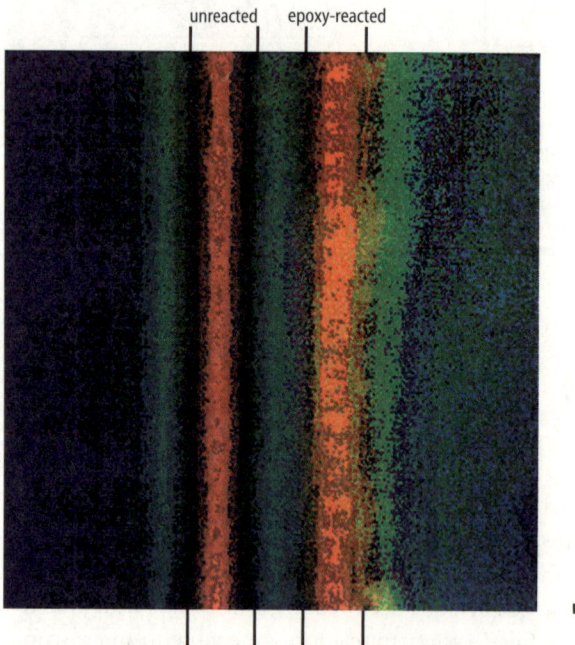

Fig. 8. FTIR maps of the reacted and unreacted fibers from the NH stretching vibration at 3326 cm^{-1} (from Ref. [91])

tation information could be obtained in a case where nitrile reaction, conjugated C=C formation and oxidation occurred at the same time to possibly affect orientation.

2.4
Summary

The single detector element microspectrometer was demonstrated over 50 years ago. Systems today employ FTIR spectrometers, sensitive and stable detectors and high-speed computers with large storage capability, which allows for the mapping of millimeter size spatial areas. However, due to the sequential nature of the mapping process, the experimental time is large. Hence, for spatial distribution information of chemical species, FTIR imaging is becoming popular (vide infra). However, the single detector element systems are not very expensive and remain the method of choice when discrete measurements from a number of points have to be made, experimental time is small compared to the time scale of changes in the sample, high spatial resolution is not important, or a spectrum from a small sample is required. These qualities have made the single element microscopy spectrometer a useful tool for industrial applications while providing unprecedented information for research activities.

3
FTIR Imaging

3.1
Introduction

The state of the art in FTIR microspectroscopic instrumentation today is the combination of a Focal Plane Array (FPA) detector and a step-scan interferometer [96, 97]. FPA detectors consist of thousands of individual detectors laid out in a grid pattern. Hence, the interferometer signal is simultaneously sampled by each detector leading to a multiplexed spectral and multichannel detection advantage. Spectral discrimination is provided by the interferometer and spatial resolution by discrete detector elements on the FPA. Each FPA element corresponds to a specific spatial region on the sample and no apertures are required to restrict radiation. A schematic of the setup may be seen in Fig. 9. This configuration allows for spatially resolved imaging of a wide field of view in a data collection time a little more than for a single pixel. From the results of the imaging experiment, a spectrum from a small region may be examined or spectroscopic signal from a spectral feature plotted for the field of view, providing simultaneous examination along spatial and spectral dimensions. This technique considerably reduces experimental time for imaging static samples and has enabled the examination of many dynamic processes [98].

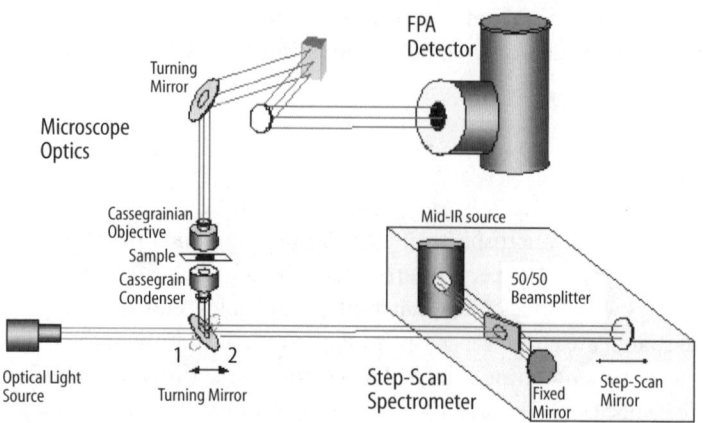

Fig. 9. Schematic Diagram of an FTIR Imaging Spectrometer

3.1.1
Mapping Versus Imaging

A comparison of the mapping and imaging techniques has been carried out [99]. As explained above, the sample is imaged onto the FPA using the magnification of auxiliary optics such as a microscope. The optics determine magnification and hence, the nominal spatial resolution afforded by the size of the individual detectors in the array. No apertures are required to limit the sample area examined in an imaging experiment and the sample does not need to be moved as a given field of view is imaged in a single collection experiment. It may be immediately seen that the collection time is decreased by a factor of n^2, where n is the number of spatial resolution elements in one direction of a square sample area imaged. Analogously, n is the number of steps in a mapping experiment. Neglecting the effects of resolution, a mapping step may be considered equivalent to a pixel in the imaging experiment. However, the practical spatial resolution limit in the mapping technique is usually ~15 µm × 15 µm and additional time is required to move the sample from position to position. For imaging, the resolution is essentially wavelength limited but additional time may be required to process the large data sets that result.

As an example to compare the two techniques, a 64×64 element array imaging a 500 µm × 500 µm area will provide reasonable quality spectral information in less than a minute. The same measurement at a resolution of ~15 µm will take approx. 10 hours using a point by point mapping approach. FTIR imaging has allowed the collection of images in faster time with higher resolution. Moreover, no stray light problems that compromise spatial discrimination are involved (no aperturing), the spatial resolution is hardware/wavelength limited and the collection time is decoupled from the two for the imaged field of view. Typically, it is tougher to fabricate miniaturized detection systems and due to unique hardware issues, the performance of an FPA pixel typically lags behind single element mapping detectors that may be ten times larger. Thus, spectral (and image) quality desired usually determines the time for data acquisition.

The optical and infrared microscopic capabilities of an imaging instrument do not have to be matched. Thus, a major requirement of the mapping spectrometer that the optical and infrared paths be parfocal and collinear, is not necessary for imaging. The focusing in the sample plane can be independently carried out using a real time "bright field" IR image of the FPA. In practice, the optical path is not even required but an imaging spectrometer equipped with a field of view larger than the FPA camera helps to localize regions for imaging and examine neighboring regions before deciding to image them. Hence, most imaging instrumentation is equipped with a relatively cheap CCD visible camera with a field of view larger than that afforded by the more expensive FPA.

3.2
Instrumentation

At the core of most FTIR micro-imaging spectrometers is the step-scan interferometer [100]. The beam from the interferometer is diverted through standard microscope optics with the FPA at the end of the optical train. Cassegranian optics are used to focus radiation in the sample plane. Initially, a refractive setup consisting of an (usually ZnSe, BaF$_2$ or CaF$_2$) imaging lens was used to form the image onto the FPA. This has given way to relatively aberration free reflective optics. The microscope setup is similar to that used in mapping experiments. A diffuser is sometimes used to increase spatial homogeneity in incident intensity [106]. The diffuser also reduces total light flux reaching the detector, thus preventing saturation of pixels. A bandpass filter can also be used to restrict wavelengths of light incident on the detector, which limits infrared flux to wavelengths of interest and eliminates background radiation outside it. An optimally sized cold shield inside the detector housing can be used to maximize available dynamic range by rejecting stray radiation arising outside the acceptance angle of the optics [101]. While the nominal resolution is determined by the system optics and the detector characteristics, the resolution limit is usually determined by the diffraction limit of the wavelength of interest.

A step-scan interferometer provides a means to maintain constant optical retardation for an arbitrarily large time. This large time is required when the FPA co-adds collected frames at each optical retardation point to give reasonable signal to noise ratio images, which is similar to in-scan co-addition for regular step-scan spectroscopy. An electric pulse triggers each interferometer step, which may also

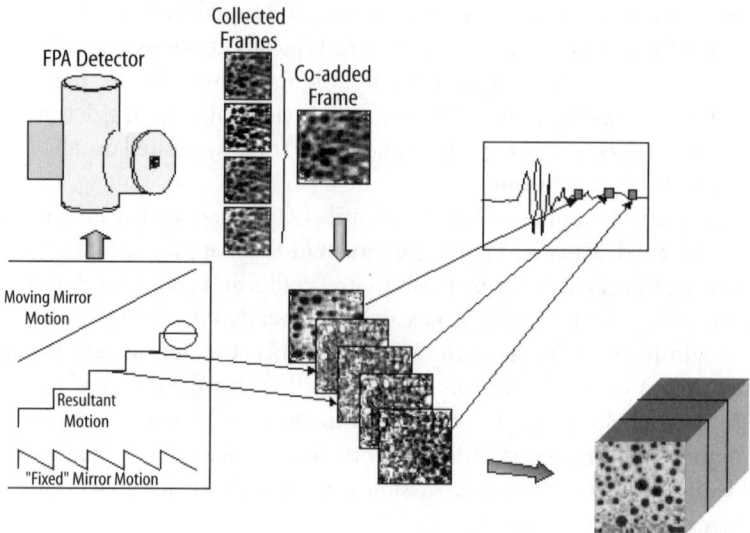

Fig. 10. Spectrometer Steps modulate the IR signal. At each step, the FPA collects frames and co-adds to yield a single frame for each step. Frames (equal to the number of steps) can be associated with a unique point on the interferogram. This interferogram data can be Fourier transformed, ratioed and truncated to yield an absorbance data cube

be used to initiate data acquisition by the FPA. A small delay to allow for mirror stabilization is allowed at the onset of the step and frames are then collected and some time is left for signal readout from the FPA and storage to a computer. The readout format may be sequential (common in MCT arrays) or single-shot (common in MCT and InSb arrays) in the focal plane. Details about solid state arrays [102] can be found elsewhere. The signal is integrated for a fraction of the time required for collection of each frame. The integration time, number of frames coadded and number of spectrometer steps (spectral resolution) determine the time required for the experiment and the data quality. This process is schematically shown in Fig. 10.

Another imaging configuration that utilizes a rapid scan interferometer has been proposed [103]. This configuration is essentially similar to a fast step-scan experiment in that the mirror speed is very low. Frames collected by the FPA are in a small enough time to be as close to instantaneous as possible on the time scale of the interferogram collection. However, the motion of the moving mirror does not allow co-addition of frames per interferometer retardation element. Hence, the process is faster but noisier. To minimize other sources of error contributing substantially to degrade the signal, the error arising from the deviation in mirror position during frame collection is sought to be made as small as possible. With current state of the art, this positional error is considerably larger than the error due to variations in mirror position during step-scan collection. The frame collection by the FPA (using a time interval based collection protocol) is triggered at the onset of the forward movement of the mirror. For maximal scanning speeds and to minimize positional errors, only one frame is collected. It is envisaged that this collected image can then be time-averaged with subsequently collected images similar to that of classical single detector element rapid scan data collection. The advantages lie in making the instrumentation cheaper than step-scan and when faster detectors are available, making data collection more efficient as the mirror stabilization time is not required [104].

3.2.1
Sampling Techniques

Sampling requirements and techniques are virtually identical to those for the single element spectrometer. However, spatial resolution achieved by using an array detector can range from less than 10 μm to hundreds of kilometers. Hence, a wide variety of optical setups may be available for imaging purposes. Most polymeric materials examined to date have generally been examined using a microscope that images areas on the order of ~500 μm × 500 μm in transmission mode as this affords spatial resolution sufficient for many polymeric composites, a convenient optical setup and sufficient field of view for most instances. However, there is no fundamental reason why other types of optical configurations may not be employed for other length scales of examination. Some reflection experiments have been successfully conducted in our laboratories. However, reports on any studies utilizing reflectance configurations are yet to appear in the literature. Attenuated total reflection (ATR) imaging [105] and micro-ATR imaging are also being attempted by researchers. Improvements in the SNR attained in short experimental times by the evolving detector technology will likely benefit applications using these novel sampling techniques.

3.2.2
Sources of Error

FTIR imaging suffers from low signal to noise compared to other types of FTIR spectroscopy and microspectroscopy [106, 107]. However, the major sources of error in an imaging system arise from detector noise and not from optical components of the imaging system (for example, apertures). The major factors in this regard are:

3.2.2.1
Interferometer Noise

Noise contributions from the interferometer arise for a variety of reasons. However, research grade spectrometers usually incorporate interferometers that have excellent stability and low noise. Thus, the noise from the interferometer is usually small, especially when compared to noise from the detector. For step scan imaging instruments, noise due to the interferometer is almost three orders of magnitude lower than the noise due to the FPA.

3.2.2.2
Detector Noise

The small size of individual FPA detectors and their relatively early stage of development result in higher noise than usually observed in infrared detectors employed for spectroscopy. Read Noise in an FPA results from random charge movement in the array and its associated electronics. This is independent of the signal magnitude and hence, can be reduced by increased photon flux by an increase in signal or integration time. Read noise is usually reduced in detectors by using correlated double sampling, multiple pixel reads and binning [102].

Dark current noise usually forms the bulk of the detector noise. It results from accumulated charge in the detector. Impinging photons are not necessary to observe this noise and it can be detected as a signal in the FPA when radiation is blocked. It is usually due to thermal excitation of electrons to the valence band producing charges. Cooling the detectors considerably reduces this noise. Charge arising from dark currents increase linearly with time to reduce full well capacity resulting in poor dynamic range. However, it must be remembered that cooling also reduces the charge transfer efficiency. Given the desire for short experimental times and performance of IR sensitive FPAs, interferometer noise is usually orders of magnitude smaller than FPA noise and the noise arising in an imaging experiment can be considered entirely random.

3.2.2.3
Failing Bump Bonds

Most IR-sensitive arrays employed to examine the "fingerprint" region of the IR spectrum of polymers are hybrid arrays – an IR sensitive material is coated on top of an underlying FPA comprised of sophisticated readout circuitry. The first generations of IR sensitive arrays were made by attaching an IR sensitive material to a solid state array by conductive bumps. Repeated thermal cycling resulted in the failure of these bump bonds over time. The delaminating of pixels from their underlying substrate results in the pixels being inoperable and consequently, delaminated pixels appear as black squares on the IR image. Thus, spatial fidelity is compromised over time if a first generation array is used. Recently, IR sensitive materials have been grown using chemical techniques such as molecular beam epitaxy (MBE) on the underlying readout substrate. This has eliminated failing bonds and provided uniform arrays with a very small number of bad pixels. FPAs featuring bump-bonded substrates are unlikely to be employed for spectroscopic imaging in the future.

3.2.2.4
Intensity Distribution, Pixel Saturation and Dynamic Range

Uneven intensity distribution leads to varying signal across the FPA field of view. While this is not a major concern when images are being ratioed and absorbance plotted, it is readily remediable. A simple array calibration (also termed flat fielding) before data collection is usually sufficient to provide a uniform field of view. However, individual pixel saturation needs to be avoided by reducing the flux at the detector, which may limit the signal at pixels with lower intensity. Pixel-to-pixel non-uniformity in the detector array is typically ~5 %. Flat fielding prior to data collection also helps to reduce apparent heterogeneity due to manufacturing differences. For absorbance imaging applications, which are typical for polymeric systems, signal non-uniformity can be compensated for by ratioing against a suitable reference. The maximum dynamic range is determined by the A/D converter and is typically 14–16 bits for arrays in use today.

3.2.2.5
Sample

The source of the most unpredictable (and in most cases, un-correctable) noise pattern is the sample itself. Differences in refractive index at phase boundaries contribute to scattering at some pixels. This scattering at the interface results in an intensity loss at all frequencies leading to an offset in the baseline compared to other sample areas when sample single beams are ratioed to substrate single beams. While this deviation of the baseline from zero can be used to locate the interface without using a component specific absorbance frequency, it may cause problems in quantification [108]. The sample also reduces signal to noise ratio by reducing incident infrared intensity on the detector.

3.2.3
Post-Collection Operations

The major difference between all other forms of FTIR spectroscopy and imaging is the large volume of data generated in a short time period. Thousands of spectra with a specific location relative to each other need to be handled. This large amount of information may be organized in the form of a data cuboid, i.e., a three-dimensional matrix. The first two dimensions denote the spatial location (i.e., specific pixel) while the third dimension stores the infrared spectrum of the sample imaged onto that pixel. Since the spectrum is digitized, the data cube may also be imaged to be a stack of two-dimensional arrays. The image corresponding to a particular data point (wavenumber), z_i, may then be extracted and the spatial distribution of the species with specific absorbance at that data point may be seen in the spatial area denoted by $E(x, y, z_i)$. The extracted spectrum from a pixel at (x_i, y_i)

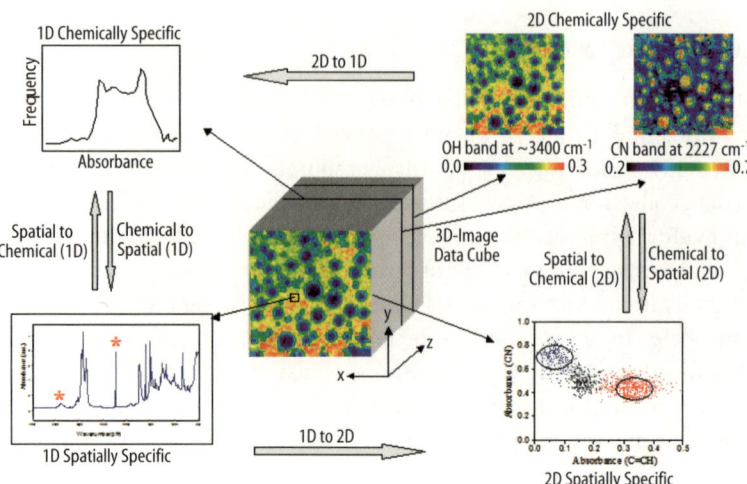

Fig. 11. Various visualizations possible with data extracted from the image data cube

may then be denoted as $E(x_i, y_j, z)$. The absorbance at a particular wavenumber at that pixel, z_j, can then be represented by $E(x_i, y_j, z_j)$.

The extraction of the image plane corresponding to a component specific wavenumber is the most popular method to visualize data. Images may be thresholded for maximum contrast between phases or between values for comparison across a set of images. Simple thresholding is affected by non-uniform thickness effects, which may lead to erroneous conclusions about a species' distribution. Peak height (or area) ratios are used to normalize thickness effects and relate the observed absorbance to species' concentration. The relative concentration of two species may be further seen using 2d plots of the absorbance of one component against the other. The various visualizations possible are shown in Fig. 11. Clearly, the simplest form of representation to comprehend is a single number (zero dimensions) or a simple distribution plot (1 dimension). However, the information content is also reduced in such representations. Further, the whole image cube clearly contains all the informational content but is extremely difficult to visualize in its entirety and such a representation may not be completely relevant to the task. Hence, some form of trade-off between loss of relevant data and ease of representation is usually found in the analyses of an imaging experiment. The chart in Fig. 11 serves as a useful guide to some of the possibilities in such an endeavor.

The low SNR in FTIR images has been a major concern since the inception of the technique. The first effective approach had been to co-add as many frames as possible during the available experimental time. While this approach results in significant gains initially, the SNR as a function of co-added frames achieves a plateau [106]. Thus, after a number of frames, the benefits are relatively minor compared

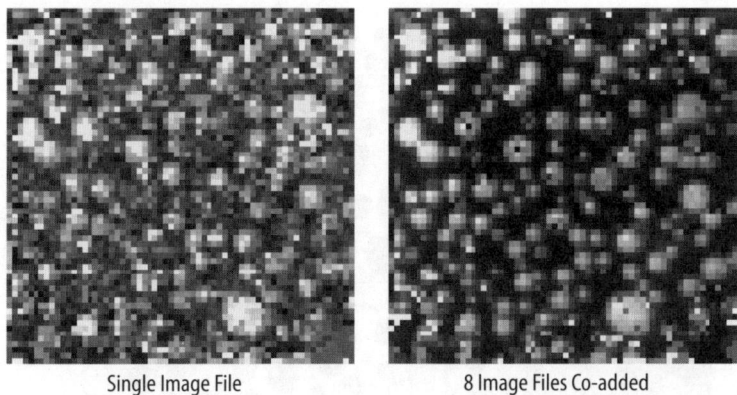

Single Image File   8 Image Files Co-added

Fig. 12. Noise reduction by co-addition as applied to a polymer dispersed liquid crystal sample. Co-addition results in a reduction of random salt and pepper noise allowing for better morphology visualization

to the large increase in experimental time. An optimum number may be derived based on the SNR as a function of co-added frames. Once the optimum number of co-added frames is determined, other techniques may be used to increase the SNR. One such technique has been to co-add single beam images of the sample and ratio them to similarly co-added single beam images used for background [106]. The SNR scales as $\sqrt{n}$, where n is the number of sample single beam images co-added and ratioed against n co-added background single beam images. The effects of this noise reduction procedure can be seen in Fig. 12. Random salt and pepper noise is vastly reduced allowing for better morphology visualization.

While image co-addition results in improved SNR characteristics, the process involves a $2n^2$ increase in collection time for a SNR improvement factor of n. However, image co-addition is not suitable for application to real-time phenomena or fast examination of large numbers of samples. Another proposed idea was to ratio a single sample single beam image to multiple background images [109]. The resultant absorbance images are then co-added. This method, termed pseudo co-addition, is also limited in the benefits achievable, presumably because noise from the sample single beam starts to dominate the resultant. In many cases, the sample geometry will allow for the co-addition of many pixels with the same true absorbance [109]. This may be termed geometric co-addition (when co-added regions can be reproduced geometrically on the image) or phase co-addition (when pixels from a phase are selected with regard to the sample morphology). Spectra from these pixels can be extracted and co-added to yield a lower noise spectrum. The improvement in SNR scales as $\sqrt{n}$, where n is the number of co-added pixels.

The benefits of pseudo co-addition are limited and the phase and geometric co-additions do not result in improved image quality. Mathematical noise reduction

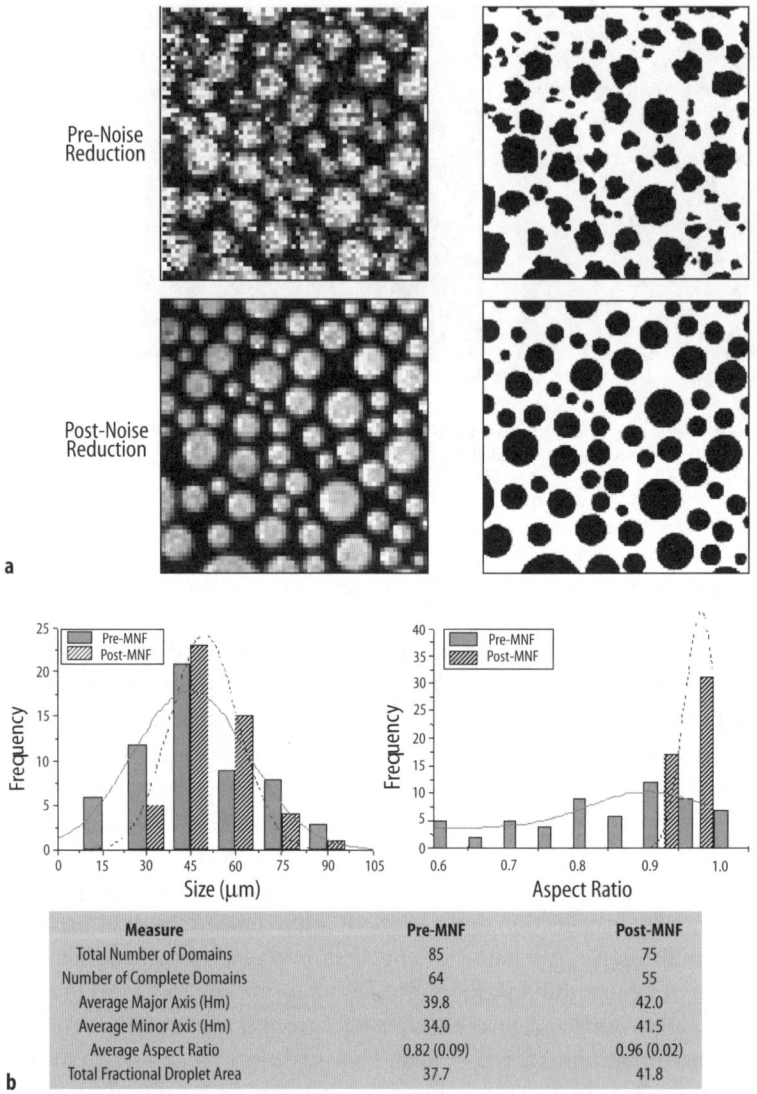

Fig. 13. (**a**) Binarized Images (*right*) for image analysis derived from (*top, left*) the original data, and (*bottom, left*) noise reduced data after application of the MNF transform to reduce noise (from Bhargava R, Wang S-Q, Koenig JL (2000) Appl. Spectrosc. 54: 486). (**b**) Morphology characteristic quantities derived from the binarized images for after image analysis (from Bhargava R, Wang S-Q, Koenig JL (2000) Appl. Spectrosc. 54: 486)

is an alternate pathway to higher fidelity images [110]. Mathematical techniques operating on the entire data set are computationally intensive but do not result in loss of the image or spectral content and do not affect collection time. A mathematical transform (the Minimum Noise Fraction [MNF] transform) that re-orders spectral data points into decreasing order of SNR was employed to transform images. Re-transforming the ordered data set using only a few relevant data points reduced noise. This approach is shown to result in significant gains in terms of image SNR (see Fig. 13). The actual gains were found to depend on the SNR characteristics of the original data. Noise is reduced by a factor greater than 5 if the noise in the initial data is sufficiently low. For a moderate absorbance level of 0.5 a.u., the achievable SNR by reducing noise is greater than 100 for a collection time of ~3.5 minutes. This noise reduction was shown to provide high quality images for accurate morphological analysis. The careful coupling of mathematical transformation techniques with spectroscopic FTIR imaging has tremendous potential to obtain high fidelity images without increasing collection time or drastically modifying hardware.

3.3
Applications

3.3.1
Contaminants/Defects in Polymers

The generally poor SNR characteristics of FTIR imaging limit its sensitivity to small concentrations. Recently, the SNR levels achievable for data collected in a matter of minutes have reached close to 1000 in a few minutes, implying quantification levels of a percent and detection levels down to a fraction of a percent in the same experiment. With the application of other co-addition and post-collection techniques, FTIR Imaging has the potential of being even more sensitive. Defects (such as air bubbles) in thin films may be easily visualized. An area where this technique is expected to be popular is the analysis of polymer inclusions and defects in the semi-conductor processing industry. The imaging of products will help determine the presence of contaminants and help in optimization of the process. With wide field imaging, even defects that have similar refractive indices with the examine material can be readily detected and separate visible imaging is not required.

There has been considerable controversy over the failure of breast implants and the subsequent contamination of breast tissue. Implants typically consist of an elastomeric shell with a soft filling inside. Over time and due to mechanical failure, the implant filling leaks into the breast tissue. This contamination may lead to various complications including capsular contracture, calcification, and some connective tissue disorders. Controversial aspects of the results of malfunction have not prevented numerous lawsuits. However, to assess any histopathological changes, it is first necessary to confirm the presence of material from the implant in tissue. The presence of silicone could be readily detected [111] in a tissue section using FTIR imaging. Si-CH$_3$ characteristic vibrations were used to provide chemical contrast between the tissue and silicone inclusions as small as ~10 μm. Inclusions could be readily found even in cases where optical microscopy contrast was poor.

3.3.2
Amorphous Polymer Blends

PVC is a useful polymer that is becoming less popular due to its degradation behavior. However, blends of PVC with some other polymers have the capability to yield composites with greater stability. In one such case, images of phase separated PVC/PMMA blends are reported [112]. Phase separation and concentration quantification can be carried out as a function of starting polymer properties and process conditions. However, the dispersion sizes examinable by infrared imaging are limited to tens of microns. In many polymeric systems, the dispersed phase is smaller than this dimension and the composites are not amenable to FTIR imaging analyses. Analyzing scattering [113] and relative concentrations of species in a pixel may yield qualitative clues to the morphology of phase separated material but probably cannot be used to obtain quantitative morphological information. We anticipate that spatial distribution of specific interactions in phase separated polymeric systems could also lead to such information and studies examining these aspects could appear in the near future.

3.3.3
Semi-Crystalline Polymers and Their Blends

FTIR imaging allows for the examination of semi-crystalline polymers via their dichroism [98, 114] The spherulitic structure, as visible in polarized optical microscopy, can be reproduced based on the orientation of transition dipole moments of functional groups in the sample. With an infrared polarizer in the beam path, the degree of orientation for each vibrational mode was determined by the generation of spatially resolved dichroic ratio images. When a melt-miscible polymer system is analyzed at temperatures lower than the crystallization temperature of one constituent, a phase separated or single crystalline phase structure might result. When the T_g of the non-crystalline component is lower than the crystallization temperature, the component phase separates as it is rejected from the crystal structure. The extent of phase separation may be easily visualized using IR imaging. PVAc/PEO and High M_w/Low M_w PEO blends have been imaged to examine the degree of segregation (Fig. 14) [114].

3.3.4
Polymer Laminate Films

The first reported application of FTIR imaging to polymeric materials was the imaging of a cross-section of a laminated polymer film. The adhesive layer between

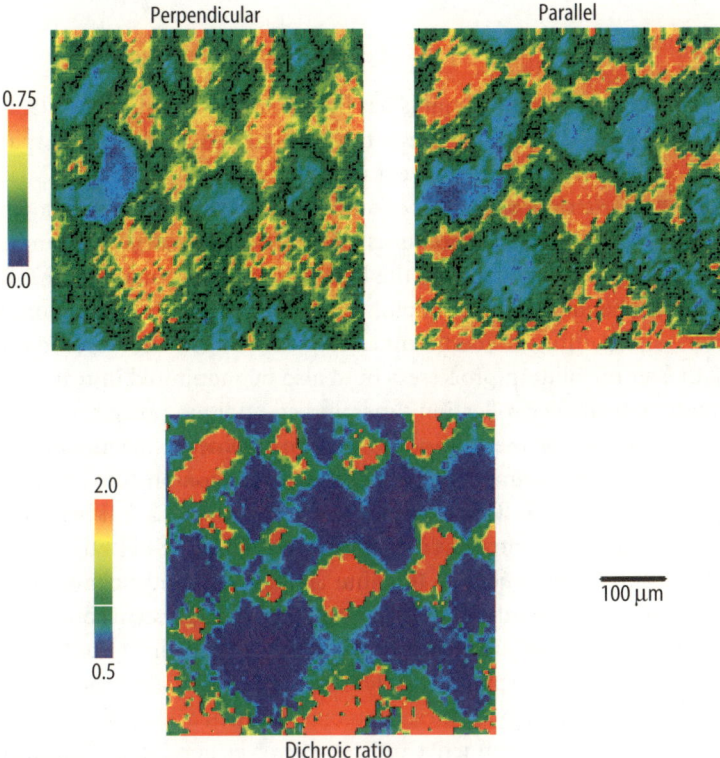

Fig. 14. Infrared images of spherulites using perpendicular (*left*, *top*) and parallel (*right*, *top*) polarized radiation obtained from PEG (M_v=35 K), quenched from 100 °C to room temperature. The images were obtained by plotting the 1343 cm^{-1}: CH$_2$ wag vibrational mode of the PEG chain. The image at the *bottom* is a dichroic ratio image obtained by pixel-by-pixel division of the upper two. (reproduced from Snively CM, Koenig JL (1999) Spectroscopy)

two polymers was identified. Thin layers of PET, EVA and LDPE could be readily identified based on absorbance images from their characteristic vibrational peaks [115]. Since many properties (for example, barrier properties) of polymeric laminates depend on the structure of the composite, FTIR imaging presents a unique method to simultaneously examine the morphology and chemical composition of laminates for quality control. The morphology, composition and chemical interaction in the interphases can also be employed to correlate to properties and design better composite materials. High throughput chemical imaging as provided by FTIR imaging will considerably speed up quality checks and the characterization component of polymer engineering.

3.3.5
Polymer-Liquid Systems

The dissolution of polymer films by liquids has been a subject of much investigation in polymer science. FTIR imaging presents a method to determine both spatial and spectral content of a polymer-solvent interphase. By monitoring the spatial distribution of concentration as a function of time from an initially known state, the diffusion of polymer and solvents as well as the dissolution rate can be determined. This experiment was among the first reported applications of FTIR imaging to real-time phenomena. The dissolution of a PMMA film was monitored and the polymer-solvent diffusion coefficient calculated [116]. In another study, it was shown that faster dissolution processes could also be monitored in real time using a combination of smaller collection times and co-addition processes [109]. The dissolution rate of the polymer film could then be obtained using its concentration profile as a function of time. The nature of the diffusion profiles can provide clues to the nature of the diffusion process. For example, a Case II behavior was found for MIBK diffusing into poly(α-methyl styrene) (PAMS) [109].

Not only can dynamic phenomena in time but changes in liquid-polymer systems due to reaction can be measured readily. A polymer-liquid system consisting of phase-separated mixtures of uncured poly(Butadiene) and diallyl phthalate were studied to characterize morphology differences before and after the curing process [117]. Optical microscopy of these systems is particularly challenging as the resultant phases have similar refractive indices. However, good image contrast was achieved by FTIR imaging due to the inherent chemical differences manifest in their infrared spectra. The morphological changes were characterized over a period of time and the post-cure sample exhibited homogeneity at the resolution level of the instrument.

3.3.6
Embedded Polymer Systems

An area of interest for polymer applications is in the encapsulation of biologically active proteins in sustained release devices. Among these, microparticles composed of poly(lactic-*co*-glycolic acid)(PLGA) have been studied extensively [118]. As the capsule of biologic containing polymer is prepared, it is subjected to a number of manufacturing processes that may denature the biological molecule. In such a case, it is advantageous to have a method available to perform a spectroscopic analysis on the final product without the need to extract the biologic. In one such case, PLGA containing egg-lysozyme was analyzed using FTIR imaging [119]. Protein (Fig. 15) was found to be present inside the microspheres and distributed evenly on the scale of examination. Little protein was found on the surface and this result was confirmed by subsequent ATR and PAS measurements. This helped sug-

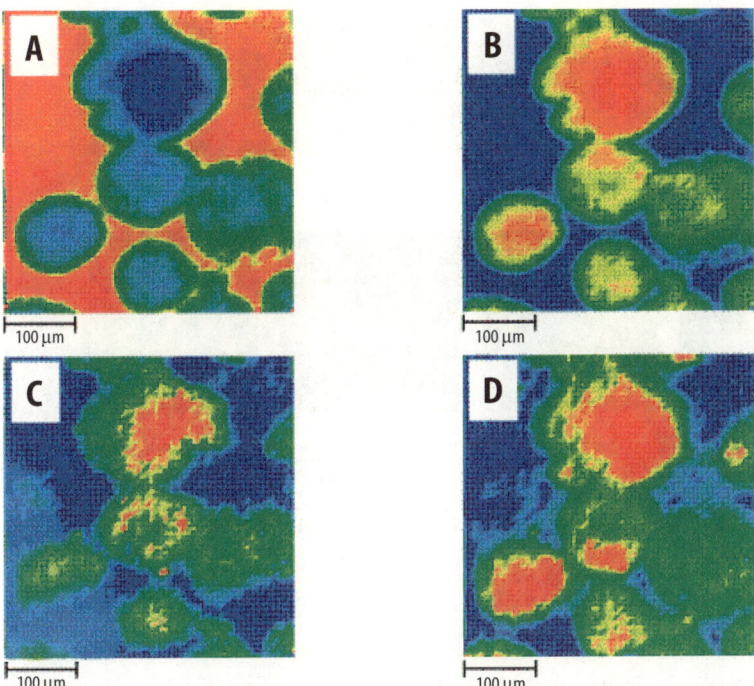

Fig. 15. False color infrared images of sliced lysozyme-loaded microspheres: blue indicates low intensity, green intermediate intensity, red high intensity. Intensity distribution of **A** embedding material, 1720 cm^{-1}; **B** PLGA, 1756 cm^{-1}; **C** protein Amide I, 1650 cm^{-1}; **D** protein Amide II, 1550 cm^{-1}. (reproduced from Ref. [119])

gest that an alternative mechanism of release from that proposed earlier [120, 121] may be at work. The formation and subsequent characterization of such composites can be carried out to not only examine the spatial distribution of chemical species afforded by examining but also obtain all the structurally sensitive information afforded by infrared spectroscopy. The structural changes and/or morphology development can then be characterized to the formation process providing a rapid analysis of the efficacy of the process.

3.3.7
Polymer-Liquid Crystal Systems

Polymer Dispersed Liquid Crystals (PDLCs) are widely researched materials of commercial importance in the electro-optical device industry as their light transmission can be changed by the application of an electric field. The refractive index of these rod-shaped liquid crystals along and perpendicular to their major axis is

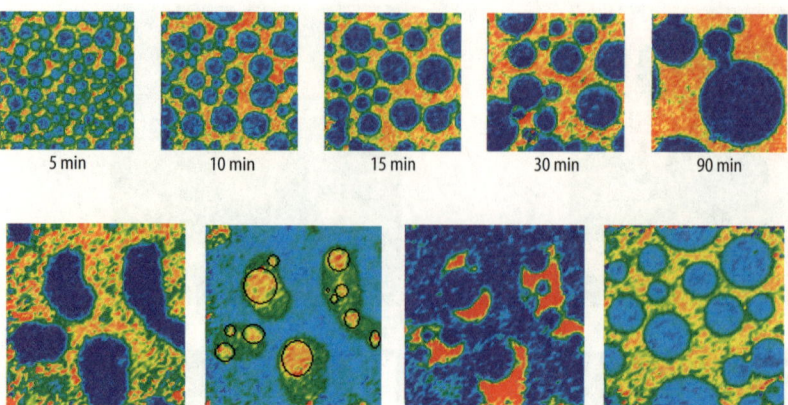

Fig. 16. The increase in droplet size can be seen as a function of time for a poly(butadiene) – liquid crystal composite (*top*). The distribution of the polymer and the liquid crystal can be seen in the first two images (*bottom*), segregated as before. The surfactant (*middle*, *right*) can be seen to be distributed around the droplets and segregated from the matrix. A sample without surfactant

different. One of the refractive indices is chosen to correspond to that of the polymeric matrix in which they are to be imbedded. The preferred method for PDLC formation is to induce phase separation between the (growing) polymer and liquid crystal from their homogeneous solution. This method results in a liquid crystal phase with some matrix solubility and a matrix phase with high liquid crystal solubility. FTIR imaging of samples from a method used to lower this mutual solubility was carried out [122]. Since liquid crystals display phase sensitive IR responses [113], a calibration curve corresponding to the achieved concentrations was first constructed and the imaging data was subsequently used to determine concentrations. The concentration of each component in the other was calculated to a high degree of accuracy and the error estimated using statistical methods. The results from the imaging experiment agreed well with those from DSC and optical examination. FTIR imaging can also be used to visualize the distribution of surfactants in such systems [123]. The addition of surfactant was shown to retard droplet growth and was found to be close to the dispersed phase (liquid crystal) in the sample (Fig. 16).

The diffusion of liquid crystals into polymers is particularly intriguing given that the liquid crystal may be in an organized state or an isotropic state depending on the temperature. The diffusion of a liquid crystal (5CB) into a PBMA matrix was studied by using the contact method to prepare a gradient [114]. Concentration profiles were obtained as a function of time and temperature. The presence of an anomalous diffusion process was detected. It was shown that fast FTIR was able to correctly identify the diffusion process as anomalous. As opposed to this, a bulk mass uptake analysis would have led to the conclusion that the process proceeded according to Fick's second law. An abrupt change in the diffusion behavior was observed as a function of temperature. This change corresponded to the T_{NI} of the LC.

3.3.8
Phase Separation in Polymers

A tedious process is currently used to determine the phase diagram of binary polymeric systems. A single concentration is cycled through a temperature range and the onset of a refractive index difference (scattering or microscopy) is observed. Hence, a number of measurements are required at small temperature differences and accuracy depends on time spent at the experiment. This procedure is then repeated for many different concentrations to determine the phase diagram. Small refractive index (which may be temperature dependent) differences between the components may complicate analysis. Finally, the phase diagram may correspond to transitions in one (or both) components leading to a complicated situation. The basic issue in characterizing the phase separation behavior of a system is the determination of phase composition as a function of temperature. Any starting composition in the single phase region is capable of yielding information about two phases in the two-phase region if it phase separates upon changing the temperature to a region where the blend is non-homogeneous. If the chemical composition of the phases is directly determined, say by FTIR imaging, then the phase diagram can be obtained.

FTIR imaging coupled to thermal control of the sample [124] has been employed to determine phase compositions. We next report a study where the phase diagram of a liquid crystal mixture, E7, in a polymer (PBMA) was obtained. A single composition (80% LC) is held in the single phase region and quenched to different temperatures in the two phase region. Once hydrodynamic and coarsening equilibrium is attained, the sample is imaged (Fig. 17a). We acquired images continuously over time and used data from an image only when the spatial features were larger than ~20 μm and the structures of two consecutive images were found to not change. The characteristics absorbances are determined and phase composition obtained from a previously determined calibration curve. Thus, the two points of the co-existence curve at the temperature of quench are simultaneously determined. In this manner, the compositions at each temperature are determined. The phase diagram (Fig. 17b) obtained corresponds well with the phase diagram obtained using optical microscopy following the classical methods of obtaining phase diagrams. Using the described approach, phase diagrams may be readily obtained in a fast, straightforward manner that is decoupled from the kinetics of the phase separation process. However, there is a discrepancy between the optical and infrared predictions, which needs to be further explored.

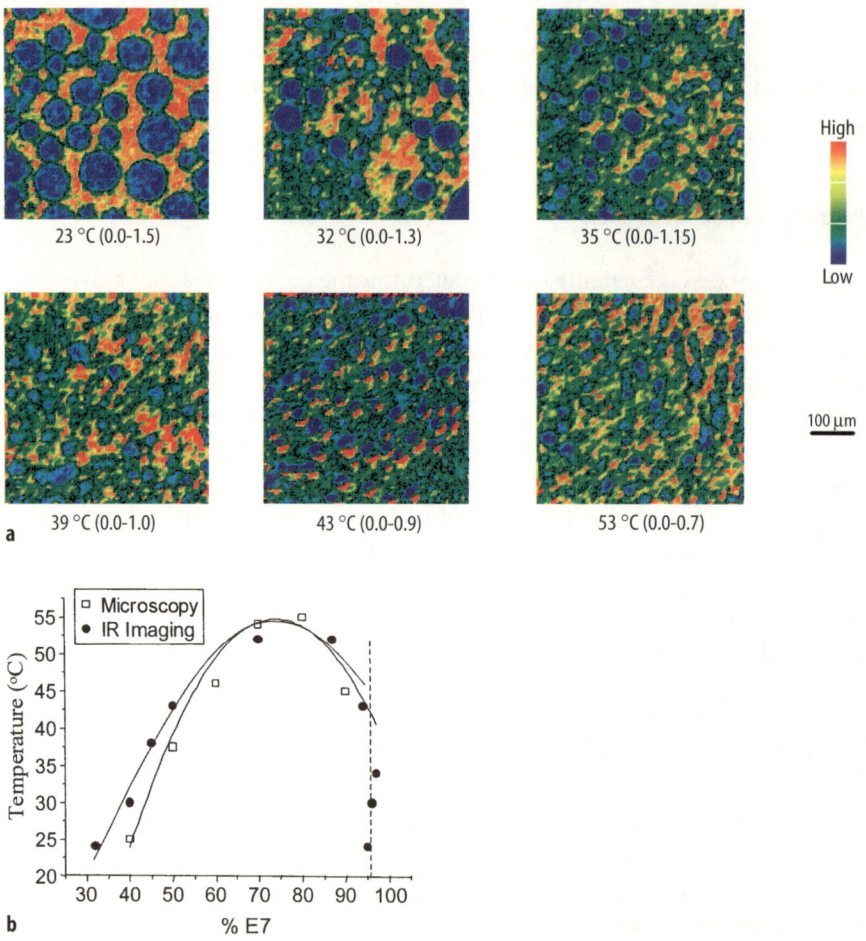

Fig. 17. (**a**) A liquid crystal mixture, E7, dispersed in PBMA (80% liquid crystal by weight) imaged after quenching from the single phase to various temperatures. The figures in parenthesis indicate the low and high limits the images are plotted between as shown by the color bar on the right. The characteristics infrared spectroscopic absorbance from the two phases can be determined to yield concentrations from a previously obtained calibration curve (not shown). (**b**) The phase diagram of the liquid crystal mixture, E7, dispersed in PBMA imaged in Fig. 16(a) compared to the phase diagram obtained by optical microscopy reported in Ref. [84]

3.4
Summary

FTIR imaging using an FPA coupled to an interferometer has been demonstrated as an excellent technique to achieve non-invasive spatially specific information. It combines the benefits of an IR spectrometer with those of an optical microscope providing unprecedented amount of information and level of detail in both the spectral and spatial domains. Some unique considerations are required in the collection and processing of the large amount of data from an imaging spectrometer. In a short time, FTIR imaging has been applied to a number of multicomponent polymer samples with good results. The techniques used for and applications to multicomponent polymers are expected to increase exponentially in the foreseeable future as the cost of instrumentation reduces and the SNR characteristics of available detectors improve.

4
Other Infrared Microspectroscopy Approaches

4.1
Hadamard Transform Infrared Microscopy

Hadamard transform infrared microscopy involves the use of encoding masks [125] to achieve selective light throughput from the optical train. By using masks of known patterns or encoding, a large spatial area could be mapped to provide an effective multichannel advantage. Spatial multiplexing of the field of view can also be accomplished by the use of masks while retaining the advantages of the Fourier transform [126]. The process is schematically shown in Fig. 18. While this tech-

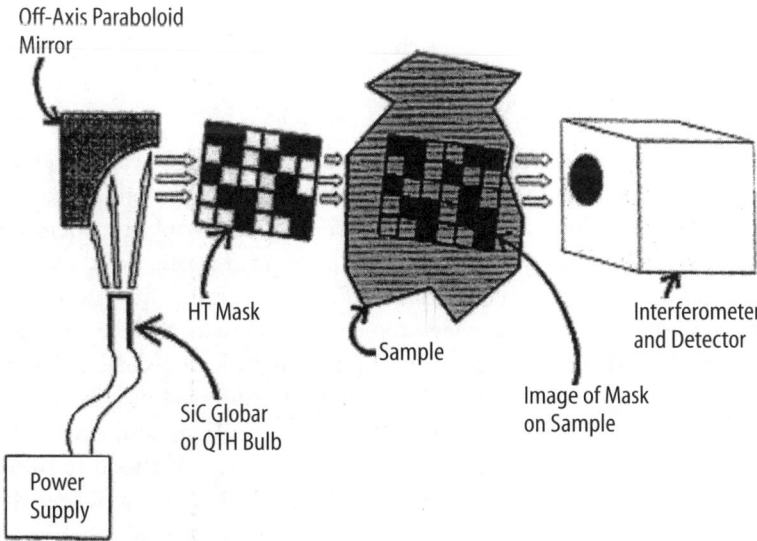

Fig. 18. Schematic of the chemical mapping instrumentation in the Mid- and Near-IR Spectral Regions by Hadamard Transform/FT-IR Spectrometry. (Reproduced from [126])

nique is not new, it has also not achieved as widespread applicability as the single aperture microscopy technique. The primary limitation of the technique arises from its need for opaque masks, which need to be reproducibly and rapidly positioned. Recently, programmable micro-mirror arrays have been used to achieve large multichannel advantages compared to raster scan approaches [127]. Wide applicability of this route to conduct microspectroscopy will determine on the advances in optical masking that will probably evolve from optical communications applications in the future.

4.2
Synchrotron Infrared Microscopy

The large collection times and poor SNR for high spatial resolutions in IR microscopy are largely a result of relatively weak light intensity. A synchrotron source provides for very high throughput and it's coupling to an infrared microscope results in high quality data. Not only is the spatial resolution of an infrared microscope using a synchrotron source superior and SNR higher (~100 times) compared to a standard globar source, it also provides an extended spectral range. Hence, data collection over wide spectral ranges is considerably faster and sensitivity for thin samples is excellent. Apertures as low as 3×3 µm^2 have been used [128]. However, the major drawback arises from the unique component of the technique: a synchrotron source is required. Thus, reports in literature examining polymeric materials have been limited [129] as the instrument is not widely available.

4.3
Solid State Focal Plane Array Imaging

It is also possible to use infrared acousto-optic tunable Filters (AOTFs) [130] to restrict radiation on the array to be of a certain wavelength and sequentially sample different wavelengths using multichannel detection by employing array detectors [131]. Fig. 19 demonstrates an application [132], in which plastic waste was classified using numerical methods from data collected using remote and on-line near-infrared measurements on a macroscopic scale using a focal plane array. In a variation of the approach of using filters in conjunction with array detectors [133], an acousto-optic tunable filter (AOTF) was employed to diffract unpolarized incident light into two diffracted beams with orthogonal polarization. One of the beams was directed to an array detector sensitive in the visible region (a silicon camera) while the other is detected to an infrared sensitive FPA. Near-IR spectroscopy could be carried out while the visible image was simultaneously observed. A curing epoxy-amine system was found to exhibit spatial inhomogeneities when examined by the instrument. Such an arrangement, however, introduces polariza-

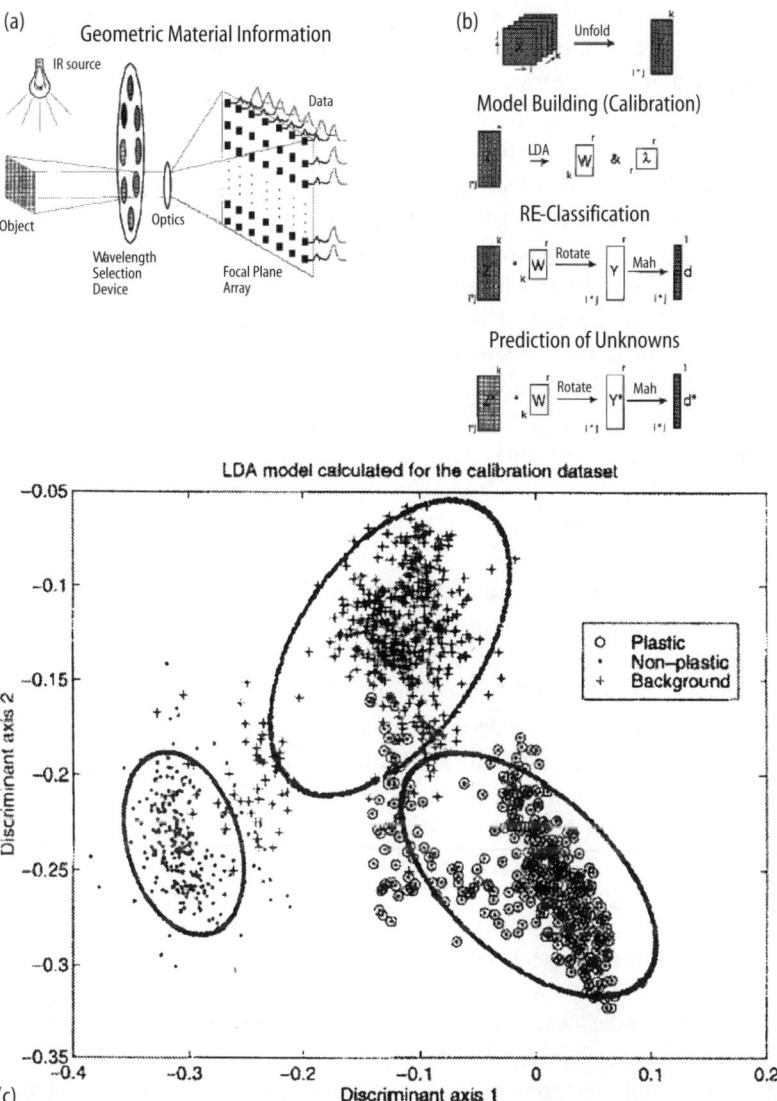

Fig. 19. (**a**) Schematic view of the measurement principles. Reflected radiation from an object is measured for several wavelength regions. Each individual detector pixel integrates possible reflected photons in the particular wavelength range. (**b**) Schematic representation of the LDA classification procedure. The three-dimensional calibration dataset **X** is unfolded to a two-dimensional matrix **Z**, which is used to calculate weight matrix **W**. Next, matrix **Z** is transformed to **Y** where for each class (nc) the covariance matrix and corresponding means are calculated, according to the Mahalanobis measure. An unknown image **Z*** is transformed to **Y*** and classified to **d*** with the use of the calculated covariance and mean data. (Reproduced from Ref. [132]). (**c**) Visualization of the calibration dataset using two linear discriminant components for three classes: plastic, nonplastic, and background mini-spectra. The ellipses indicate the classes modeled by LDA. In this manner, waste is discriminated

tion effects in the spectra. In many oriented polymer systems, these polarization effects may be useful to determined relative organization of sub-molecular species, especially for thick polymer films.

Yet another approach to spatially resolved infrared spectroscopy using a focal plane array can be visualized where one of the dimensions of the focal plane is employed to detect spectrally discriminated data. The combination of a device that disperses different wavelengths to different spatial locations and an FPA can be employed to image a line perpendicular to the wavelength dispersion direction [134]. Alternately, the same can be achieved by means of a linear variable filter (LVF). In either case, the spatial position and wavelength discrimination are linked. Hence, an integrated system needs to be designed where the sample is moved such that every pixel samples a different set of a wavelength and a spatial component to yield the entire imaging data set. These approaches have not been employed for imaging yet, but can be expected to be available in the near future for at least some special analysis cases. While the optical components in these configurations are fixed, resulting in a cheaper and more robust system, the sample would need to be moved.

Clearly, solid state approaches such as these provide an alternative where high spatial fidelity is required, a high rate of data acquisition is desirable and a rugged system required. The optical arrangement determines the spatial resolution of the system, which can range from the usual "macroscopic" configuration where the FPA is employed for multichannel detection or a combination of the optical system with a sophisticated data analysis protocol is for spatially resolved data collection. While solid state approaches are useful in such settings and provide flexibility and rapid data acquisition, interferometer based systems provide much higher sensitivity, broad spectral coverage, and make possible much higher spectral resolutions.

4.4
Scanning Probe Microspectroscopy or Photothermal Imaging

A new technique is under development combining scanning probe microscopy at sub-micron spatial resolution with FTIR spectroscopy. The idea springs from a recently developed technique termed Micro-Thermal Analysis [135, 136]. This technique allows localized thermal analysis to be combined with surface and sub-surface imaging by means of scanning thermal microscopy. Chemical fingerprinting is thereby achieved, giving a microscopic version of two well-known thermal analysis techniques, TMA and modulated-temperature DSC. It also provides for combining the advantages of the thermal characterization techniques with infrared spectroscopy.

Modulated radiation from a standard spectrometer is directed onto a sample and detected by a microprobe (see Fig. 20) [137, 138]. The IR imager's thermal probe is the same as used in micro-thermal techniques. It detects photo-thermal response of a region on the specimen heated by exposure to the beam from a FTIR

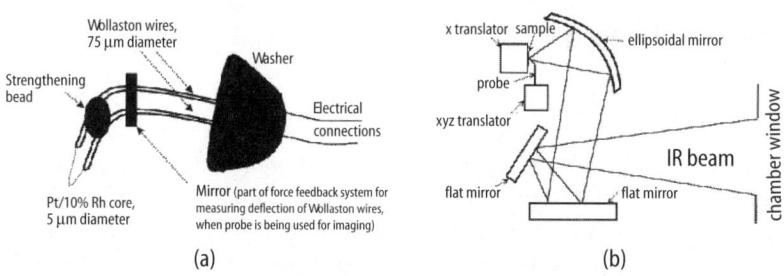

Fig. 20. a Schematic Diagram for a thermal probe and **b** schematic diagram of the probe and sample arrangement layout inside the spectrometer. (Reproduced from Ref. [137])

spectrometer. The signal from this probe measures the resulting temperature fluctuations, thus provides an interferogram that replaces the interferogram normally obtained by means of direct detection of the IR transmitted by a sample. This enables IR microscopy at a spatial resolution below diffraction limits as the resolution now depends on the probe and not system optics. This resolution could be as low as tens of nanometers. However, the interaction of the probe with the sample may not be desirable and may limit the actual resolution achieved.

4.5
Near Field Infrared Mapping

Near-field infrared microscopy allows examination of spatial regions at a spatial resolution better than the diffraction limit imposed by the wavelength of the probing light. One approach employs the use of a standard FTIR interferometer-microscope combination [21]. The experimental setup was such that near-field conditions were possible but the effective spatial resolution attained was not clear. The In another implementation, the diffraction limit is overcome by employing specialized radiation collection optics for the infrared region [139] that permit partial conversion of the evanescent fields into propagating waves. The amount of light propagating from a sub-wavelength aperture through a flat substrate strongly increases, as the distance between the light collection tip and the sample is decreased. Hence, for a sample, topographic artifacts often dominate near-field images, which are sought to be eliminated by flat sample preparation techniques [140]. In addition, the transmitted power is dependent on the refraction index of the sample, resulting in deviations between the near-field and far-field spectra. The most useful applications for the apertureless near-field IR spectroscopic techniques will involve ideal sample preparation, multiplexing of signal, complementary information from mathematical modeling and better probe design [141]. Near-field systems are expected to be especially useful for imaging phase-separated high polymer blends, sub-microscopic self-organization, thin films and aqueous polymer systems.

4.6
Infrared Diode-Based Imaging

An alternative approach to using interferometers for imaging is to use another source that provides spectral discrimination directly. One such source is an infrared diode laser. A laser coupled to an infrared microscope provides a means to rapidly obtain spatial resolution approaching the diffraction limit [142]. A multi-layer polymer composite contaminated by migration of a volatile additive from an exogenous source was imaged using this system. The distribution of this additive in the layered structure was shown to correlate with specific layers (Fig. 21). A con-

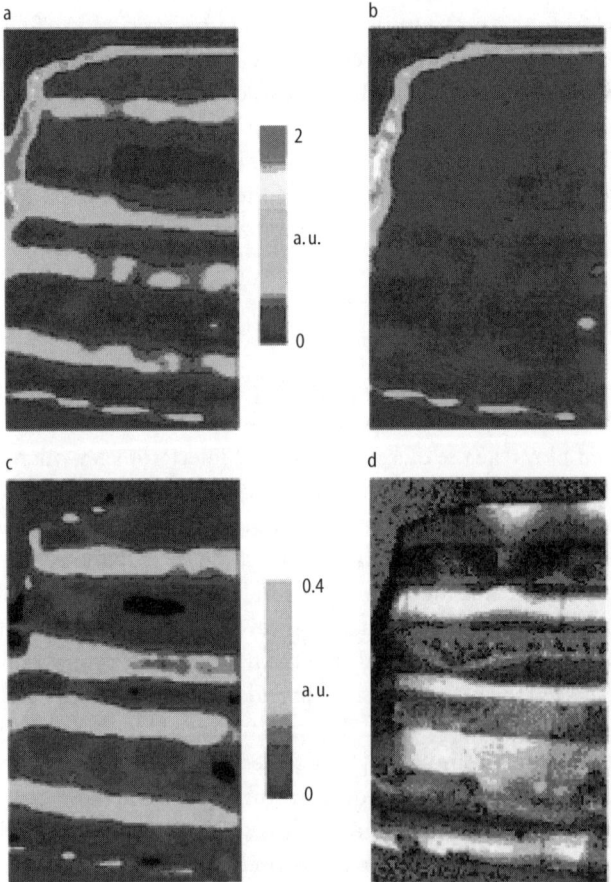

Fig. 21. Diode laser false-color images of an NP-exposed sample: **a** absorbance at 1568 cm^{-1}, **b** absorbance at 1530 cm^{-1}, and **c** difference between (**a**) and (**b**). A white-light image is shown in (**d**). The absorbance intensity maps to the color scale adjacent to each image. Each image is 300 µm_500 µm (Reproduced from Ref. [142])

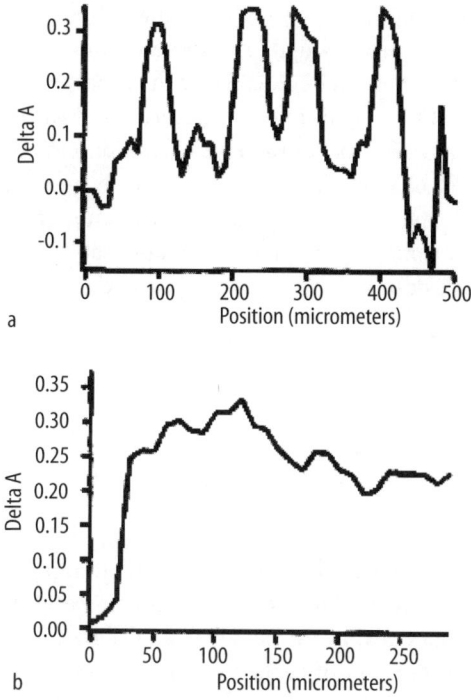

Fig. 22. Absorbance profiles from diode laser images in Fig. 21. **a** 100 µm from left edge of image (perpendicular to layer structure); **b** 100 µm from top edge of image (parallel to layer structure following an epoxy-binder layer)

centration gradient suggesting a diffusive mechanism of additive migration parallel to the layered structure was observed (Fig. 22). A coupling of the laser to a focal plane array could provide a multichannel advantage, considerably increasing the scope of such instrumentation.

4.7
Summary

There are a number of approaches to allow infrared microspectroscopic investigations of polymers. While interferometer based approaches dominate the instrumentation in general, other approaches may be effective in specific cases or when dedicated instrumentation examining a limited range of problems is desired. FPAs, employed to achieve spatial discrimination as in FTIR imaging or spatial and spectral discrimination as in dispersive and filter approaches, provide flexibility and faster spectroscopy due to their multichannel detection advantage. Many new approaches to microspectroscopy are dedicated to achieving spatial resolution better than limited by diffraction. While the progress to that end is remarkable, most techniques are far from developed to a mature enough level to be routinely applied. It is expected that the maturation of these technologies and the resulting instrumentation will be particularly useful for examining the interfacial regions in polymeric samples or cases where domains are smaller than the resolution limit of current techniques. However, the SNR characteristics would probably require large collection times for a single image and may result in the instruments being used for specialized cases where current imaging methods are inapplicable. The development of new imaging capabilities can only mean greater progress for the scientist studying multicomponent polymers.

5
Concluding Remarks

5.1
Instrumental Directions

The trend towards more stable spectroscopic systems with high spatial resolution and lower spectral noise data acquired in smaller time periods is going to be a feature of instrumental development for many years to come. This new generation of instruments awaits the development of more sensitive, faster response and lower noise detectors as well as better miniaturized electronics. As problems in polymer science become more sophisticated, the need to analyze samples with greater precision, track changes in a sample more minutely and carry these out faster is going to be a primary need for the spectroscopist. New accessories continue to be developed for old techniques and several specialized accessories will no doubt be available for the emerging techniques in the near future. New instrumentation consisting of motionless spectrometers, ultrafast spectrometers and sub-micron resolution imaging spectrometers will probably develop in the foreseeable future. In particular, FTIR imaging techniques have reached maturity levels where they can be used for routine analysis and we expect to see a large increase in the applications reported. Development of larger format arrays with expanded wavelength sensitivity and higher efficiency is one direction, which will spur such growth. Emergence of faster and more sophisticated computers continues to help in data acquisition and analysis.

5.2
Applications

We believe that much growth in applications will be spurred on by the desire of scientist to observe dynamic phenomena and to analyze large numbers of samples in a short time period. Static, spatially resolved measurements have been routinely carried out for over 30 years. The last decade has seen a major initiative towards examining dynamic phenomena ranging from simple changes like polymerization to complicated spatially resolved interactions. Even physical phenomena ranging from orientation to phase separation to temperature rise continue to be explored. Such demanding/specialized needs will contribute to further development. A major area where spatially resolved methods will become popular is in the analysis of combinatorial libraries [143]. The multichannel detection advantage of the imaging techniques available today is well suited to the analysis of high throughput reaction sequences and is expected to be widely applied in this field in the future.

5.3
Processing Techniques and Strategies

FTIR spectroscopy has been applied to identify many multicomponent systems using techniques such as spectral subtraction and factor analysis. The high degree of reproducibility and accuracy of IR measurements have made this possible. One of the largest reasons for that has been the use of digital computers. Indeed the development of spectroscopy has gone hand in hand with that of faster, cheaper and better computing power. High spectrometer fidelity and reproducibility is possible not only because of better hardware but numerous checks and control loops for fine corrections that can be made using sophisticated computers. The very existence of mapping and imaging technologies is due to the ability to handle large data sets made possible by computers.

It may be envisioned that, in the coming years, appropriate software will play an increasing role in the collection, post-collection data extraction and manipulation, identification of species and processing of results. Multivariate analysis coupled to automated detection schemes and fuzzy logic identification protocols based on spectroscopic data will be increasingly used. Automated Artificial Neural Networks coupled to FTIR instrumentation [144] can then be used to monitor and control the formation and properties of multicomponent polymers. To make that aim possible, numerical methods for the analysis and characterization of polymers [145] are becoming increasingly important. The use of imaging techniques in synthesis, processing and quality control applications is expected to grow with the development of stable and ruggedly usable instrumentation.

5.4
Summary

FTIR microspectroscopy has made remarkable progress in understanding multicomponent polymer systems. A range of issues – from static morphology to submolecular molecular dynamics can now be analyzed by the versatile set of tools this technique affords. The last 5–10 years have been a period of intense development on both instrumental and data interpretation fronts. For applications to polymers, a lot of effort has been devoted to understand interactions between polymers, dynamics of polymers and microscopic characterization of polymers. These analyses have been carried out faster and with greater accuracy than ever before. The future holds good promise for the development in spatially resolved infrared spectroscopic techniques, especially imaging, which will greatly assist polymer scientists in structure-property relationship elucidations.

References

1. Painter PC, Coleman MM, Koenig JL (1982) The theory of vibrational spectroscopy and its application to polymeric materials. John Wiley, New York
2. Koenig JL (1992) Spectroscopy of polymers. American Chemical Society (ACS), Washington DC, USA
3. Chalmers JM, Everall NJ, Hewitson K, Chesters MA, Pearson M, Grady A, Ruzicka B (1998) Analyst 123:579
4. Smith MJ, Carl RT (1989) Appl. Spectrosc. 43:865
5. Barer R, Cole ARH, Thompson HW (1949) Nature 163:198
6. A general discussion of sampling techniques for microscopy may be found in Allen TJ (1992) Vib. Spectrosc. 3:217
7. Augerson CC (1998) Appl. Spectrosc. 52:1353
8. Tungol MW, Bartick EG, Montaser A (1993) Appl. Spectrosc. 47:1655
9. Lang PL, Katon JE, Schiering DW, O'keefe JF (1986) Polym . Mat. Sci. Eng. 54:381
10. Kaito A, Nakayama K (1992) Macromolecules 25:4882
11. Harrick, NJ, Milosevic M, Berets SL (1991) Appl. Spectrosc. 45:944
12. Gentner JM, Wentrup-Byrne E (1999) Spectrochim. Acta. A-Mol. Bio. 55:2281
13. Lewis L., Sommer AJ (1999) Appl. Spectrosc. 53:375
14. Mirabella FM Jr, (1987) Applications of microscopic Fourier transform infrared spectrophotometry sampling techniques for the analysis of polymer systems. In: Rousch PB (ed) The design, sample handling and applications of infrared microscopes. American Society for Testing and Materials, Philadelphia, PA, USA, p 74
15. Ferrer N (1997) Mikrochim Acta 329
16. Reffner JA; Ressler G; Schiering DW; Wihlborg WT (1997) Mikrochim Acta 333–337
17. Blair D.S., Ward KJ (1991) Part. Surf. 3:123
18. Sommer AJ, Katon, JE (1991) Appl. Spectrosc. 45:1663
19. Chase B (1998) In: Messerschmidt RG, Hartcock MA (eds) Infrared microspectroscopy. Marcel Dekker, New York, USA, p 97
20. Messerschmidt RG (1987) In: Rousch PB (ed) The design, sample handling and applications of infrared microscopes. American Society for Testing and Materials, Philadelphia, PA, USA p 12
21. Sahlin JJ. Peppas NA (1997) J. Appl. Polym. Sci. 63:103
22. Born M, Wolf E (1980) Principles of optics, 6th edn. Pergamon, Elmsford NY, USA
23. Heimann PA, Urstadt R (1990) Appl. Opt. 29:495
24. Koenig JL (1992) Microspectroscopy of polymers. ACS, Washington, DC, USA
25. Katon JE (1994) Vib. Spectrosc 7:201
26. Rintoul L, Panayiotou H, Kokot S, George G, Cash G, Frost R, Bui T, Fredericks P (1998) Analyst 123:571
27. Cournoyer R, Shearer JC, Anderson DH (1977) Anal Chem 49:2275
28. Hartcock MA (1987) In: Rousch PB (ed) The design, sample handling and applications of infrared microscopes. ASTM STP 949. American Society for Testing and Materials, Philadelphia, PA, USA, p 85

29. Shearer JC, Peters DC (1987) In: Rousch PB (ed) The design, sample handling and applications of infrared microscopes. American Society for Testing and Materials, Philadelphia, PA, USA, p 27
30. Weesner FJ, Carl RT, Boyle RM (1991) Proc SPIE Int Soc Opt Eng, 1575:486
31. Ward KJ (1989) Proc SPIE Int Soc Opt Eng 1145:212
32. Garcia D (1997) J Vinyl Additive Tech. 3:126
33. Hartcock MA, Lentz LA, Davis BL, Krishnan K (1986) Appl Spectrosc 40:
34. Joshi NB, Hirt DE (1999) Appl Spectrosc 53:11
35. Gavrila DE, Gosse B (1994) J Rad Nuc Chem 185:311
36. Jouan X, Gardette JL (1987) Polym Commun 28:239
37. Miseo EV, Guilmette LW (1987) Industrial problem solving by microscopic Fourier transform infrared spectrophotometry. In: Rousch PB (ed) The design, sample handling and applications of infrared microscopes. ASTM STP 949. American Society for Testing and Materials, Philadelphia, PA, USA, p 97
38. Wilhelm P (1996) Micron 27:341
39. Ahlblad G, Forsstrom D, Stenberg B, Terselius B, Reitberger T, Svensson LG (1997) Polym Degrad Stabil 55:287
40. Tungol MW, Bartick EG, Montaser A (1990) 44:543
41. Grieve MC; Griffin RME; Malone R (1998) Sci Justice 38:27
42. Bouffard SP, Sommer AJ, Katon JE, Godber S (1994) Appl Spectrosc 48:1387
43. Katon JE, Lang PL, Schiering DW, O'Keefe JF (1987) Instrumental and sampling factors in infrared microspectroscopy. In: Rousch PB (ed) The design, sample handling and applications of infrared microscopes. ASTM STP 949. American Society for Testing and Materials, Philadelphia, PA, USA, p 4
44. Bartick EG(1987) Considerations for fiber sampling with infrared microspectroscopy. In: Rousch PB (ed) The design, sample handling and applications of infrared microscopes. ASTM STP 949. American Society for Testing and Materials, Philadelphia, PA, USA, p 64
45. Cho LL, Reffner JA, Gatewood BM, Wetzel DL (2001) J Forensic Sci 46:1309
46. Chase DB (1987) Infrared microscopy: a single fiber technique. In: Rousch PB (ed) The design, sample handling and applications of infrared microscopes. ASTM STP 949. American Society for Testing and Materials, Philadelphia, PA, USA, p 4
47. Hirschfeld T (1985) Appl Spectrosc 39:424
48. see for example, Cho LL, Reffner JA, Wetzel DL (1999) J Forensic Sci 44:283
49. Vermette P, Thibault J, Levesque S, Laroche G (1999) J Biomed Mater Res 48:660
50. Costa L, Luda MP, Trossarelli L, del Prever EMB, Crova M, Gallinaro P (1998) Biomaterials 19:659
51. Kakoulides EP, Smart JD, Tsibouklis J (1998) J Control Release 54:95
52. Kressler J, Schafer R, Thomann R (1998) Appl Spectrosc 52:1269
53. Benedetti E, Catanorchi S, D'Alessio A, Vergamini P, Ciardelli F, Pracella M (1998) Polym Int 45:373
54. Ludvigsson M, Lindgren J, Tegenfeldt J (2000) J Electrochem Soc 147:1303
55. Withey RE, Hay JN (1999) Polymer 40:5147
56. Karamancheva I, Stefov V, Soptrajanov B, Danev G, Spasova E, Assa J (1999) Vib Spectrosc 40:369
57. Camacho NP, Rinnerthaler S, Paschalis EP, Mendelsohn R, Boskey AL, Fratzl P (1999) Bone 25:287
58. Benedetti E; Catanorchi S; D'Alessio A; Vergamini P; Ciardelli F; Pracella M (1998) Polym Int 45:373
59. Dibbern-Brunelli D, Atvars TDZ, Joekes I, Brabosa VC (1998) J Appl Polym Sci 69:645
60. Schafer R, Zimmermann J, Kressler J, Mulhaupt R (1997) Polymer 38:3745
61. Coleman PB, Ramamurthy AC (1999) Appl Spectrosc 53:150
62. Gal T, Toth P (1992) Can J Appl Spectrosc 37:55
63. Kellner R, Fischböck G, Minich C (1986) Mikrochim Acta 1:271
64. Nishioka T, Nakano T, Teramae N (1992) Appl Spectrosc 46:1904

65. Pell RJ, Mckelvy ML, Hartcock MA (1993) Appl Spectrosc 47:634
66. Guilment J, Markel S, Windig W (1994) Appl Spectrosc 48:320
67. High MS, Painter PC, Coleman MM (1992) Macromolecules 25:797
68. Sahlin JJ, Peppas NA (1997) J Biomat Sci Polym Ed 8:421
69. Sheu K, Huang S J, Johnson JF (1989) Polym Eng Sci 29:77
70. Hsu SC, Lin-Vien D, French RN (1992) Appl Spectrosc 46:225
71. Cameron RE, Jalil MA, Donald AM (1994) Macromolecules 27:2713
72. Riquet AM, Wolff N, Laoubi S, Vergnaud JM, Feigenbaum A (1998) Food Addit Contam 15:690
73. Schafer R, Kressler J, Neuber R, Mulhaupt R (1995) Macromolecules 28:5037
74. Iijima M, Ukishima S, Iida K, Takahashi Y, Fukada E (1995) Jap J Appl Phys Lett 34:L65
75. Kaito A, Kyotani M, Nakayama K (1992) Polymer 33:2672
76. Glime J; Koenig JL (2000) Rubber Chem Technol 73:47
77. Kaito A, Kyotani M, Nakayama K (1993) J Polym Sci B Poly Phys 31:1099
78. Kaito A, Kyotani M, Nakayama K (1993) J Appl Polym Sci 48:2147
79. Fondeur F, Koenig, JL (1993) Appl Spectrosc 47:1
80. Fondeur F, Koenig, JL (1993) J Adhes 43:263
81. Palinko I, Torok B, Prakash GKS, Olah GA (1998) Appl Catal A-Gen 174:147
82. Ohtake Y, Kobayashi T, Asabe H, Murakami N (1998) Polym Degrad Stabil 60:79
83. Challa SR, Wang S-Q, Koenig JL (1996) Appl Spectrosc 50:1339
84. Challa SR, Wang S-Q, Koenig JL (1997) Appl Spectrosc 51:297
85. Wall BG, Koenig JL (1998) Appl Spectrosc 52:1377
86. Challa SR, Wang S-Q, Koenig JL (1997) Appl Spectrosc 51:10
87. McFarland CA, Koenig JL, West JL (1993) Appl Spectrosc 47:321
88. McFarland CA, Koenig JL, West JL (1993) Appl Spectrosc 47:598
89. Kohri S, Kobayashi J, Tahata S, Kita S, Karino I, Yokoyama T (1993) Appl Spectrosc 47:1367
90. Markwell RD, Butler IS, Gao JP, Shaver A (1992) Appl Org Chem 6:693
91. Maverich AM, Ishida H, Koenig JL (1995) Appl Spectrosc 49:354
92. Arvanitopoulos C, Koenig JL (1996) Appl Spectrosc 50:1
93. Arvanitopoulos C, Koenig JL (1996) Appl Spectrosc 50:11
94. Noobut W, Koenig JL (1999) Polym Composite 20:38
95. Dalton S, Heatley F, Budd PM (1999) Polymer 40:5531
96. Lewis EN, Treado PJ, Reeder RC, Story GM, Dowrey AE, Marcott C, Levin IW (1995) Anal Chem 67:3377
97. Colarusso P, Kidder LH, Levin IW, Fraser JC, Arens JF, Lewis EN (1998) Appl Spectrosc 52:106A
98. Koenig JL, Snively CM (1998) Spectroscopy 13:22
99. Bhargava R, Wall BG, Koenig JL (2000) Appl Spectrosc 54:470
100. Lewis EN, Kidder LH, Arens JF, Peck MC, Levin IW (1997) Appl Spectrosc 51:563
101. Bhargava R, Fernandez DC, Schaeberle MD, Levin IW (2000) Appl Spectrosc 54:1743
102. Harnly JM, Fields RE (1997) Appl Spectrosc 51:334A
103. Snively CM, Katzenberger S, Oskarsdottir G, Lauterbach J (1999) 24:1841
104. Huffman SW, Bhargava R, Levin IW (2002) Appl Spectrosc 56:965
105. Sommer AJ, Tisinger LG, Marcott C, Story GM (2001) Appl Spectrosc 3:252
106. Snively CM, Koenig JL (1999) Appl Spectrosc 53:170
107. Bhargava R, Levin IW (2001) Anal Chem 73:5157
108. Bhargava R, Wang S-Q, Koenig JL (1998) Appl Spectrosc 52:323
109. Bhargava R, Ribar T, Koenig JL (1999) Appl Spectrosc 53:1313
110. Bhargava R, Wang S-Q, Koenig JL (2000) Appl Spectrosc 54:1690
111. Kidder LH, Kalasinsky VF, Luke JL, Levin IW, Lewis EN (1997) Nature Med 3:235
112. Artyushkova K, Wall BG, Koenig JL, Fulghum JE (2000) Appl Spectrosc 54:1549
113. Bhargava R, Wang S-Q, Koenig JL (2000) Macromolecules 32:8989
114. Snively CM, Koenig JL (1999) J Poly Sci Pol Phys 37:2353

115. Marcott C, Story GM, Dowrey AE, Reeder RC, Noda I (1997) Mikrochim Acta Suppl 14:157
116. Snively CM, Koenig JL (1998) Macromolecules 31:3753
117. Oh SJ, Koenig JL (1998) Anal Chem 70:1768
118. van de Weert M, Hennink WE, Jiskoot W (2000) Pharma Res 17:1159
119. van der Weert M, van't Hof R, van de Weerd J, Heeren RMA, Posthuma G, Hennink WE, Crommelin DJA (2000) J Control Release 68:31
120. Cohen S, Yoshioka T, Lucarelli M, Hwang LH, Langer R (1991) Pharm Res 8:713
121. Takahata H, Lavelle EC, Coombes AGA, Davis SS (1998) J Control Release 50:237
122. Bhargava R, Wang S-Q, Koenig JL (1999) Macromolecules 32:2748
123. Koenig JL, Wang S-Q, Bhargava R (2000) Inst Phys Conf Ser 165:43
124. Koenig JL, Wang S-Q, Bhargava R (2001) Anal Chem 73:351A
125. Zhang F, Gu T (1990) Proc SPIE 1205:150
126. Bellamy MK, Mortensen AN, Hammaker RM, Fateley WG (1997) Appl Spectrosc 51:477
127. DeVerse RA, Hammaker RM, Fateley WG (2000) Appl Spectrosc 54:1751
128. Guilhaumou N, Dumas P, Carr GL, Williams GP (1998) Appl Spectrosc 52:1029
129. Ocola LE, Cerrina F, May T (1997) Appl Phys Lett 71:847
130. Bennet JM (1976) Appl Opt 15:2705
131. Lewis EN, Levin IW (1995) J Microscopy Soc Am 1:35
132. Van den Broek WHAM, Wienke D, Melssen WJ, Feldhoff R, HuthFehre T, Kantimm T, Buydens LMC (1997) Appl Spectrosc 51:856
133. Tran CD (2000) J Near Infrared Spectrosc 8:87
134. Elmore DL, Tsao MW, Frisk S, Chase DB, Rabolt JF (2002) Appl Spectrosc 56:145
135. Hammiche A, Reading M, Pollock HM, Song M, Hourston DJ (1996) Rev Sci Instrum 67:4268
136. Pollock HM, Hammiche A, Song M, Hourston DJ, Reading M (1998) J Adhesion 67:217
137. Hammiche A, Pollock HM, Reading M, Claybourn M, Turner PH, Jewkes K (1999) Appl Spectrosc 53:810
138. Anderson MS (2000) Appl Spectrosc 54:349
139. Piednoir A, Licoppe C, Creuzet F (1996) Optics Commun 129:414
140. Palanker DV, Simanovskii DM, Huie P, Smith TI (2000) J Appl Phys 88:6808
141. Dragnea B, Leone SR (2001) Int Rev Phys Chem 20:59
142. Bailey JA, Dyer RB, Graff DK, Schoonover JR (2000) Appl Spectrosc 54:159
143. Snively CM, Oskarsdottir G, Lauterbach J (2000) J Comb Chem 2:243
144. Dolmatova L, Ruckebusch C, Dupuy N, Huvenne J-P, Legrand P (1998) Appl Spectrosc 2:329
145. Chalmers JM, Everall NJ (1996) Trends Anal Chem 15:18

Editor: J. E. McGrath
Received: June 2002

Microwave Assisted Synthesis, Crosslinking, and Processing of Polymeric Materials

Dariusz Bogdal*[1], Piotr Penczek[2], Jan Pielichowski[1], Aleksander Prociak[1]

* E-mail: pcbogdal@cyf-kr.edu.pl
[1] Department of Chemistry, Politechnika Krakowska, 31-155 Kraków, ul. Warszawska 24, Poland
[2] Industrial Chemistry Research Institute (ICRI), 01-793 Warszawa, ul. Rydygiera 8, Poland

Abstract The purpose of this review is to provide useful details concerning the application of microwave irradiation to polymer chemistry. Research in this area has shown some potential advantages of microwaves in the ability not only to drive chemical processes but to perform them in reduced time scale. In some cases the afforded products exhibited properties that may not be possible using conventional thermal treatments. This chapter exposes and discusses the microwave-assisted polymer syntheses, crosslinking, and processing with the stress on chemistry of those processes. A short description of the nature of microwaves as well as their interactions with different matter, in particular with organic substances, is given to provide a brief review of those topics. Then the equipment necessary to carry out microwave experiments is described with links to commercially available systems. Eventually, the free-radical polymerization reaction of various unsaturated compounds is presented together with dental applications followed by step-growth polymerization reactions (i.e., polyamides, polyimides, polyesters, polyethers, epoxy resins, and polyurethanes), polymer modification, reaction on polymer matrices, and recycling.

Keywords Microwave irradiation · Polymerization · Polymer cure · Composite · Polymer matrices

1	Introduction	196
2	Nature of Microwave Irradiation	197
2.1	Equipment for Microwave Experiments	198
2.1.1	Microwave Generators	198
2.1.2	Microwave Applicators	201
2.1.3	Microwave Reactors	203
2.1.4	Temperature Measurement	204
2.2	Microwave Techniques Applied to Chemical Transformations	204
3	Synthesis and Processing of Polymers Under Microwave Conditions	205
3.1	Free-Radical Polymerization	205
3.1.1	Preparation of Dental Materials	214
3.2	Step-Growth Polymerization	217

© Springer-Verlag Berlin Heidelberg 2003

3.2.1 Polyamides and Polyimides 217
3.2.2 Polyesters and Polyethers 226
3.2.3 Epoxy Resins 233
3.2.4 Polyurethanes 246
3.3 Polymer Modification 248
3.4 Reactions on Polymer Matrices (Immobilized Reagents) 250
3.5 Rubber Vulcanization 254
3.6 Recycling 256
3.7 Other Applications 257

4 Conclusions 259

References 260

List of Symbols and Abbreviations

2-MI	2-Methylimidazole
3,4'-ODA	3,4'-Oxydianiline
3DCM	4,4'-Diamino-3,3'-dimethyldicyclohexylmethane
ACA	ε-Aminocaproic amid
AIBN	2,2'-Azobis(2-methylpropanenitrile)
BCMO	3,3-Bis(chloromethyl)oxetane
Bis-GMA	2,2-Bis[4-(3-methacryloyloxy-2-hydroxypropoxy)phenyl]
BTDA	3,3',4,4'-Benzophenonetetracarboxylic acid dianhydride
BTDE	3,3',4,4'-Benzophenonetetracarboxylic acid methyl ester
BWT	Backward wave tubes
CF	Carbon fiber
CTBN	Carboxy terminated butadiene-acrylonitrile random copolymer
DBTM	Dibutyltin maleate
DCM	Methylene dichloride
DDM	4,4'-Methylenedianiline
DDS	4,4'-Diaminodiphenyl sulfone
DDSA	2-Dodecen-1-ylsuccinic acid
DEG	Diethylene glycol
DGEBA	Diglycidyl ether of bisphenol A
DGEBF	Diglycidyl ether of bisphenol F
DMA	Dynamical mechanical analysis
DME	Ethylene glycol dimethyl ether
EDA	Ethylenediamine
ELNR	Epoxidized liquid natural rubber
E-M	Elevated-molecular-weight epoxy resins
ERL-4221	3,4-Epoxycyclohexylmethyl-3,4-epoxycyclohexanoate

ERL-4234	2-(3,4-Epoxycyclohexyl-5,5-spiro-3,4-epoxy)cyclohexane-1,3-dioxane
ERL-4299	Bis-(3,4-epoxycyclohexyl)adipate
ETBN	Epoxy terminated butadiene-acrylonitrile random copolymer
GF	Glass fiber
HDPE	High density polyethylene
HEMA	2-Hydroxyethyl methacrylate
HHPA	Hexahydrophthalic anhydride
ICP	Inherently conductive polymers
KPS	Potassium persulfate
LA	D,L-Lactic acid
LDPE	Low density polyethylene
LLS	Laser light scattering
L-M	Low molecular weight epoxy resins
l-MDI	Liquid mixture of bis(isocyanatophenyl)methane isomers
MA	Methyl acrylate
MDA	Methylene dianiline
Me-HHPA	4-Methyl-hexahydrophthalic anhydride
MMA	Methyl methacrylate
MMT	Microwave Materials Technologies
MORE	Microwave-organic reaction enhancement
mPDA	m-Phenylenediamine
NAA	α-Naphthyl acetic acid
NE	5-Norbornene-2,3-dicarboxylic acid monomethyl ester
NMP	N-Mcthyl 2 pyrrolidone
NVC	N-Vinylcarbazole
PA-6	Polyamide-6
PANI	Polyaniline
PE	Polyethylene
PEPA-3,4'-ODA	3,4'-Bis[(4-phenylethynyl)phthalimido]diphenyl ether
PET	Poly(ethylene terephthalate)
PETI-5	Phenylethynyl-terminated imide oligomer
PETI-5/IM7	Carbon fiber reinforced phenylethynyl-terminated polyimide composites
PG	Propylene glycol
PGE	Phenyl glycidyl ether
PMDA	Pyromellitic acid dianhydride
PMMA	Poly(methyl methacrylate)
PS	Polystyrene
PTC	Phase-transfer catalysis
PTFE	Polytetrafluoroethylene
PVC	Poly(vinyl chloride)

PVK	Poly(*N*-vinylcarbazole)
RP-46	Polyimide resin
RTM	Resin transfer molding
SA	Succinic anhydride
SEP	Poly(styrene)-*block*-poly(ethene-*alt*-propene)
SPOS	Solid-phase organic synthesis
TBAB	Tetrabutylammonium bromide
TBAI	Tetrabutylammonium iodide
TE	Transverse electric wave
TEBA	Triethylbenzylammonium chloride
TEGDMA	Triethylene glycol dimethacrylate
TEM	Transverse electromagnetic wave
TM	Transverse magnetic
TPP	Triphenyl phosphate
TWT	Traveling wave tube
VFM	Variable frequency microwave

1
Introduction

Microwave irradiation is a rapid way of heating materials for domestic, industrial and medical purposes. Microwaves offer a number of advantages over conventional heating such as non-contact heating (reduction of overheating of material surfaces), energy transfer instead of heat transfer (penetrative radiation), material selective and volumetric heating, fast start-up and stopping, and, last but not least, reverse thermal effect, i.e., heat starts from the interior of material body.

In the last decade, microwaves have attracted the attention of chemists who have begun to apply this unconventional technique of material heating as a routine in their practice. The reduced time of processing under microwave conditions found for a great number of chemical reactions was the main reason that microwave techniques became so attractive for chemists. Increase of productivity, improved product characteristics, uniform processing, less manufacturing space required, and controllability of the process are the advantages that microwave processing of material can provide. Microwave processing seems to be easily scaled up from a small scale batch process to a continuous process employing a conveyor.

Besides ceramic processing, polymer chemistry is probably the largest single discipline in microwave technology, and the methods and procedures used therein are certainly among the most developed. Although a great number of papers have been published on the processing and curing of polymers and polymeric materials under microwave irradiation and the application of microwaves to the organic synthesis is still growing; there exists only a limited number of reports on the microwave-assisted polymer synthesis.

The purpose of this chapter is to provide useful details concerning the application of microwave irradiation to polymer chemistry and technology. A survey of the past achievements in polymer synthesis and polymer composites can be found in review papers [1–5]. However, in this chapter a short description of the nature of microwaves as well as their interactions with matter, in particular with organic substances, is given to provide a brief review of those topics; the most extensive explanation can be found in the cited books and papers [6–9].

Fundamentals of electromagnetic heating and processing of polymers, resins, and related composites were summarized by Parodi [10]. An overview of application of microwaves in organic synthesis together with reactions applied in polymer preparation can be found in recently published reviews [11–16] and in a comprehensive review with over 600 references cited in a tabular format [17]. The influence of microwave irradiation on the rate of chemical reactions is still under debate and some researchers have mentioned the existence of so-called non-thermal microwave effect, i.e., an effect that seems to be inadequate to the observed reaction temperatures in sudden acceleration of reaction rates. A useful review summarizing and evaluating the latest theories on the existence of non-thermal or specific microwave effects according to the reaction medium and reaction mechanism was recently published by Perreux and Loupy [18].

2
Nature of Microwave Irradiation

Microwaves are the electromagnetic radiation which can be placed between infrared radiation and radio frequencies with wavelengths of 1 mm to 1 m, which corresponds to the frequencies of 300 GHz to 300 MHz, respectively. The extensive application of microwaves in the field of telecommunication has resulted in only specially assigned frequencies being allowed to be allocated for industrial, scientific or medical applications (e.g., most of the wavelengths between 1 and 25 cm is used for mobile phones, radar, and radio-line transmissions). Currently, in order to avoid interferences with telecommunication devices, household and industrial microwave ovens (applicators) are operated at either 12.2 cm (2.45 GHz) or 33.3 cm (900 MHz). However, some other frequencies are also available for heating [5]. Most common domestic microwave ovens utilize the frequency of 2.45 GHz, and this may be one reason why almost all commercially available microwave reactors for chemical use operate at the same frequency.

When a piece of material is exposed to microwave irradiation, microwaves can be reflected from its surface (when the surface is conducting as in metals, graphite, etc.), can penetrate the material without absorption (in the case of good insulators), and can be absorbed by the material (lossy dielectrics). Thus, heating in microwave ovens is based upon the ability of some liquids and solids to absorb and to transform electromagnetic energy into heat. Microwave radiation – as every radiation of electromagnetic nature – consists of two components: magnetic and electric fields. The electric field component is responsible for dielectric heating mechanisms since it can cause molecular motion either by migration of ionic species (conduction mechanism) or rotation of dipolar species (dipolar polarization mechanism). In a microwave field, the electric field component oscillates very quickly (at 2.45 GHz the field oscillates 4.9×10^9 times per second), and the strong agitation, provided by cyclic reorientation of molecules, can result in an intense internal heating which can lead to heating rates in excess of 10 °C per second when microwave radiation of a 1 kW-capacity source are used [6].

From the point of view of microwave processing, the material is characterized by dielectric properties among which of greatest importance are electric permittivity sometimes called dielectric constant (ε) and the loss tangent ($\tan\delta$). The complex permittivity ($\varepsilon=\varepsilon'-j\varepsilon''$) is a measure of the ability of dielectric materials to absorb and to store electric potential energy, with the real permittivity, ε', characterizing the propagation of microwaves into the material and the loss factor, ε'', indicating ability of the material to dissipate the energy. The most important property in microwave processing is $\tan\delta=\varepsilon''/\varepsilon'$, indicating the ability of the material to convert the absorbed energy into heat. For optimum coupling, a balanced combination of moderate ε', to permit adequate penetration, and high loss (high ε'' and $\tan\delta$) are required [19].

2.1
Equipment for Microwave Experiments

In general, microwave systems consist of a microwave source (generator), a section of transmission line which delivers microwaves from the generator into applicator, and microwave applicator which can efficiently transfer the energy to heated materials. Also a reliable measuring system is necessary to monitor temperature of the process. For processes running at elevated pressures appropriate pressure and temperature control may be required.

2.1.1
Microwave Generators

Magnetrons are the most widely applied generators for laboratory and industrial microwave processing such as those currently used in domestic microwave ovens. Magnetrons are mass produced and the cheapest available sources of microwaves. They are classified as diodes having an anode and cathode. The anode is kept at high potential compared to the cathode. As the cathode is heated, electrons are removed from the cathode and are accelerated toward the anode by the electric field. However, there is a constant magnetic field superimposed orthogonal to the electric field between anode and cathode. External magnets mounted around the mag-

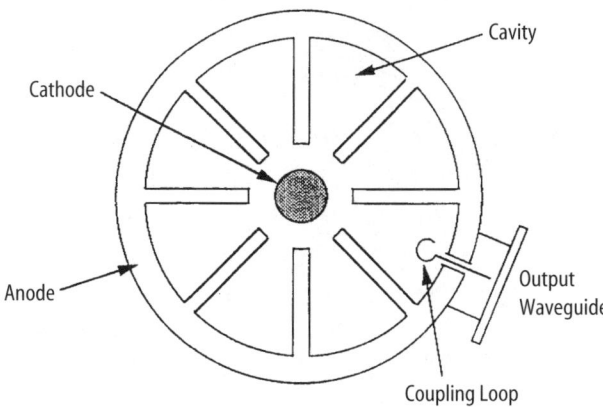

Fig. 1. Schematic diagram of a magnetron shown in cross-section. Reprinted from Microwave Processing of Materials (1994), National Academy Press, Washington DC [19] with permission

netron or a coil wound around the anode block create the magnetic field which is always parallel to the axis of the cathode. The magnetic field enforces the electrons to rotate in the magnetron before they can reach the anode. The anode consists of even number of small cavities that form a series of microwave circuits which are tuned to oscillate at specified frequency. The frequency of the oscillations depends on the size of the cavities. The rotating electrons form a rotor moving around the cathode synchronously in a way that they decelerate and thus transform their energy into microwave oscillations in the cavities. The electromagnetic energy is transferred from one of the resonant cavities to the output transmission line, usually terminated with an antenna (Fig. 1).

In fact, the magnetron belongs to a wider class of microwave vacuum devices which are called traveling wave tubes (TWT) with such a difference that in the magnetron there is no input for microwave energy and therefore the magnetron can only be used as a generator while a TWT can also be used as an amplifier. It is so because in TWT the electron beam is moving straight along the so-called slow wave or traveling wave structure which can retard propagation of electromagnetic waves so that they can conveniently (from the point of view of energy exchange) interact with and gain energy from the electrons moving in the same direction. Consequently, there can be yet another type of microwave amplifier in which the electrons and the wave travel in opposite directions. These are called Backward Wave Tubes (BWT) which can be even more energy efficient as power sources in the sub-millimeter range of microwaves.

The extensive description of microwave sources like traveling wave tubes (TWT), klystrons, disk-seal triodes, and power grid tubes can be found in a number of books on industrial microwave applications [19].

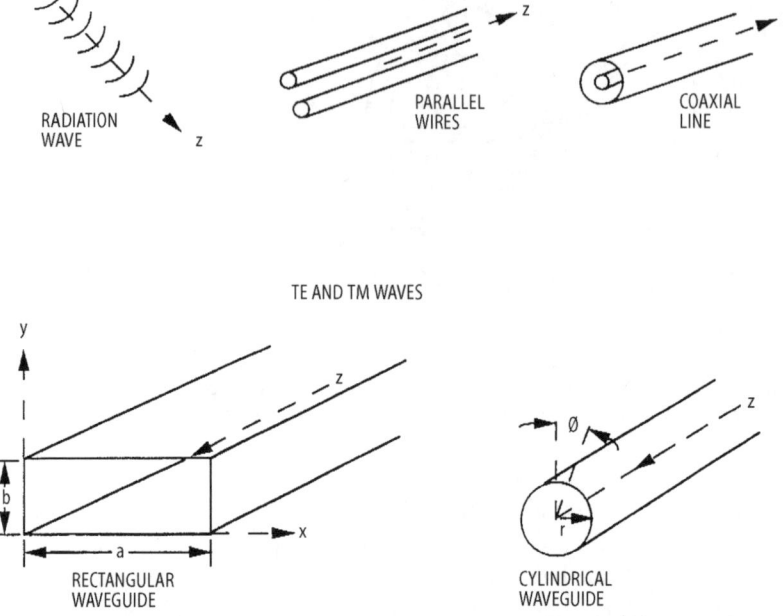

Fig. 2. Waveguides for propagating transverse electromagnetic (TEM), transverse magnetic (TM), and transverse electric (TE) waves. Reprinted from Microwave Processing of Materials (1994), National Academy Press, Washington DC [19] with permission

Microwaves can be transmitted through various media without much loss. Therefore, an applicator can be remote from the power source. Various types of transmission lines can efficiently couple the energy of the microwave source to the applicator. Within transmission lines the microwaves can propagate using three modes. For the so-called transverse electromagnetic wave (TEM), all the components (magnetic and electric) are transverse which is an approximation of the radiation in free space. Both electric and magnetic components in the propagation direction (z) are missing. TEM waves can propagate between two parallel wires, two parallel plates, or in a coaxial line (Fig. 2), and they are applied in general to low power systems (<10 kW).

The transverse electric (TE) and transverse magnetic (TM) are typical modes propagating inside metallic waveguides, which are basically hollow conducting pipes having either a rectangular or a circular cross section. In the TE wave and TM wave, the component of the electric field and the component of the magnetic field in the propagation direction are missing, respectively. In general, every electromagnetic wave in rectangular or cylindrical waveguides can be described as a linear

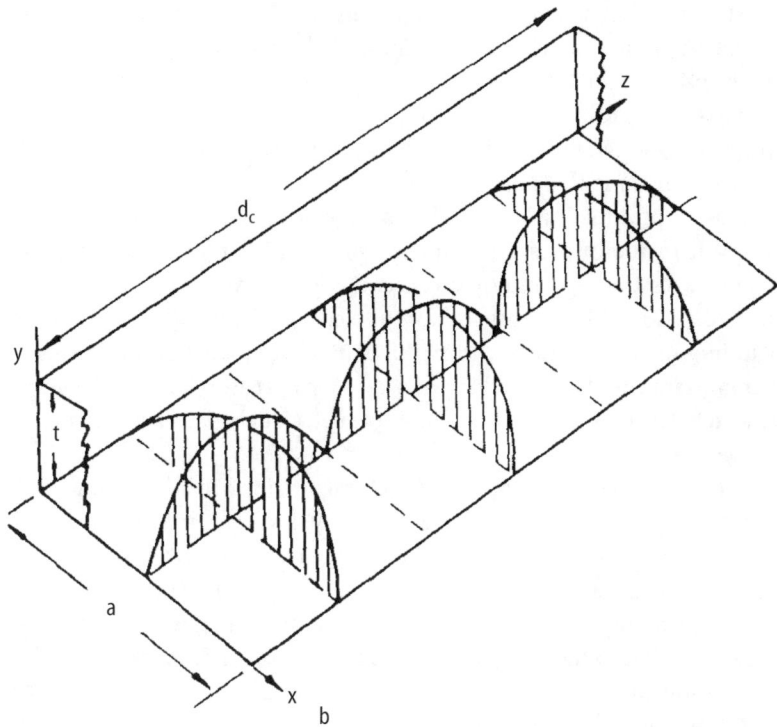

Fig. 3. Development of standing waves: TE_{103} resonant cavity electric field pattern. Reprinted from Industrial Microwave Heating, Peter Perigrinus, London (1983) [6] with permission

combination of the TE and TM modes. The indexes in propagation modes (e.g., TE_{10}) describe particular eigenvalues and define wave variation in a given direction. For example, a TE_{10} mode in rectangular waveguide specifies fundamental TE mode with one half-wavelength in the x-direction and zero half-wavelengths in the y-direction [6]. If the waveguide section forms a cavity a third index describes how many of half-wavelengths would fit along this section. For instance, TE_{103} is a cavity having one half-wavelength along the x-direction, three half-wavelengths along the z-direction, and zero variation of electric field in the y-direction (Fig. 3).

2.1.2
Microwave Applicators

Microwave applicators may appear in many different shapes and dimensions and in fact their design is critical to microwave processing since within applicators microwave energy is transferred (coupled) to processed material. Most common microwave applicators belong to three main types: multi-mode cavity, single-mode cavity including TEM structures, and traveling wave applicators.

It is worth stressing that an important criterion characterizing applicators is the electrical field strength attainable in the sample. While it is possible to map the field strength distribution within an oven by moving, for example, a small water load around it and recording the rate of temperature rise, the results of such an exercise apply only to water. Once the target material is placed in the oven, the field distribution is changed self-consistently.

The simplest applicator is a rectangular metal box that can accommodate the target load. When microwaves are launched into such a box via a waveguide launcher, the waves undergo multiple reflections from the walls. The reflected waves interfere and establish a distribution of electrical field within the internal space (including the load) which, within the band of frequencies, correspond to many different stable modes of oscillations. And for this reason it is called a multi-mode applicator. The most common example of such a device is a cavity in domestic microwave oven.

The field distribution within the load which is contained in a multi-mode applicator depends not only on dielectric permeability or dielectric loss, but also on size and location of the load within the applicator. In this respect, a multi-mode device is best suited to very lossy loads occupying relatively large volumes (more than 50%) of the applicator. For low and medium loss materials occupying less than about 20% of the applicator's volume, the temperature rise in the material, at best, will be a non-uniform one and, at worst, potentially damaging "hotspots" will occur as the result of extremely high local fields.

By incorporating a mode stirrer (a rotating reflector) and by continuously rotating the load placed on a turntable, temperature uniformity within the load can readily be improved. A major criticism of the use of a multi-mode oven for scientific studies is that, since the spatial distribution of field strength is unknown, the feasibility to generalize the results from a particular investigation is compromised, making it very difficult to attempt reliable scaling-up.

By far the most efficient applicator, particularly for filamentary materials, is a single mode resonant cavity. Within such a cavity, only one mode of propagation is permitted and hence the field pattern is defined in space and the target load can be positioned accordingly to it. A single mode cavity may be both cylindrical or rectangular. The simplest (and physically smallest) single mode cavity operates in the TM_{010} mode, in which the electrical field strength is constant along the main axis without variation of angle and is only a function of radius. The radius of the loaded cavity can be determined by solving the electric and magnetic field equations within the cavity under assumption of dimensions and particular boundary conditions. The resonance condition so derived defines radius in terms of the dielectric constants of the load.

A rectangular single-mode cavity consists of a section of waveguide which is terminated with a non-contact plunger which can tune the effective cavity length (resonant frequency). The mode of operation of such a device is, typically, TE_{10n}

with the target load positioned in a region of high field strength. The major limitation of this device is that the length of the load must be less than half a wavelength, otherwise the wave standing along the load would release heat in a periodic manner leading to periodic heating pattern along the target. This can be overcome by moving the plunger in a reciprocating fashion, so that the time-averaged field seen by the load is smoothed.

A traveling wave applicator consists of a section of a slow-wave structure with traveling wave propagating along it and coupled to the load in a spatially distributed manner. The wave is traveling which makes possible efficient coupling also to loads having bigger volumes or large surfaces. As microwaves pass along the slow-wave structure they are absorbed by the load exponentially with distance and according to the dielectric properties and size of load. As a precaution of proper operation, a dummy water load is attached to the load-end of the waveguide to absorb residue of microwave energy which was not absorbed by the load. Provided that the load can move at a constant speed along the structure, each part of it would experience the same total field strength once steady state conditions were achieved, i.e., after at least one length of material has already passed throughout the structure.

2.1.3
Microwave Reactors

Most microwave-assisted reactions in organic and polymer synthesis have been carried out in multi-mode household microwave ovens; however, more recently, the use of microwave reactors for chemical syntheses has become more advocated. The main advantage of household microwave ovens is their low cost and possibility to adjust such ovens to chemical synthesis by drilling shielded outlets in the oven walls for assembling appropriate elements like an upright condenser, stirrer, dropping funnel, etc. The disadvantage is a pulsating-mode duty cycle – adjustable power regulation in such cheap microwave ovens in which variable power is always obtained by switching the magnetron off and on.

Specialized microwave reactors for chemical synthesis are commercially available from such companies as CEM [20], Lambda Technologies [21], Microwave Materials Technologies (MMT), Milestone [22], PersonalChemistry [23], and Plazmatronika [24] which are mostly adjusted from microwave systems for digestion and ashing of analytical samples [25]. They are equipped with built-in magnetic stirrers and direct temperature control by means of an IR pyrometer, shielded thermocouple or fiber-optical temperature sensor, and continuous power feedback control, which enable one to heat reaction mixture to a desired temperature without thermal runaways. In some cases, it is possible to work under reduced pressure or in pressurized conditions within cavity or reaction vessels.

2.1.4
Temperature Measurement

Measurement of reaction temperature during microwave irradiation is a major problem one faces during microwave-enhanced processing of materials. Fiber-optic thermometers can be applied up to 300 °C, but are too fragile for industrial application. In turn, optical pyrometers and thermocouples can be used, but pyrometers only measure surface temperature that can be lower than the interior temperature of the material, while application of thermocouples which are metallic probes can results in arcing between the thermocouple shields and cavity walls leading to failures in thermocouple performances. The temperature of microwave irradiated samples can also be measured by inserting either a thermocouple or thermometer into the hot material immediately after turning off the microwave power.

2.2
Microwave Techniques Applied to Chemical Transformations

The simplest method for conducting microwave-assisted reactions involves irradiation of reactants in an open vessel. Such a method, termed "microwave-organic reaction enhancement (MORE)", was developed by Bose at al. [26]. During the reaction, reactants are heated by microwave irradiation in polar, high-boiling solvent so that the temperature of reaction mixture does not reach the boiling point of a solvent. Despite the convenience, a disadvantage of the MORE technique consists in limitation to high-boiling polar solvents such as DMSO, DMF, *N*-methylmorpholine, diglyme, etc. The approach has been adapted to lower-boiling solvents (e.g., toluene) [27], but it generates a potentially serious fire hazard.

For reactions at reflux, domestic microwave ovens have been modified by making a shielded opening to prevent leakage, and through which the reaction vessel has been connected to a condenser [8].

The pressurized conditions for microwave reactions first reported by the groups of Gedye [28] and Giguere [29] have also been developed. Gedye et al. used a domestic microwave oven, and commercially available screw-up pressure vessels made of PET and PTFE (both being microwave transparent). The "bomb" strategy has been successfully applied to a number of syntheses, but it always generates a risk of hazardous explosions. Recently, Majetich and Hicks [30] reported 45 different reactions with a commercial microwave oven and PET vessels designed for acid digestion.

Microwave heating has been proven to be of benefit particularly for the reactions under "dry" media in open vessel systems (i.e., in the absence of a solvent, on solid support with or without catalysts) [31]. Reactions under "dry" conditions were originally developed in the late 1980s [32], but solventless systems under microwave irradiation offer several advantages. The absence of solvent reduces the risk of explosions when the reaction takes place in a closed vessel. Moreover, aprotic dipolar solvents with high boiling points are expensive and difficult to remove from the reaction mixtures. During microwave induction of reactions under dry conditions, the reactants adsorbed on the surface of alumina, silica gel, clay, and other mineral supports absorb microwaves whereas the support does not, and transmission of microwaves is not restricted. Consequently, such supported reagents efficiently induce reactions under safe and simple conditions with domestic microwave ovens instead of expensive specialized commercial microwave systems.

Finally, the mixture of neat reagents in an open vessel can lead to a reaction under microwave conditions provided that one of the reagents is liquid or a low melting solid that couples well with microwaves.

3
Synthesis and Processing of Polymers Under Microwave Conditions

The effect of microwave irradiation on chemical reactions is generally evaluated by comparing the time needed to obtain a desired yield of final products with respect to conventional thermal heating. Research in the area of chemical synthesis has shown some potential advantages in the ability not only to drive chemical reactions but to perform them in reduced time scale. In some cases the products exhibited properties that may not be possible using conventional thermal treatments. In this chapter the microwave-assisted polymer syntheses, crosslinking, and processing with the stress on chemistry of those processes is discussed. For this purpose, syntheses run under microwave conditions were compared with conventional heating methods. In most examples reported in the literature, the amount of reagents varied from a few milligrams to several grams. Moreover, the shape and size of the reaction vessel are important factors for the processing of material under microwave irradiation as well as applied microwave system (i.e., applicator and temperature detection). Therefore, in each case we have tried to describe briefly microwave systems that were used by different research groups together with amounts of reagents to give readers a deeper feeling about the microwave experiments.

3.1
Free-Radical Polymerization

Microwave irradiation has been applied to free-radical polymerization and copolymerization reactions of various unsaturated monomers.

In 1983, Teffal and Gourdene investigated the bulk polymerization of 2-hydroxyethyl methacrylate (HEMA) under conventional and microwave conditions [33]. Since HEMA bears ester as well as alcohol functions, it is a polar species

Δ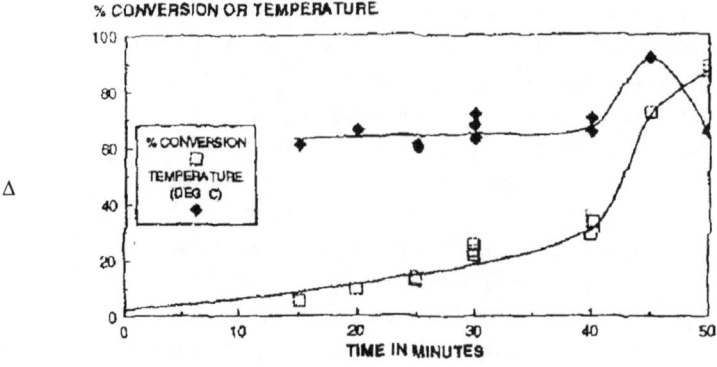

MW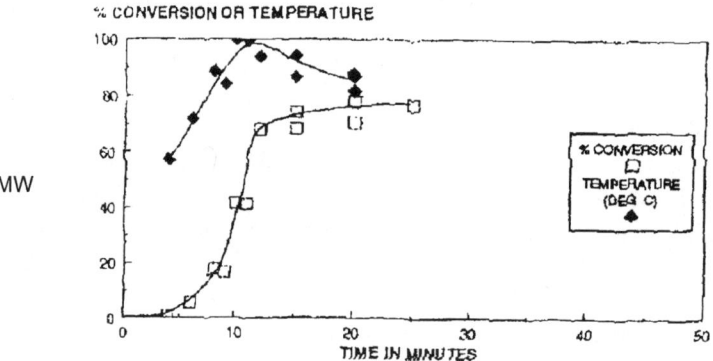

Fig. 4. Thermal profiles and conversions for conventional and microwave polymerization of methyl methacrylate (MMA). Reprinted from (1995) J Mat Proc Techn 48:445 [35] with permission

capable of interacting and absorbing microwaves. Therefore, the reactions were carried out without any radical initiator, and the liquid monomer polymerized to form a solid material that was insoluble in all usual solvents but swelled in water. Thus, it was demonstrated that the radical polymerization can be achieved by microwave irradiation of neat reagents. In a similar example, Palacios et al. showed that in the case of bulk copolymerization of HEMA with methyl methacrylate (MMA), microwave-assisted polymerizations gave copolymers with molecular weight twice as high and of narrower molecular weight distribution in comparison with copolymers obtained under conventional thermal conditions [34].

Using variable power of microwave irradiation, Boey et al. showed that bulk polymerization of methyl methacrylate (MMA) was faster by ca. 130–150% compared with the conventional methods (Fig. 4) [35]. Also, the limiting conversion of the reaction varied for the thermal and microwave polymerization. The thermal polymerization of MMA at temperatures of 69, 78, and 88 °C showed a limiting conversion of about 90%, while limiting conversion of microwave polymerization decreased in following series: 200 W, 88%>300 W, 84%>500 W, 78% [36]. The reactions were run in a sample vial with a narrow tube of 10 mm diameter and 2 ml capacity, in which 4 mg of AIBN was placed together with 0.5 ml of MMA. The microwave polymerization was conducted in a multimode cavity equipped with a rotating platform to prevent formation of hotspots due to non-uniform heating. NMR analysis showed that the tacticity of the polymers for thermal and microwave polymerization are similar, suggesting that the polymerization mechanisms are the same.

Under similar conditions, polymerization of methyl acrylate (MA) with AIBN as an initiator under microwave irradiation was carried out by Chia et al. [36]. To prepare samples for polymerization, 4.1 mg of AIBN (0.85 wt%) was taken in a 4-ml sample vial of 15 mm diameter together with 0.5 ml (478 mg) of MA. The reaction rate enhancement of microwave polymerization compared to thermal method was found to be as follows: 500 W, 275%; 300 W, 220%; and 200 W 138%. Even though the comparable temperature at variable power was the same, 52 °C, the reaction rate enhancement increased with increase in microwave power indicating a significant correlation between the reaction rate enhancements and the level of microwave power used [36].

On the other hand, by free radical suspension polymerization of MMA in *n*-heptane solution in the presence of poly(styrene)-*block*-poly(ethene-*alt*-propene) (SEP) as a dispersing agent, PMMA samples were prepared with similar molecular weights and polydispersity under both conventional and microwave conditions [37]. The reactions were run for 1 h at 70 °C with different monomer (9.0–28.3 vol.%), SEP (21.7–5.4 wt%), and AIBN (1.0–0.27 wt%) concentrations (Table 1). In a typical experiment, 30 ml of the reaction mixture was fed into a 50-

Table 1. Thermal (Δ) and microwave-induced (MW) non-aqueous free-radical dispersion polymerization of MMA in the presence of SEP dispersing agent. Reprinted from (1996) Acta Polym 47:74 [37] with permission

Method of activation	*n*-Heptane [vol.%]	MMA [vol.%]	SEP[a] [vol.%]	AIBN[a] [wt%]	Conversion [%]	Diameter[b] [nm]
Δ	91	9	21.7	1	42	67±16
MW	91	9	21.7	1	50	92±53
Δ	83.5	16.5	10.8	0.53	73	144±23
MW	83.5	16.5	10.8	0.53	71	134±42
Δ	71.7	28.3	5.4	0.27	75	1330±133
MW	71.7	28.3	5.4	0.27	73	1370±107

[a] Wt% with respect to MMA
[b] Average diameter of the particles with standard deviation

ml reaction vessel that was equipped with a stirrer (200 rpm), fiber-optic thermometer, and gas inlet tube to avoid air in the reaction mixture. Then the reaction vessel was placed into a microwave cavity equipped with a waveguide with TE_{10} mode. Under such conditions, no significant effect of microwave irradiation was detected, and molecular weights, polydispersities, stereoregularities, particle size and their distribution were similar for both activation methods.

The bulk polymerization of styrene at two different power levels – 300 and 500 W – conducted in a multimode microwave cavity was investigated by Chia et al. [38]. The reactions were run in 2-ml sample vials of 10 mm diameter, in which 23.0 mg of AIBN was placed together with 455 mg of styrene. The conversion profiles of the microwave polymerization were significantly different from that of the thermal cure at the same temperature of 80 °C. The thermal cure was characteristic of a gradual gel effect at 30–50% conversion while, with the microwave cure at 300 W and 500 W, a sharp and large gel effect was recorded at conversions 20–69% and 20–65%, respectively (Fig. 5). Moreover, the comparison of thermal and microwave polymerization under similar conditions showed a reaction rate enhancement of 190% for 500 W and 120% for 300 W. Similar to the microwave polymerization of MMA [35], the limiting conversion of styrene decreased from 72% for conventional thermal conditions down to 69% at 300 W and 65% at 500 W of microwave irradiation power. Finally, it was stated that comparison of kinetic results of microwave induced reactions should consider the temperature as well as the power of microwave irradiation due to different energy supplied to the reaction system [38].

Emulsion polymerization of styrene under microwave irradiation in the presence of potassium persulfate (KPS) as an initiator was reported by Palacios and Velverde [39]. The reactions were carried out in glass tubes of 20 ml volume while 15 ml of the reaction mixture with different initiator concentration (0.2–31.1 mmol/l) (Table 2) was added. During the experiments, each tube was placed in a 2-l glass beaker filled with vermiculite. The beakers were irradiated in a multimode microwave with 389 W power in the magnetron, which was turned on and off twice per minute to maintain a constant temperature (50 °C).

In order to reach a conversion of styrene of 70%, constant heating in an oil bath for as long as 6 h was required in comparison with only 8.3 min in the microwave oven. Calculated values of R_p (mol/l·s) for both processes showed that the rate of microwave-assisted emulsion polymerization of styrene was higher by a factor of 26 than the reaction activated by conventional heating (Fig. 6) [39]. Molecular weight distribution and polydispersities of the polystyrene samples obtained under microwave and conventional conditions were similar for both activation methods and depended on the initiator concentration.

More recently, using a bench-scale microwave polymerization reactor (Fig. 7), Correa et al. reported that emulsion polymerization of styrene could be carried out more rapidly with significant saving of energy and time when compared to conventional methods [40]. For all the experiments, the concentration range of mate-

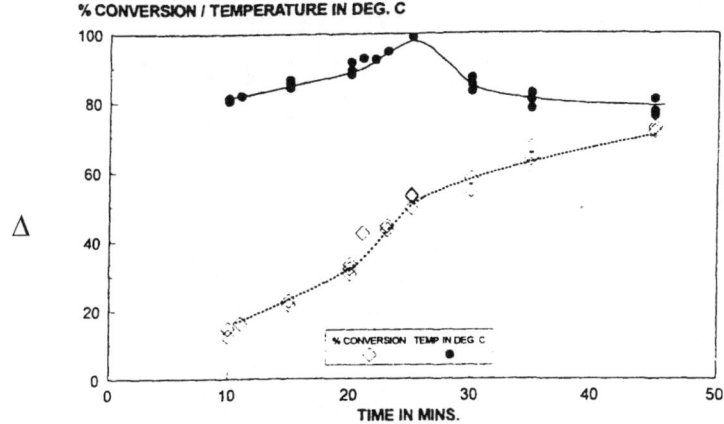

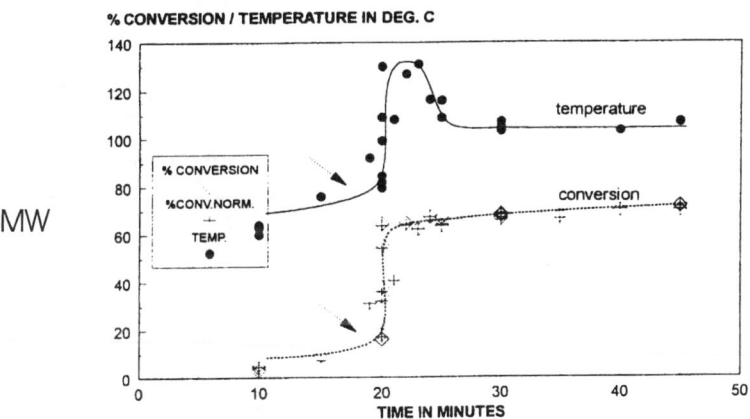

Fig. 5. Thermal profiles and conversions for conventional and microwave polymerization of styrene. Reprinted from (1996) J Polym Sci A, Polym Chem 34:2087 [38] with permission

rials was as follows: KPS (initiator) 0.04–0.16 wt%, sodium lauryl sulfate (emulsifier) 1.24%, and monomer to water ratio 1:3, 1:16, and 2:3. The reaction temperature for both methods was 70 °C while two power levels – 175 W and 800 W – of microwave irradiation were applied.

The reaction time for the conventional heating method was longer by a factor of 70 than for the microwave irradiation method. Moreover, it was observed that the molecular weight of polystyrene samples prepared by microwave irradiation (M_w=350,000) was higher by a factor of 1.2 than that of polystyrene obtained by conventional heating. As it could be expected, the requested period of time under microwave irradiation depended on the applied microwave power and was longer

Table 2. Emulsion polymerization of styrene initiated by microwaves. Reprinted from (1996) New Polym Mat 5:93 [39] with permission

Experiment no.	KPS initiator concentration (mol/l) × 10^4	Polymer yield (%)[b]	$M_n \times 10^{-3}$ (g/gmol)	M_w/M_n	R_p (mol/l×s) × 10^3	Time required for conversion of 50% (s)
1	0	5.0			0	
2	2.90	52.0	385.1	3.23	3.57	1100
3	9.70	73.0			6.20	800
4	31.15	80.0	131.0	2.51	7.88	830
5	124.6	91.0	79.7	3.35	9.65	350
6	192.6	97.0	74.4	2.38	13.63	240
7	269.6	95.0	10.8	3.52	13.65	245
8	311.5	66.0			13.75	550
9[a]	192.6	69.0	70.1	2.41	51.84	18500

[a] 6 h, conductive heating. T=50 °C
[b] Time=33 min. R_p – polymerization rate

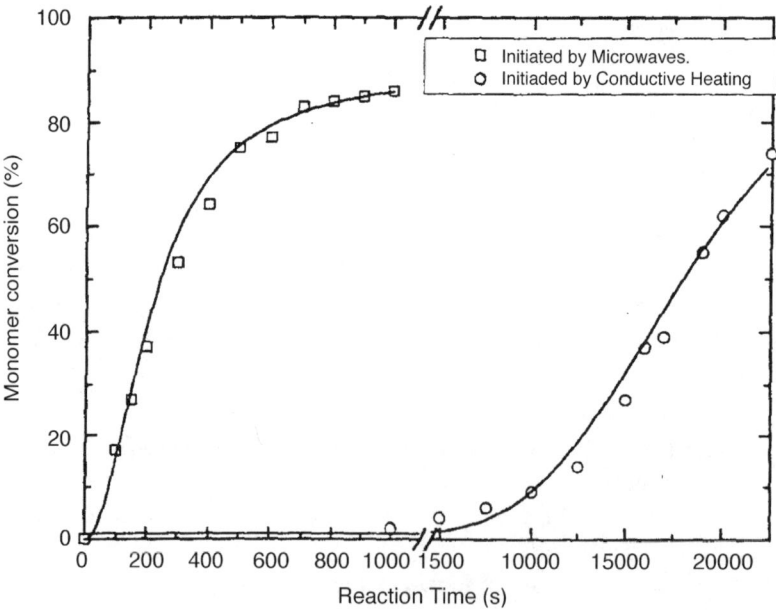

Fig. 6. Comparison between conventional and microwave initiated emulsion polymerization of styrene. Reprinted from (1996) New Polym Mat 5:93 [39] with permission

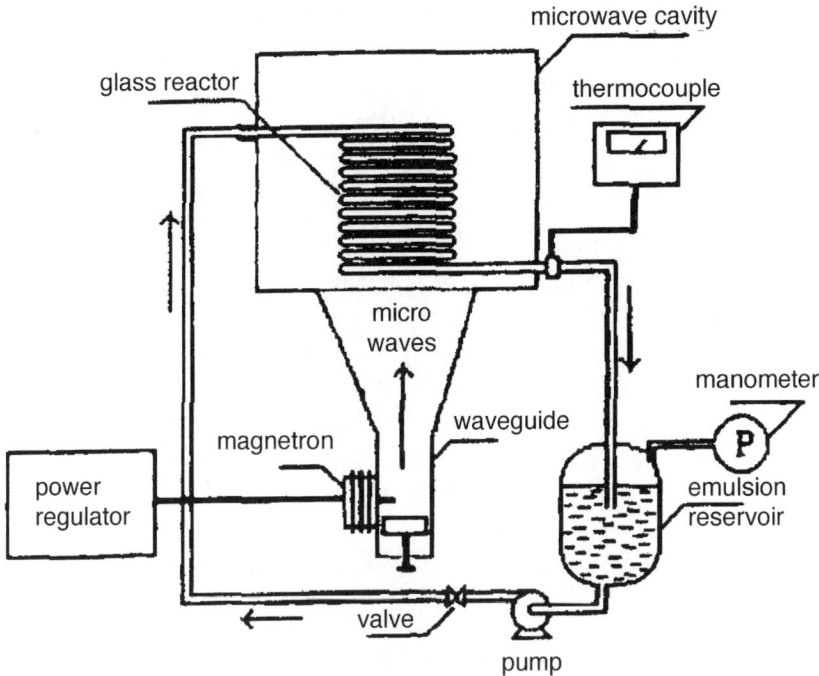

Fig. 7. Experimental set-up for discontinuous operation of the microwave reactor system. Reprinted from (1998) Polymer 39:1471 [40] with permission

for lower power level [40]. In fact, microwave power was used in turn on/off cycles of 20 and 600 s, respectively, while the reaction time was recorded only during turn-on cycles. In our opinion, both turn-on and turn-off cycles had to be counted, which means that the total reaction time under microwave irradiation was comparable with the reaction time under conventional conditions.

Wu et al. reported that surfactant-free stable polystyrene nanoparticles could be prepared by applying microwave irradiation with KPS as an initiator in water solutions [41]. In comparison with conventional heating, this method can shorten the reaction time by a factor of 20, results in narrowly distributed nanoparticles, and leads to moderately distributed polymer chains inside the nanoparticles (Table 3). The reactions were conducted in a multimode microwave cavity with maximum output of 900 W, wherein a reaction flask equipped with a stirrer and reflux condenser and charged with 250 ml of the reaction mixture was irradiated. Typically, under reduced microwave irradiation power (80 W), the reaction temperature was maintained at 70 °C, which led to 98% conversion of styrene within 40 min [42].

Even in the absence of surfactant, the resultant nanoparticles were still narrowly distributed and stable for months. The application of conventional heating meth-

Table 3. Summary of LLS results of the polystyrene nanospheres in water and the polystyrene chains in toluene at 25 °C. Reprinted from (1997) Macromolecules 30:6388 [42] with permission

Sample no.	C_m (g/ml) × 10^2	C_i (g/ml) × 10^4	Polystyrene nanospheres		Polymer chains	
			R_h (nm)	$1+\mu_2/D^2$	$M_w \times 10^{-5}$ g/mol	M_w/M_n
1	0.0544	3.02	35.2	1.10		
2	0.181	3.02	39.2	1.10		
3	0.362	3.02	56.6	1.01		
4	1.13	3.02	82.1	1.01	1.06	1.8
5	1.45	3.02	87.9	1.03		
6	2.26	3.02	105	1.01	2.11	1.6
7	3.04	3.02	113	1.01		
8	1.13	1.39	73.4	1.01		
9	1.13	6.08	91.2	1.01	1.01	1.6
10	1.13	11.7	117	1.01	0.71	2.0

C_m – monomer concentration, C_i – initiator concentration, R_h – hydrodynamic radius of the nanoparticles, $1+\mu_2/D^2$ – the small values of this coefficient indicate that the nanospheres are narrowly distributed

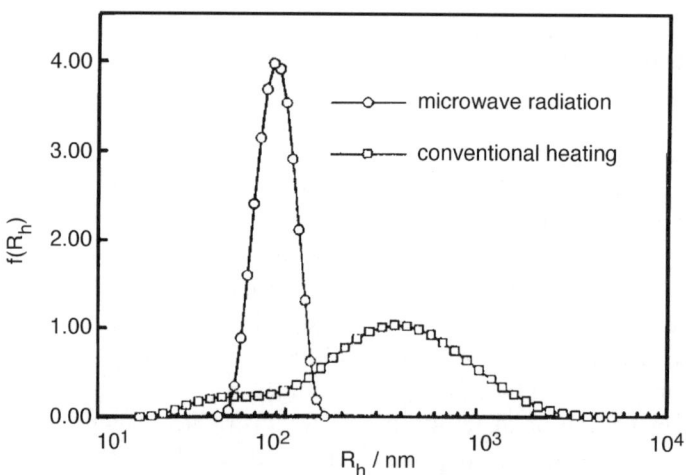

Fig. 8. Comparison of the typical hydrodynamic radius distribution $f(R_h)$ of the polystyrene nanospheres prepared microwave irradiation and conventional heating method. Reprinted from (1997) Macromolecules 30:6388 [42] with permission

od to polymerize the same reaction mixture under similar conditions resulted in broader particles size distribution and took a much longer time (10 h) to reach the same extent of conversion (Fig. 8) [42].

Eventually, it was reported that it was possible to prepare a very concentrated (up to 40 wt%) uniform polystyrene nanoparticles (down to 60 nm) in the presence of only 1.82 wt% of anionic surfactant, sodium dodecyl sulfate [41], if microwave irradiation was applied.

Applying microwave irradiation, Cai et al. demonstrated that N-vinylcarbazole (NVC) could be polymerized in bulk in the presence of fullerene-C60 as a charge-transfer initiator [43]. Microwave polymerizations were carried out in sealed argon-filled thick-walled Pyrex tubes filled with NVC (from 0.19 to 3.22 g) together with C60 (0.02 g) and placed in a commercially available multimode microwave oven for 10 min. In comparison with polymerization under conventional conditions (water bath, 70 °C), microwave polymerization was found to be advantageous due to decrease in the reaction time and considerable improvement of yield of poly(N-vinylcarbazole) (PVK) – 70% as opposed to 8% after 10 min of conventional heating. In fact, during microwave experiments, temperature was neither measured nor evaluated, so that it is hard to make any yield comparison between these two techniques. In spite of different heating methods, the molecular weights and polydispersities of the resultant polymers were similar.

The copolymerization reactions of maleic anhydride with allylthiourea [44] and dibenzyl maleate [45] were studied by Lu at al. The reactions were carried out in the solid state without any initiator in 10-ml vials, which were irradiated in a domestic microwave oven for 20–70 s. In the case of allylthiourea, the conversion of monomers reached ca. 20% after 70 s at 430 W, while yield of the copolymer obtained from dibenzyl maleate was 94% after 32 s of irradiation at 700 W. Under conventional conditions, the copolymer was produced with 89% yield after 2 h at 45 °C [45]. The copolymers were used as a metal complexing agent [44] and suspending agent for the synthesis of superabsorbent oil resins [45].

More recently, the microwave copolymerization of dibutyltin maleate (DBTM) and allyl thiourea was investigated, under conditions specified above to produce organo-tin copolymers [46]. Under microwave irradiation, the copolymers could be synthesized without and in the presence of free radical initiator; however, in the presence of initiator the conversion of monomers was slightly higher to reach ca. 35%. Without using any initiator, the monomers did not copolymerize by conventional heating within 7 h at all. The copolymers were applied for thermal stabilization of PVC. The thermal stability was increased from 180 °C for neat PVC to above 240 °C for PVC modified with the organo-tin copolymers [46].

Methathesis polymerization of phenylacetylenes under microwave irradiation was achieved by Sundararajan and Dhanalakshmi [47], applying in situ generated (arene)M(CO)$_3$ complexes (Fig. 9).

Microwave irradiation was carried out in a domestic microwave oven under oxygen free dry nitrogen atmosphere in a specially designed long-necked round bottom flask which could withstand elevated pressure. The catalyst/monomer ratio was kept at a 1:50 level. For example, W(CO)$_6$ (0.199 mmol) was taken together

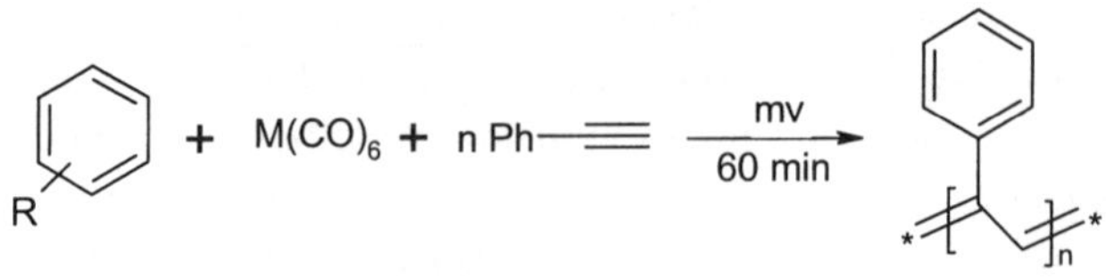

Fig. 9. Polymerization of phenylacetylenes under microwave conditions (M=Cr, Mo, and W). Reprinted from (2000) Polymer 41:7877 [47] with permission

with phenylacetylene (5.975 mmol) and toluene (0.5 ml) in a 1,2-dichloroethylene solution and irradiated for 5 min at time intervals of 10 min to cool down the reaction mixture. The reaction time was reduced to 1 h in contrast to refluxing conditions of 24 h. Among the group VIB metal carbonyl, $W(CO)_6$ was found to be the best catalyst precursor, while phenol showed the highest activity among arenes. Under such conditions, poly(phenylacetylene) was obtained in 67–75% yield.

3.1.1
Preparation of Dental Materials

In the preparation of dental base material, poly(methyl methacrylate) (PMMA) is the most commonly used acrylic resin. It is well-mixed with the second component, liquid methyl methacrylate (MMA), prior to use [48]. The prepared in such a way polymer dough is placed in a mould and polymerized in a water bath. The initiator (e.g., benzoyl peroxide) that is usually added to the PMMA powder by a manufacturer decomposed at an elevated temperature and started polymerization and crosslinking.

The use of microwave irradiation to polymerize dental materials was reported by Nishii [49], who was able to obtain porosity-free acrylic resin specimens of similar physical and mechanical properties to the water bath cured specimens. Recently, research in this area was continued by Kimura [50, 51] as well as Reitz et al. [52, 53]. It was shown that specimens of 2.5 mm thick processing for 2.5 min on each side at 400 W of microwave irradiation were not statistically different in respect of porosity, hardness, or strength from those specimens cured in a water bath at 74 °C for 8 h. However, in thicker sections of specimen (i.e., ca. 10 mm) it was difficult to eliminate porosity even by increasing the exposure time to microwaves.

More recently, Al Doori et al. studied microwave and conventional polymerization of denture base acrylic materials in respect of their molecular weights, conversion of monomer and porosity [54]. For this purpose, four commercially available resins that contained benzoyl peroxide as an initiator were compared while a household microwave oven was applied as a microwave source. It was found that molecular weight values of polymerized materials cured under microwave irradiation and in a water bath were essentially the same. Moreover, the conversion of monomer under microwave conditions was substantial, but minimum residue

Table 4. Viscosity-average molecular weight of cured PMMA samples prepared under microwave and conventional conditions. Reprinted from (1999) J Appl Polym Sci 74:2971 [55] with permission

Curing method	Curing period (min)	Average viscosity [η]	Molecular weight (g/mol)×10^{-6}
MW	1	2.9174	0.88
MW	2	4.3253	1.4
MW	3	7.2927	2.7
Δ	5	4.3798	1.4
Δ	10	3.9662	1.2
Δ	15	3.0972	0.94
Δ	20	3.3594	1.0
Δ	25	3.5871	1.1
Δ	30	3.9808	1.2

monomer levels attainable with the water bath systems were not achieved. However, microwave curing at 70 W for 25 min minimized porosity problems associated with rapid heating of dough; porosity-free material was guaranteed in specimens not thicker that 3 mm.

In a similar study, Usanmaz et al. polymerized commercially available acrylic resins under both conventional (water bath) and microwave conditions (multimode cavity) [55]. In a typical experiment, 100 g of PMMA powder was mixed together with 43 ml of liquid monomer (MMA) to make homogenous dough that was then placed into a mould and polymerized at 70 °C for 30 min. The curing was done in boiling water for 5–35 min or under microwave irradiation for 1–3 min. It was found that viscosity-average molecular weight values were larger for microwave curing and increased with curing time (Table 4), which is an important advantage of microwave curing in dental applications.

As could be expected, the glass transition temperature of different samples increased with curing periods. However, the values were close to each other for both curing methods. Dynamical mechanical analysis (DMA) showed that crosslinking density of the samples increased with an increase in cure time but the changes were more noticeable under microwave irradiation compared to conventional thermal curing [55].

On the other hand, a comparison of some properties of denture base polymers done by Blagojevic et al. under both microwave and water bath conditions demonstrated that water bath polymerization with a long curing cycle (14 h at 70 °C) and a terminal boil (3 h) could produce, in general, superior properties [56]. Although all the tested materials did not behave uniformly, the general trend suggested that the water bath method enhanced the degree of polymerization resulting in a lower level of residual monomer and increased glass transition temperature, which was attributed to the long curing time and boil period. Similarly, denture materials processed with a new polymerization system in which a pressure of about 3.5 MPa was applied until the polymerization was complete exhibited significantly smaller discrepancies compared with dentures obtained by microwave Method [57]. The dentures were prepared from commercially available resins by mixing 20 g of powder with 30 ml of liquid monomer. Thus, under microwave conditions, the samples were irradiated in a multi-mode microwave oven for 3 min, whereas under pressurized conditions the samples were kept for 5 min. It was found that the level of water absorption of the resins disks processed in the pressurized system was less than of the disks made by microwaves, and the former exhibited significantly better adaptation to the cast [57].

Urabe et al. [58] used microwave polymerization methods to make composite resin inlays, applying standard light-curing posterior teeth composition such as 2,2-bis[4-(3-methacryloyloxy-2-hydroxypropoxy)phenyl] propane (Bis-GMA) and triethylene glycol dimethacrylate (TEGDMA) [59]. The monomers were mixed in the weight ratio of 6:4 (Bis-GMA:TEGDMA) together with benzoyl peroxide as an initiator in the amount of 0.1–0.9 wt%. The samples of 6×3 mm and 3×6 mm were irradiated in Teflon molds in a conventional microwave oven for 5 min. The results showed that the degree of conversion increased with increasing concentration of benzoyl peroxide and reached its maximum of 75% at 0.5 wt% of initiator concentration. Then there were no significant differences at the level of 0.5–0.9 wt%. Compression strength, diametral strength, and the Knoop hardness showed a similar tendency as the degree of conversion. No significant differences were recorded in hardness between the top and bottom surfaces of the samples, which suggested that more Uniform polymerization in the cured samples occurred compared with light curing [58]. In fact, at the present stage of investigations, it is a little difficult to imagine a direct application of microwaves in an oral cavity.

In conclusion, the porosity, hardness, and strength of microwave cured acrylic resins that are processed for less than 5 min exhibited no significant differences in properties when compared with conventionally polymerized resins [60–63]. However, the investigation of color stability of several commercially available heat-cured, quick heat-cured denture and microwave cured denture base resins (tested exclusively under microwave conditions) revealed that the materials demonstrated differences in color stability, but the standard heat-cured materials treated under microwave irradiation exhibited color changes that were negligible [64].

The influence of microwave irradiation on the polymerization process of silicone rubber denture soft lining material to PMMA, and its properties were studied by Wright et al. [65]. Three groups of specimens were polymerized under microwave irradiation at 650 W for 3, 5, and 10 min in a household microwave oven. The fourth group of specimens was polymerized using conventional thermal methods, i.e., by boiling in a water bath for 2 h. The study demonstrated that microwave energy activation of polymerization did not compromise the adhesion of a silicone soft liner to PMMA. Control and test groups exhibited cohesive rather than adhesive failure in the peeling test. Both groups polymerized by microwave and conventional methods tore instead of peeling from the PMMA. That study also suggested the use of 3 min of irradiation at 650 W power level for processing a silicone soft lining material [65].

3.2
Step-Growth Polymerization

3.2.1
Polyamides and Polyimides

The kinetic study of solution imidization under microwave irradiation of polyamic acid prepared from 3,3′,4,4′-benzophenonetetracarboxylic acid dianhydride (BTDA) and DDS was performed by Lewis et al. (Fig. 10) [66].

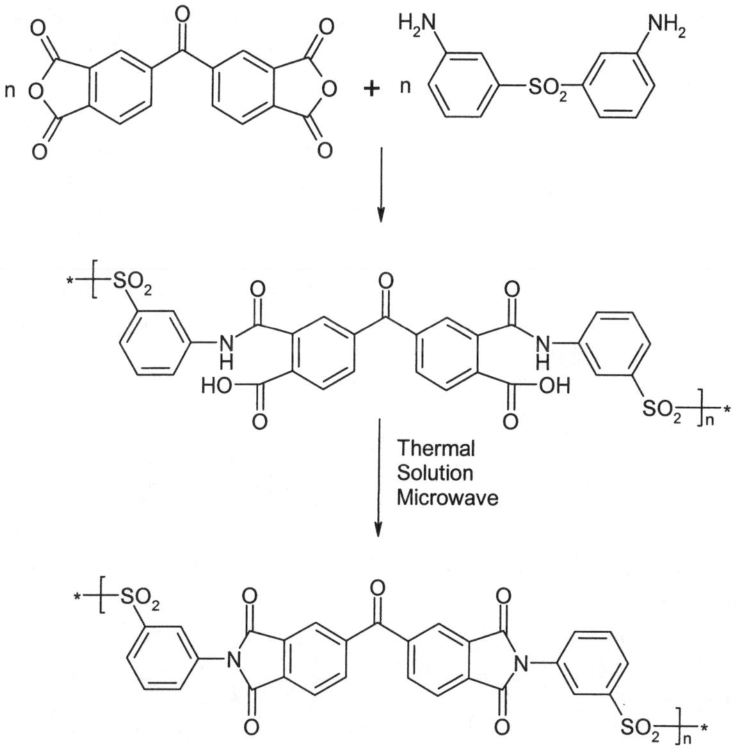

Fig. 10. Chemical structures and reaction scheme for 3,3′,4,4′-benzophenontetracarboxylic acid dianhydride (BTDA) and 4,4′-diaminodiphenyl sulfone (DDS) polyamic acid. Reprinted from (1992) J Polym Sci A Polym Chem 30:1647 [66] with permission

The microwave equipment consisted of a microwave generator (85 W) and a tunable cavity operating in the TE_{111} mode, while temperature was monitored using a fiber-optic temperature sensor. The samples were maintained in a Teflon vessel of 1.5 cm diameter hole and 1.5 cm deep. During a typical run, 25 W of microwave power was required to heat the sample to the desired temperature over 80–200 s. It was demonstrated that microwave irradiation increased the rate of solution imidization over that obtained for conventional treatment by a factor of 20–34, depending on the reaction temperature. The apparent activation energy for this imidization, determined from an Arrhenius analysis, was reduced from 105 to 55 kJ/mol when microwave activation was utilized rather than conventional thermal processing.

In the series of papers, synthesis of aliphatic polyamides and polyimides under microwave irradiation was described by Imai et al. [67–71]. The reactions were carried out in a modified domestic microwave oven with a small hole on the top of the oven so that nitrogen was introduced to a 30 ml wide-mouth vial adapted as a reaction vessel. In the case of polyamides synthesis, they were prepared of both ω-amino acids and nylon salt type monomers while polyimides were obtained from the salt form of monomers composed of aliphatic diamines and pyromellitic acid or its diethyl ester in the presence of a small amount of a polar organic medium (Fig. 11).

In a typical experiment, a monomer or its salt (2 g) in a polar high boiling solvent (1–2 ml) was irradiated under nitrogen atmosphere. The microwave assisted polycondensation proceeded rapidly and was completed within 5 min (inherent viscosity around 0.5 dL/g) for the polyamide synthesis [68] and within 2 min for the polyimides (inherent viscosity above 0.5 dL/g) [69]. The rate of polycondensation of the salt monomers under various conditions decreased in the following series: the microwave induced polycondensation>solid-state thermal polymerization>high-pressure thermal polycondensation [71].

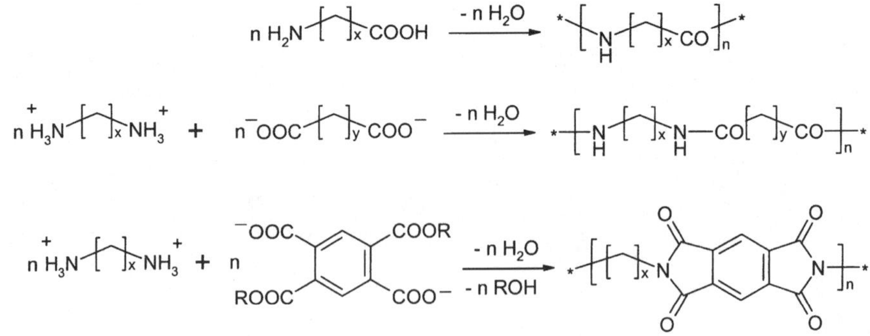

Fig. 11. Microwave-assisted synthesis of various polyamides from ω-amino acids and nylon salts. Reprinted from (1996) React Funct Polym 30:3 [70] with permission

In a similar way, the synthesis of aromatic polyamides from aromatic diamines m-phenylenediamine, p-phenylenediamine, bis(4-aminophenyl)methane, and bis(4-aminophenyl)ether and dicarboxylic acids such as isophthalic and terephthalic acid was performed in a household microwave oven [72]. The polycondensation was carried out in an N-methyl-2-pyrrolidone (NMP) solution in the presence of triphenyl phosphite (TPP), pyridine, and lithium chloride as condensing agents to produce a series of polyamides with moderate inherent viscosities of 0.21–0.92 dL/g within 30–50 s. However, no marked differences in molecular weight distribution and inherent viscosities between the polyamides produced by conventional (60 s, 220 °C) and microwave methods were found [72].

The polyether-ester polyamic acid imidization process in a solid state under microwave irradiation was studied by Yu et al. [73]. The prepolymer, polyether-ester polyamic acid, was prepared by the polycondensation of poly(tetramethylene ether)glycol di-p-aminobenzoate (Polyamine-650, Polaroid, Co.) and pyromellitic acid dianhydride (PMDA) at room temperature in DMF solution. Later, the prepolymer solution was cast on polytetrafluoroethylene plates to form 200 μm thin films that were imidized under microwave irradiation in a household microwave oven at 60 °C. The temperature was measured by means of a thermocouple applied to the film surface immediately after the intervals of microwave turn off. It was found that microwave irradiation reduced both the reaction temperature and time. For example, during the solid phase thermal polymerization 68.3% polyamic acid was converted to polyimide at 155 °C, while under microwave irradiation 65% of polyamic acid was reacted at 60 °C within 3 h [73].

More recently, the synthesis of partially aromatic polyamides from linear non-aromatic dicarboxylic acids (i.e., adipic, suberic, sebacic, and fumaric acid) and aromatic diamines such as p-phenylenediamine or 2,5-bis(4-aminophenyl)-3,4-diphenylthiophene (Fig. 12) under microwave conditions was presented by Pourjavadi et al. [74].

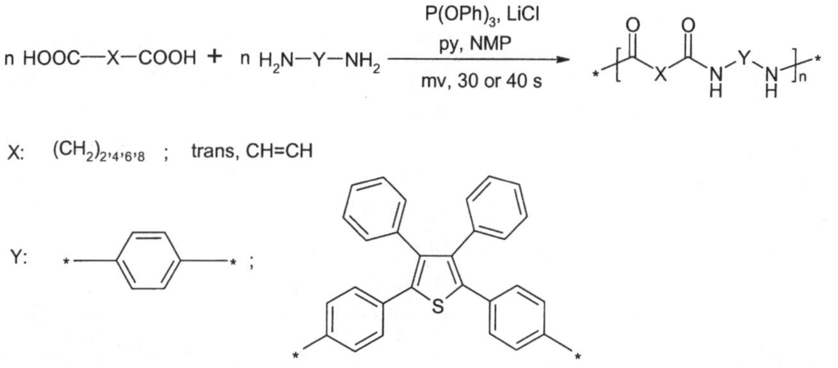

Fig. 12. Microwave-assisted preparation of polyamides via phosphorylation reaction. Reprinted from (1999) Angew Makromol Chem 269:54 [74] with permission

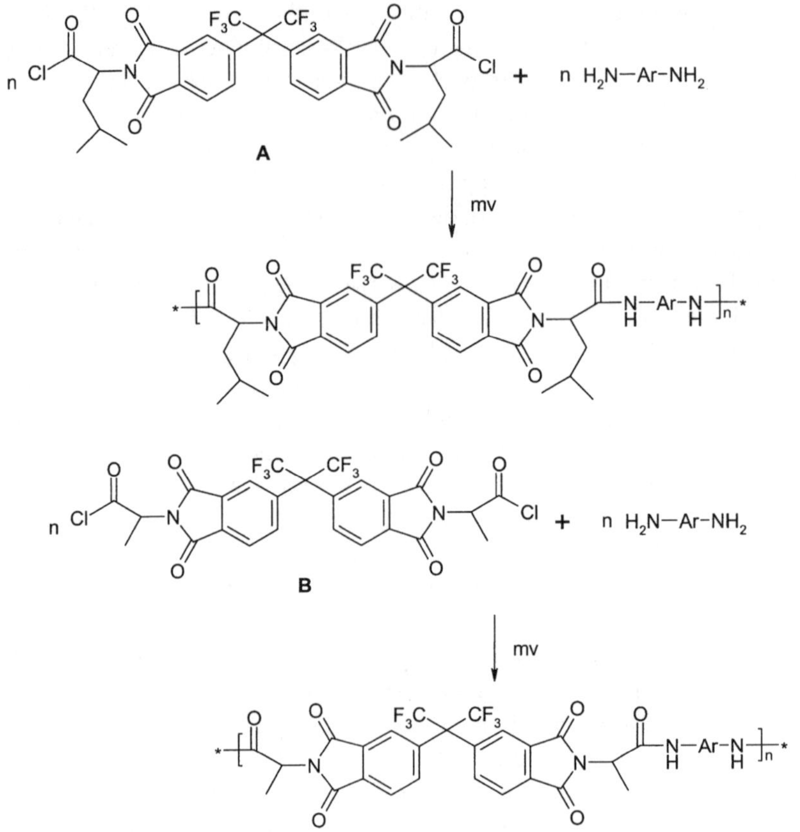

Fig. 13. Polymerization reaction of [N,N'-(4,4'-carbonyldiphthaloyl)] bisalanine diacid chloride (B) and 4,4'-(hexafluoroisopropylidene)-N,N'-bis(phthaloyl-L-leucine) diacid chloride (A) with certain aromatic amines. Reprinted from (2001) Eur Polym J 37:119[75] and (2000) J Polym Sci A Polym Chem 38:1154 [79] with permission

The reactions were carried out in a 50-ml polyethylene (HDPE) screw-capped cylinder vessel, in which aromatic acid (1.25 mmol) together with aliphatic diamine (1.25 mmol) in an NMP (3 ml) solution were irradiated in a domestic microwave oven (30–40 s) in the presence of TPP (3.123 mmol), pyridine (0.75 ml) and lithium chloride (3.123 ml). The polyamides with inherent viscosity in the range of 0.1 to 0.8 dL/g were obtained in medium to high yield (60–100%) [74]. Temperature was not detected during these microwave experiments. The polyamides were characterized by thermal methods (TGA, DSC). However, no comparison to the polymers prepared by a conventional method was made.

Applying microwave irradiation, Mallakpour et al. presented a rapid method for the polycondensation of a number of diacid chlorides such as [N,N'-(4,4'-carbonyldiphthaloyl)] bisalanine diacid chloride [75–77] and 4,4'-(hexafluoroiso-

Table 5. Inherent viscosity (η_{inh}) of the polymers (differences between the solution[a] and microwave methods). Reprinted from (2000) J Polym Sci Part A: Polym Chem 38:1154 [79] with permission

Reagent	Solution η_{inh}	Microwave η_{inh}
Benzidine	0.22	1.22
4,4'-Diaminodiphenylmethane	0.29	1.32
1,5-Diaminoantraquinone	0.09	1.73
4,4'-Sulfonyldianiline	0.09	0.50
3,3'-Diaminobenzophenone	0.22	1.26
p-Phenylenediamine	0.15	1.04
2,6-Diaminopyridine	0.12	1.44

[a] One equimolar of diacid chloride and diamines was refluxed for 12 h in $CHCl_3$ and then heated for 12 h at 120 °C in dimethylacetamide

propylidene)-N,N'-bis(phthaloyl-L-leucine) diacid chloride [78, 79] with certain aromatic amines [75, 76, 79] and diphenols (Fig. 13) [77, 78].

The reactions were performed in an unmodified microwave oven. Prior to the microwave irradiation, 0.1 g of diacid chloride was ground with equimolar amount of an aromatic amine or diphenol and a small amount of a polar high boiling solvent (e.g., o-cresol, 0.05–0.45 ml) that acted as a primary microwave absorber. Under microwave irradiation, the polycondensation reactions proceeded rapidly (6–12 min) while compared with a conventional solution polymerization (reflux for 12 h in chloroform then for another 12 h in dimethylacetamide solutions [79]) to give polymers with higher inherent viscosities in the range of 0.36 to 1.93 dL/g (Table 5).

In the case of the polycondensation of 4,4'-(hexafluoroisopropylidene)-N,N'-bis(phthaloyl-L-leucine) diacid chloride with aromatic amines, the thermal stability of the resulted polymers showed no significant difference from the polymers obtained in conventional solution polymerization, but higher glass-transition temperature (by ca. 19–82 °C) were reported [79]. However, in this case, it is difficult to make a direct comparison of reactions rates for both conventional and microwave methods since the reactions were run under different reaction conditions. Moreover, the temperatures for microwave methods were not given.

The studies of the crosslinking reaction of a nadic end-capped imide, N,N'-(oxydi-3,4'-phenylene)bis(5-norbornene-2,3-dicarboximide) in thermal and microwave processes were performed by Scola at al. (Fig. 14) [80]. The investigated starting resin (RP-46) consisted of polyimide precursors, 3,3',4,4'-benzophenonetetracarboxylic acid methyl ester (BTDE), 3,4'-oxydianiline (3,4'-ODA), and end capping reagent, 5-norbornene-2,3-dicarboxylic acid monomethyl ester (NE). The cure process proceeds in two steps: imidization (1) and a thermally induced (retrodiene Diels-Alder) decomposition-recombination crosslinking step (2) (Fig. 14).

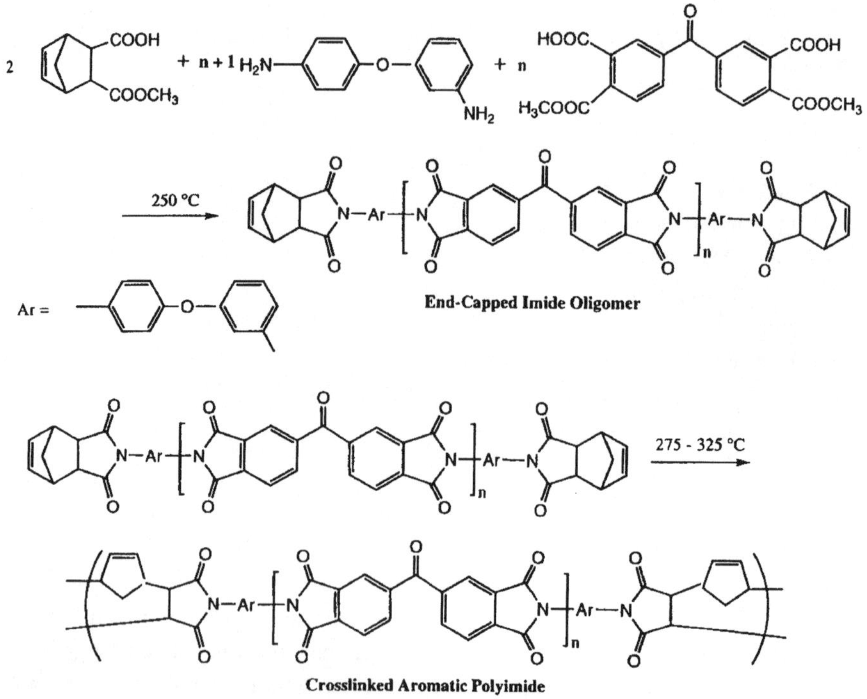

Fig. 14. Synthesis of crosslinked polyimide RP-46 under microwave conditions. Reprinted from (1998) J Polym Sci A Polym Chem 36:2653 [80] with permission

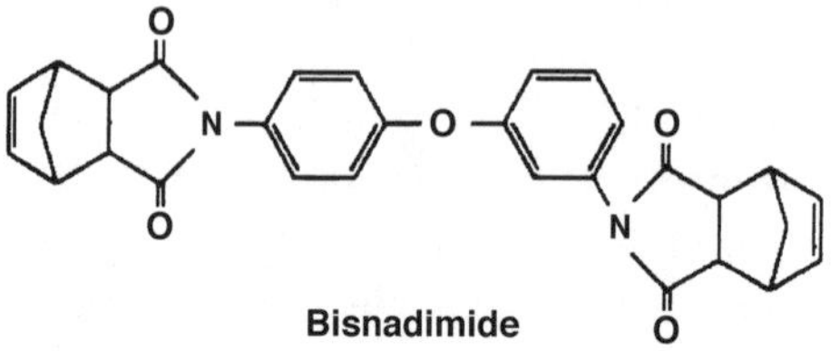

Fig. 15. Structure of the model compound, bisnadimide. Reprinted from (1998) J Polym Sci A Polym Chem 36:2653 [80] with permission

At the same time, kinetic studies of a model compound, 'bisnadimide' (Fig. 15), in thermal and microwave processes were carried out in order to simulate the crosslinking reaction of polyimide RP-46.

The reactions were run in a microwave oven (Model 10) provided by MMT, while the temperature was controlled by a thermocouple covered with a ceramic tube to avoid the influence of microwaves. The microwave cure was carried out in the temperature range of 230–325 °C, and runs at 230–280 °C were used to determine the kinetics of the reaction. At these temperatures, the microwave cure was

Table 6. Rate constants for the thermal and microwave cure processes. Reprinted from (1998) J Polym Sci Part A: Polym Chem 36:2653 [80] with permission

Temperature (°C)	Δ k (min^{-1})	MW k (min^{-1})
230	0.003	0.028
280	0.011	0.140
300	0.030	
325	0.128	
Activation energy (kJ/mol)	94	84

rapid with both isomers; conversions to crosslinked structure were about ten times faster than in case of a conventional thermal process (Table 6). The apparent activation energy for the thermal cure was estimated to be 94 kJ/mol, whereas for the microwave cure the value of activation energy fell to 74–84 kJ/mol, which suggests that the microwave process is a more efficient energy process [80].

In the next stage, microwave irradiation was investigated to cure and process nadic-end capped polyimide precursor (RP-46 resin) and glass-graphite-RP-46 composites [81]. Processing of thick section by conventional thermal processes requires slow ramp rates and a long processing time. Therefore, the composite materials containing conducting fibers could be heated by the microwave process to achieve inside to outside heating patterns and quick heat ramp rate. Additionally, microwave processes may enhance the bonding strength between resin and fiber matrix.

Both neat resin and composite with glass and graphite cloth were obtained and the effects of various parameters such as power level, mold material, and pressure were studied using a Cober Electronics microwave oven at a frequency of 2.45 GHz. It was shown that the sample size and geometry were important factors in microwave processes. For example, changing the sample size from 5 to 15 g caused an increase in temperature of 32 °C in 10 min at the same power level as shown in Fig. 16. Essentially, no coupling occurred between a sample of 5 g polyimide resin and microwave energy, proving that a critical mass was required to absorb the microwave energy with a high efficiency.

Depending on the conditions, microwave cure of glass and glass-graphite hybrid composites was accomplished in 0.6–2.16 h and the imidization of neat resin and composites was complete. Resin specimens containing only 0.057 wt% chopped graphite fibers resulted in complete imidization in 6 min. Glass and glass-graphite composites were fabricated by microwave irradiation with flexural strength and moduli equivalent from 50 to 80% of the properties of composites fabricated by conventional thermal processes [81].

The kinetic studies of the microwave cure of phenylethynyl-terminated imide oligomer (PETI-5, M_n 5000 g/mol) and a model compound, 3,4'-bis[(4-phe-

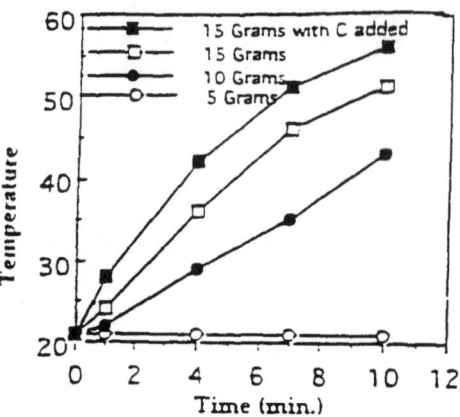

Fig. 16. The effect of sample size on the microwave absorption of undried RP-46 resin. Reprinted from (1999) J Appl Polym Sci 73:2391 [81] with permission

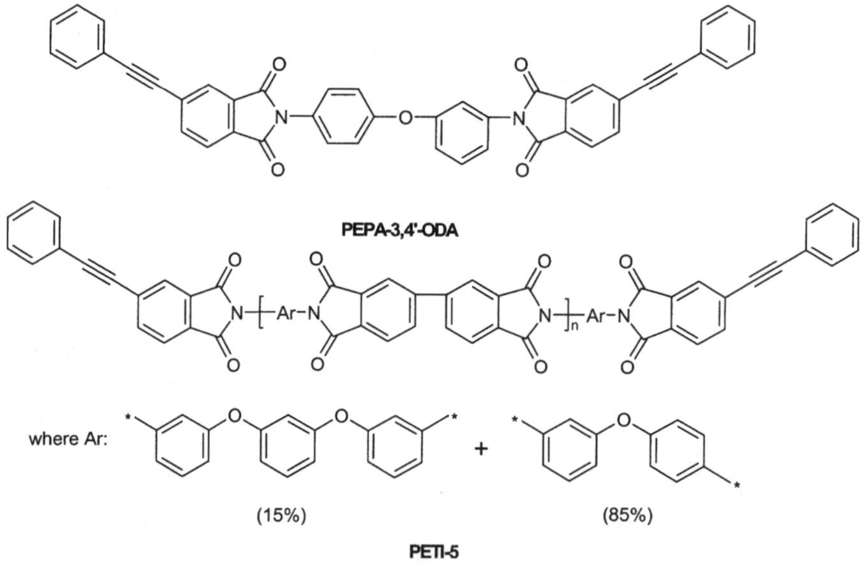

Fig. 17. Chemical structures of 3,4'-bis[(4-phenylethynyl)phthalimido]diphenyl ether (PEPA-3,4'-ODA) and phenylethynyl-terminated imide oligomer (PETI-5). Reprinted from (2000) J Polym Sci A Polym Chem 38:2526 [82] with permission

nylethynyl)phthalimido]diphenyl ether (PEPA-3,4'-ODA) were carried out (Fig. 17) by Scola et al. as well [82].

The microwave cure of PEPA-3,4'-ODA and PETI-5 oligomer was performed in a variable frequency (4.19–5.19 GHz) microwave oven LT 502 X (Lambda Technologies) in the temperature range of 300–330 °C and 350–380 °C, respectively.

Table 7. Comparison of activation energies of microwave and thermal cure reactions. Reprinted from (2000) J Polym Sci Part A: Polym Chem 38:2526 [82] with permission

Sample	MW	Δ
PEPA-3,4'-ODA	27.6 ±2.3 (kcal/mol)	40.7 ±2.7 (kcal/mol)
	1.20±0.10 (eV)	1.77±0.12 (eV)
PETI-5	17.1 ±0.7 (kcal/mol)	33.8 ±2.0 (kcal/mol)
	0.74±0.03 (eV)	1.47±0.09 (eV)

Comparing with thermal kinetic curing rate studies of PEPA-3,4'-ODA and PETI-5, microwave cure gave much higher rate constants for both. For PEPA-3,4'-ODA, the reaction followed first-order kinetics, yielding an activation energy of 27.6 kcal/mol, which was 68% that of the thermal cure. For PETI-5, the reaction followed 1.5-order, yielding activation energy of 17.1 kcal/mol, which was 51% that of the thermal cure for PETI-5 (Table 7) [82]. The kinetic data showed that maintaining the same temperature with microwave irradiation could provide a faster cure relative to conventional thermal heating.

In the next step, the application of microwave irradiation to the processing of carbon fiber reinforced phenylethynyl-terminated polyimide composites (PETI-5/IM7) was investigated and evaluated. Six different microwave cure cycles and three thermal processes were investigated. It was found that the polyimide prepreg showed no obvious difference in coupling with microwaves within a range 2.4–7.0 GHz. Higher glass transition temperatures were observed in the microwave cured composites. Thermally cured composites, fabricated from the same time-temperature cure cycles as the microwave processes, showed incomplete cure and much lower glass transition temperatures. Compared with the standard thermally cured composites, microwave cured samples exhibited higher flexural strengths and moduli and shear strength values. A microwave process was demonstrated that fabricated unidirectional polyimide-(carbon fiber) composites with superior thermal and mechanical properties relative to the thermal process in half the time required for the thermal process (Fig. 18) [83].

Microwave irradiation was also applied by Scola et al. to synthesize poly(ε-caprolactam-co-ε-caprolactone) directly from the two cyclic monomers, ε-caprolactam and ε-caprolactone, by anionic catalyzed ring opening Copolymerization [84]. The reactions were carried out using a variable frequency (0.4–3 GHz) microwave oven, programmed to set a temperature and controlled by a pulsed power on-off system. During microwave experiments, ε-caprolactam and ε-caprolactone in the molar ratio of 2 to 1 were mixed together with solid LiAl[OC(CH$_3$)$_3$]$_3$H (1–3 mol% of total reactants), and the mixture 5–10 g was irradiated for 0.5–1 h in a Teflon vessel under a nitrogen blanket. The copolymerization temperature inside the sample (140–180 °C) was measured and control via a grounded thermocouple, which was calibrated against a fiber-optic temperature probe. The same ratio of re-

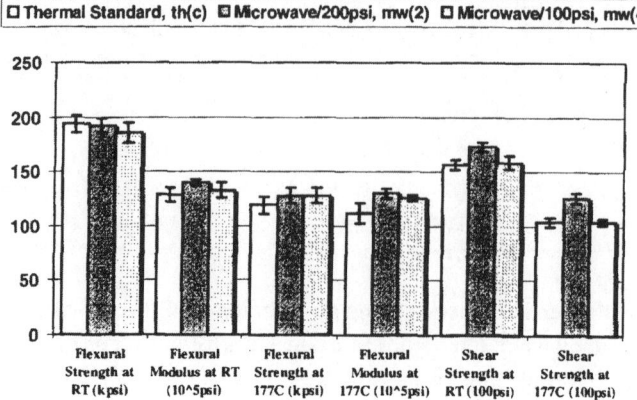

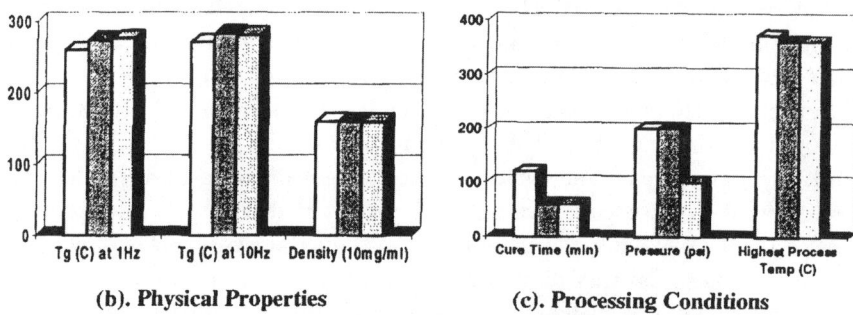

Fig. 18a–c. Comparison of processing conditions and properties between microwave and thermally cured carbon fiber reinforced phenylethynyl-terminated polyimide composites (PETI-5/IM7) composites. Reprinted from (1999) J Polym Sci A Polym Chem 37:4616 [83] with permission

actants and catalyst used in the microwave methods were used for the conventional thermal synthesis in an oil bath. Compared with the corresponding thermal products, microwave synthesized copolymers were obtained in higher yield, amide composition, glass transition temperature, and equivalent molecular weights (Table 8) [84].

3.2.2
Polyesters and Polyethers

Synthesis of a polyether by phase-transfer catalysis (PTC) under microwave irradiation was investigated by Hurduc et al. [85], who applied 3,3-bis(chloromethyl)oxetane (BCMO) and various bisphenols (i.e., bisphenol A, 4,4'-dihydroxyazobenzene, 4,4'-dihydroxybiphenyl, 4,4'dihydroxybenzophenone, and 4,4'-thiodiphenyl) to the synthesis (Fig. 19).

Table 8. Comparison of microwave and thermal activation in copolymerization reactions. Reprinted from (2000) J Polym Sci A Polym Chem 38:1379 [84] with permission

Sample	Δ-PAE, 1%[a]	Δ-PAE, 2%	Δ-PAE, 3%	MW-PAE, 1%	MW-PAE, 2%	MW-PAE, 3%
	160°C–0.5h	160°C–0.5h	160°C–0.5h	160°C–0.5h	160°C–0.5h	160°C–0.5h
Starting materials – ester:amide	1:2	1:2	1:2	1:2	1:2	1:2
Yield (%)	51.2	52.7	57.0	61.9	70.1	78.2
T_g (°C), (tan δ), DMTA (1 Hz)	−25.0	−18.5	−14.5	−14.0	−7.5	6.0
T_g (°C), from Fox equation	29	12	9	15	8	4
M_w (kg/mol), GPC	25.4	19.8	17.1	22.0	21.3	16.2
M_w/M_n	1.4	1.5	1.6	2.1	2.0	1.5

[a] Catalyst level

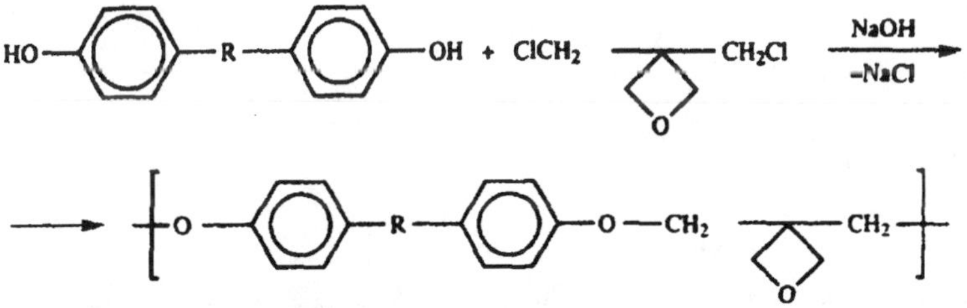

Fig. 19. Synthesis of polyethers from 3,3-bis(chloromethyl)oxetane (BCMO) and various bisphenols. Reprinted from (1997) Eur Polym J (1997) 33: 187 [85] with permission

In a typical reaction, a mixture of bisphenol (1.3 mmol), water (5 ml), NaOH (3 g), nitrobenzene (5 ml), BCMO (1.3 mmol), and TBAB (0.2 mmol) was placed in a 50-ml flask and irradiated for 1.5 h in a microwave waveguide (60 W), while temperature (95–100 °C) was monitored by a thermovision IR camera. Under conventional conditions, the reaction was carried out for 5 h at 90 °C. It was found that for semi-crystalline polymers the yields were higher under microwave conditions, whereas in the case of amorphous polymer the yields were approximately equal besides of shorter reaction time (Table 9).

Table 9. Polymer yields under classical PTC and MW conditions. Reprinted from (1997) Eur Polym J 33: 187 [85] with permission

Sample no.	Bisphenol structure	Polymer Yields (%) Classical PTC		Under MW
		A	B	
1	HO–⌬–⌬–OH	85.3	44.3	89.7
2	HO–⌬–N=N–⌬–OH	64.9	20.1	71.7
3	HO–⌬–C(CH$_3$)$_2$–⌬–OH	77.4	61.7	83.9
4	HO–⌬–C(=O)–⌬–OH	50.1	4.0	72.3
5	HO–⌬–S–⌬–OH	76.7	13.5	70.7

A – 5 hr. with stirring.
B – 1.5 hr. without stirring.

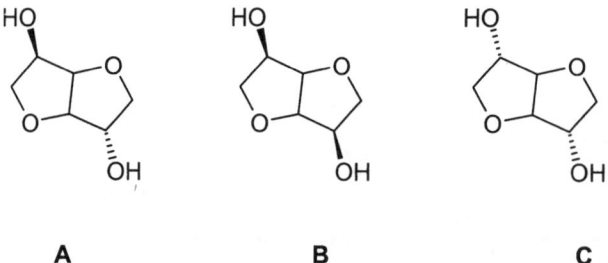

Fig. 20. A Isosorbide (1,4:3,6-dianhydro-D-sorbitol). **B** Isomannide (1,4:3,6-dianhydro-D-mannitol). **C** Isoidide 1,4:3,6-dianhydro-L-iditol (isoidide). Reprinted from (2002) Eur Polym J 38:1851 [87] with permission

The synthesis of linear polyethers either from isosorbide [86] or isoidide [87] and disubstituted alkyl bromides or methanesulfates by using microwave irradiation under solid-liquid PTC conditions was described by Loupy et al. (Fig. 20) [86, 87].

Isosorbide and isoidide are by-products of biomass obtained from sugar industry by double dehydration of starch [88]. The reactions were carried out in a single-mode microwave reactor Synthwave 402 (Prolabo) with temperature infrared detector. The reaction mixtures consisting of isosorbide or isoidide (5 mmol),

Table 10. Influence of reaction time on the yields of high molecular weight fraction (**FP MeOH**) and low molecular weight fraction (**FP Hex**) of polyethers. Reprinted from [87] with permission

Time	Mode of activation[a]	FP MeOH (%)	FP Hex (%)	Total yield (%)
30 min	MW	67	18	85
60 min	MW	71	19	90
30 min	Δ	12	81	93
1 day	Δ	64	25	89
1 week	Δ	83	5	88
1 month	Δ	91	0	91

[a] MW – microwave irradiation
Δ – conventional thermal heating

alkyl dibromide/methane disulfate (5 mmol), tetrabutylammonium bromide (TBAB, 1.25 mmol), and powdered KOH (12.5 mmol) were irradiated for 30 min to get polyethers with 70–90% yields. It was found that the use of a small amount of solvent was necessary to ensure a good temperature control and decrease in viscosity of the reaction medium.

In the case of isosorbide, the microwave-assisted synthesis proceeded more rapidly compared with conventional heating (30 min; yield 69–78%). Under conventional conditions, the polyethers were obtained with 28–30% yield within 30 min. Similar yields of the polyethers were obtained while the reaction time was extended to 24 h. These yields remained practically unchanged even though the synthesis was carried out for another seven days (Table 10)

The analysis of properties of the synthesized polyethers revealed that the structure of the products strictly depended on the activation mode (i.e., microwave or conventional activation). Under microwave conditions, the polyethers had higher molecular weight and better homogeneity. For example, within 30 min under conventional heating, the polyethers of higher molecular weight were not observed at all. Moreover, it was found that the mechanism of chain termination is different under microwave and conventional conditions. The polyethers prepared by conventional heating have shorter chains with terminal hydroxyl ends, whereas under microwave irradiation the polymer chains were longer with terminal ethylenic ends. Under microwave irradiation, terminal ethylenic ends were formed rapidly and set up a hindrance to further polymer growth. Under conventional conditions, chains were terminated essentially by hydroxyl functions [86].

The enhancement of solid state polymerization for poly(ethylene terephthalate) (PET) as well as polyamide 66 in a fluidized bed was reported by Mallon and Ray et al. [89]. For each polymerization, the reactor charged with PET or polyamide 66 pellets was placed in a multi-mode microwave cavity, purged with nitrogen, and

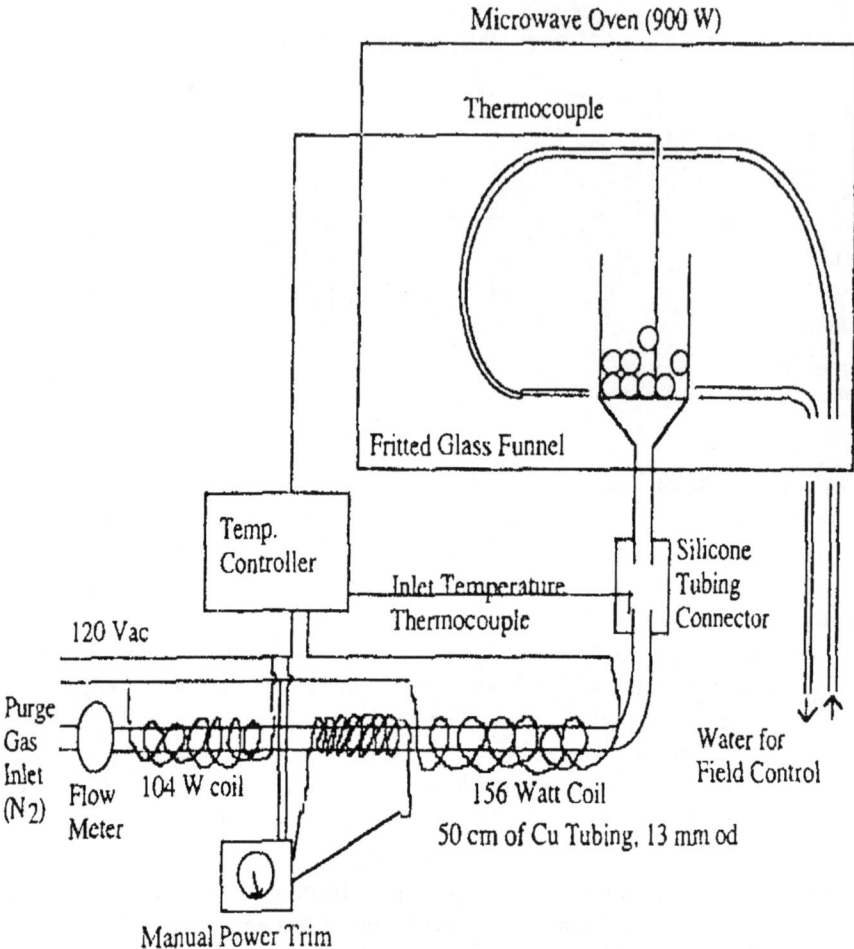

Fig. 21. Microwave reactor setup for solid state polymerization of PET and Nylon 66. Reprinted from (1998) J Appl Polym Sci 69:1203 [89] with permission

irradiated for 6–7 h (Fig. 21). The temperature was controlled by manipulating the inlet gas temperature.

It was found that the increase in polymerization rate was not due to an increase in polymerization temperature, but the effect was consistent with direct heating of the condensate leading to enhanced diffusion rates (Table 11) [89]. In the case of microwave PET polymerization, it was noticed that an increase in polymerization rate for hydroxyl and carboxyl functional groups was not monotonous; it was higher for hydroxyl end groups. In general, the increase in the solid-state reaction rate due to microwaves was about equivalent to an increase in reaction temperature of 10–15 °C.

Table 11. Influence of microwave irradiation on PET polymerization in solid state. Reprinted from (1998) J Appl Polym Sci 69:1203 [89] with permission

Conditions	Diffusion coefficient at 220°C. Both ethylene glycol and water $(cm^2/s) \times 10^6$	Activation energy (cal/mol)
Δ	1.19	16.672
MW	3.55	12.197

More recently, the bulk polycondensation of D,L-lactic acid (LA) to poly(lactic acid) under microwave irradiation was described by Zsuga et al. [90]. The reaction was carried out in a domestic microwave oven, in which 5 g of LA acid was placed in a 20-ml beaker and irradiated at 650 W power. The reaction time was varied between 10, 20 and 30 min to afford poly(lactic acid) with 96, 84 and 63% yield, respectively. The yield of polycondensation decreased with increasing irradiation time probably due to the loss of oligomers of lower polymerization degree during the polycondensation. Under conventional thermal conditions, the polymer was obtained in 94% yield after 24 h at 100 °C. According to MALDI MS analysis only linear oligomers were formed. After 20 min of microwave irradiation, oligomers with nearly the same molecular weight were obtained like those produced upon conventional heating (i.e., 100 °C, 24 h). Thus, the reaction time for preparing poly(lactic acid) could be considerably reduced. On increasing reaction time under microwave conditions, the fraction of cyclic oligomers tended to appear besides linear ones [90].

Processing of neat unsaturated polyester resins (diallyl phthalate diluted with vinyltoluene) and composites with glass fiber under microwave irradiation were studied by Hawley et al. [91]. As a microwave source, a single mode resonant cavity operated in 2.45 GHz in TE_{111} mode with constant input power of 60 W was used. An allyl phthalate polyester was chosen for the neat resin studies, vinyltoluene (30 wt%) was used as the crosslinking monomer, and benzoyl peroxide (1%) was added as an initiator. Glass fiber (70 wt%) composites with diallyl phthalate polyester as matrix material were used in the composites investigation, and the prepreg was in the form of 1.6 cm wide and 0.3 cm thick tape. A thin film technique was applied in the reaction kinetics study to avoid large temperature gradients. Samples prepared on KBr disks were isothermally microwave cured at 85, 100 and 115 °C, while the extent of cure of the samples was monitored by means of FT-IR spectroscopy. It was found that microwaves could initiate the reaction at a lower bulk temperature and shorter time than thermal heating. As the result, higher reaction rates were observed in microwave curing as compared to thermal curing. At lower polymerization temperatures, such as 85 °C, the ultimate extent of cure was higher under microwave than conventional thermal conditions.

Microwave assisted resin transfer molding was described by Rudd et al. [92, 93], who developed a specialized in-line microwave resin preheating system for such a

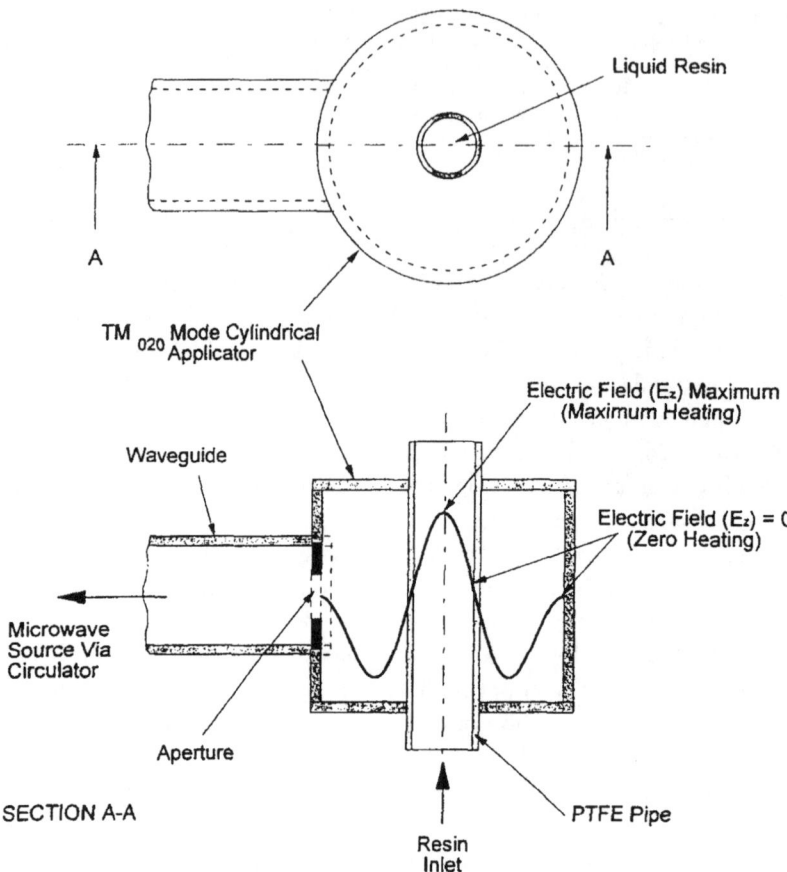

Fig. 22. Cross-sections of microwave TM_{020} mode cylindrical applicator. Reprinted from (2001) Macromol Rapid Commun 22:1063 [92] with permission

purpose. Resin transfer molding (RTM) was limited to low volume production due to protracted component cycle times. The principal cause of extended cycle time is the thermal quench near the injection gate. It occurs when cold resin enters the heated mold. The period required for the mould and resin to recover the lost heat lengthens the cycle time. One method to decrease thermal quench and to reduce the cycle time is to preheat the resin prior to injection. For this purpose, a small volume of resin, having a low thermal inertia, passes through the in-line system, affording a rapid heating response to variations in the input of microwave power. The heating profile of the TM_{020} mode cylindrical applicator is analogous to the laminar flow profile, producing zero heat at the PTFE pipe wall where the resin is stagnant, and maximum heat along the pipe axis where the velocity is the highest (Fig. 22)

Using this system, the resin can be heated accurately to a constant, pre-defined temperature. Furthermore, prescribing an analytical heating function allows profiling of the resin temperature during the injection phase. The cure sequence can be controlled using a ramping profile, with the additional benefit of a lower resin viscosity for improved flow through the mold. The molding was made with unsaturated polyester resin (Synolac 6345). As a result, impregnation and cycle times are reduced considerably. The increase in the resin temperature from 22 to 40 °C caused a 70% decrease in resin viscosity and reduced the impregnation time by 41% and the cycle time by 24% (mold temperature 60 °C).

3.2.3
Epoxy Resins

Microwave-assisted curing of epoxy resin systems was one of the first applications in polymer chemistry. It represents the most widely studied area in polymer chemistry under both continuous and pulse microwave conditions [94–97]. The epoxy-amine formulations were investigated in terms of structure, dielectric properties, toughness, mechanical strength, percentage of cure and glass transition temperature.

In the early works, it was found that the pulse method could lead to the fastest heating of epoxy resins [98] and improve their mechanical properties [99]. For example, it was shown that a computer controlled pulsed microwave processing of epoxy systems that consisted of diglycidyl ether of bisphenol A (DER 332) and 4,4'-diaminodiphenyl sulfone (DDS) in a cavity operated in TM_{012} mode could be successfully applied to eliminate the exothermic temperature peak and maintain the same cure temperature at the end of the reaction [100]. The epoxy systems under pulsed microwave irradiation were cured faster, and it was possible to cure them at higher temperatures when compared with a continuous microwave or conventional thermal processing.

Dielectric parameters, i.e., the dielectric behavior of epoxy system consisting of diglycidyl ether of bisphenol A (EPON 828 EL) and ethylenediamine (EDA) during the crosslinking process at frequency interval of 10^3–10^{10} Hz was described by Rolla et al. [101, 102]. However, these studies that confirmed the possibility to utilize dielectric quantities to obtain information on relevant parameters such as conversion, viscosity change, sol-gel transition, and glass transition temperatures were not concerned with microwave processing of epoxy systems, the basic parameters such as static dielectric constant (ε_0) and dielectric permittivity ($\varepsilon=\varepsilon'-j\varepsilon''$) being important for microwave processing can be found in these works [101, 102]. The loss factor (ε'') and permittivity (ε') increase with the reaction temperature and decrease with extent of cure, which can be attributed to the higher and lower mobility of molecular dipoles during heating and crosslinking, respectively [103, 104].

Lately, applying radio frequency (RF) radiation at the frequency range of 30–99 MHz, Whittaker et al. managed to cure an epoxy resin system consisting of diglycidyl ether of bisphenol A (DGEBA) (Araldite GY260) and diaminodiphenyl sulfone (DDS) [105]. The samples were placed in thin-walled 5- or 10-mm Pyrex tubes and cured in solenoid coils with 5 to 10 turns and diameters of 5–10 mm in the temperature range of 160–190 °C. The epoxy resin was cured rapidly (60 min) at low power levels. However, comparison of the kinetics of the RF curing with thermal curing while maintaining the same temperature revealed no differences. It was stressed that the main advantages of RF curing were much lower power levels required to achieve rapid heating, superior penetration, and possibility of design of RF coils with highly homogenous fields [96].

The microwave curing of diglycidyl ether of bisphenol A (DGEBA) with different curing agents (i.e., diaminodiphenyl sulfone (DDS) and *m*-phenylenediamine (mPDA)) in thin films were studied by DeLong et al. [106]. The samples were prepared by casting stoichiometric mixtures of DGEBA/DDS and DGEBA/mPDA onto 13 mm diameter and 1 mm thick potassium bromide disks to form approximately 10 micron thick films that were protected by other potassium bromide discs placed on the top of samples. The thin film samples were cured under microwave irradiation in a center of a cylindrical resonant cavity operating in a TE_{111} mode, while the temperature was monitored by means of fiber-optic thermometer and measured directly from the thin epoxy films.

It was found that the effects of microwave irradiation on the cure of epoxy systems depended on curing agents. Microwaves had a stronger effect on epoxy/DDS than epoxy/mPDA systems and, consequently, the magnitude of increases in glass transition temperatures was much larger for DGEBA/DDS than for DGEBA/mPDA. Similar values of glass transition temperature were obtained for microwave and thermal cure at low extents of cure while higher values were observed in microwave cure at the extents of cure after gelation. Moreover, significantly higher ultimate extents of cure and faster reaction rate were observed in the microwave cure when compared to thermal cure (Figs. 23 and 24) [106].

In contrast, Galy et al. investigated the effect of a microwave cure on the mixture of epoxy prepolymer with a cycloaliphatic diamine and compared it with a standard thermal cure [107]. The applied epoxy prepolymer was diglycidyl ether of bisphenol A (DGEBA) DER 332 from Dow, while 4,4'-diamino-3,3'-dimethyldicyclohexyl methane (3DCM, Laromin C260) from BASF was chosen as the curing agent (Fig. 25).

The epoxy prepolymer and the curing agent were mixed together prior to use, then the epoxy mixture (13 g) was poured into moulds (inside dimension 96 mm × 16 mm × 8 mm), which were irradiated in a microwave applicator with TE_{01} propagation mode. The sample temperature was measured continuously by means of an infrared pyrometer that gave the surface temperature and fiber-optic thermometer that recorded the bulk temperature. Samples cured by both thermal

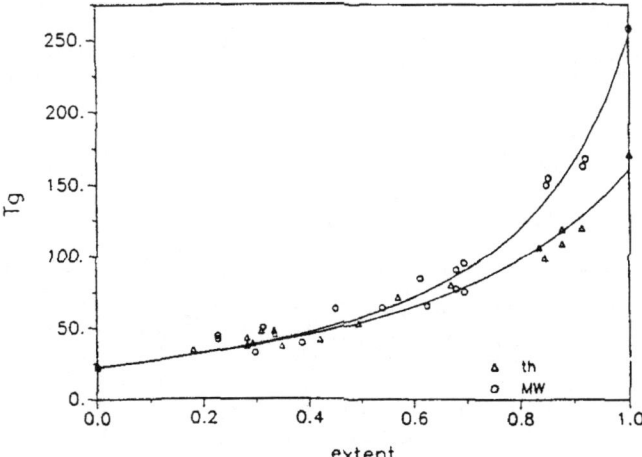

Fig. 23. Comparison of glass transition temperatures (T_g) of microwave and thermally cured epoxy resins (DGEBA/DDS). Reprinted from (1993) Polym Eng Sci 33:113 [106] with permission

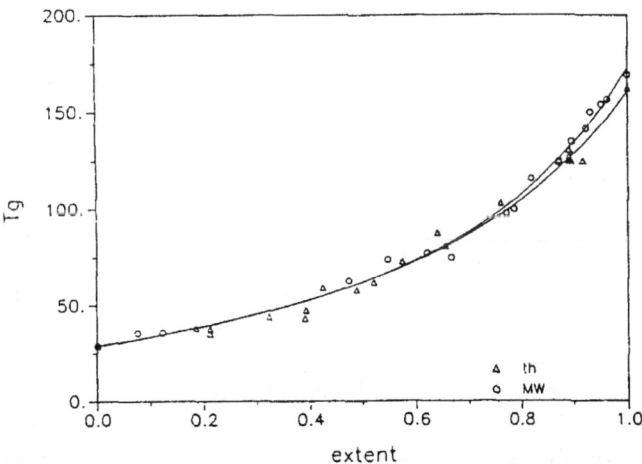

Fig. 24. Comparison of glass transition temperatures (T_g) of microwave and thermally cured epoxy resins (DGEBA/mPDA). Reprinted from (1993) Polym Eng Sci 33:113 [106] with permission

and microwave processing were characterized by dynamic and static mechanical properties and then compared with those of fully crosslinked networks. It was found that microwaves did not have direct influence on the mechanical processing of the polymer network, and the only parameter that influenced the mechanical properties is the extent of the reaction. Moreover, under microwave processing conditions, it was not possible to obtained a fully cured DGEBA/3DCM network.

DER 332 n̄ = 0,03
DOW CHEMICAL
174 g/eq

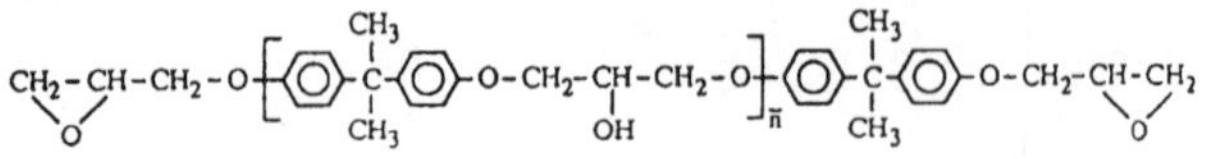

3DCM
BASF
238 g/mol

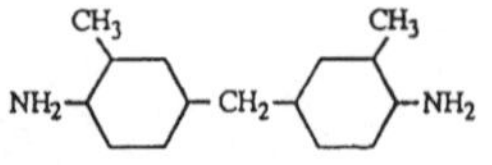

Fig. 25. Structure of the epoxy prepolymer and the diamine (3DCM) used for the synthesis of the epoxy networks. Reprinted from (1995) Polym Eng Sci 35:233 [107] with permission

Table 12. The comparison of epoxy compositions cured under both microwave and microwave/conventional conditions. Reprinted from (1995) Polym Eng Sci 35:233 [107] with permission

Curing cycle	MW	MW+Δ
Power (W)	200	200
Temperature (°C)		190
Time	15 min	15 min+14 h
Compression modulus (GPa)	3.15	2.9
Poisson's ratio	0.36	0.36
Glass transition temperature (°C)	131	186
Extent of reaction (%)	89	100

The fully cured samples were obtained either by thermal (140 °C, 1 h+190 °C, 14 h) or combined processing (microwave 200 W, 15 min+thermal 190 °C, 14 h) (Table 12) [107].

In a more recent paper, the reaction kinetics of a neat and rubber-modified epoxy formulation cured under microwave and thermal conditions were investigated [108]. The epoxy formulation consisted of DGEBA and DDS as a curing agent. The applied rubber was an epoxy terminated butadiene-acrylonitrile random copolymer (ETBN), which was prepared from CTBN with an excess of DGEBA (Fig. 26). The weight percentage of rubber introduced into the modified epoxy component amounts to 15% with respect to the total amount of the mixture. Under microwave conditions, the samples (12.5 g) were cured in a microwave applicator with the propagation mode TE_{01} at 170–190 °C for 0–3 h.

The same kinetic behavior of the epoxy-amine DGEBA-DDS and the ETBN-modified systems was observed whatever the curing process (conventional or microwave) was employed. The determined conversion volumes of the samples cured by microwave irradiation were compared with the model predictions derived for the thermal curing and found to be independent of the curing method (Fig. 27) [108].

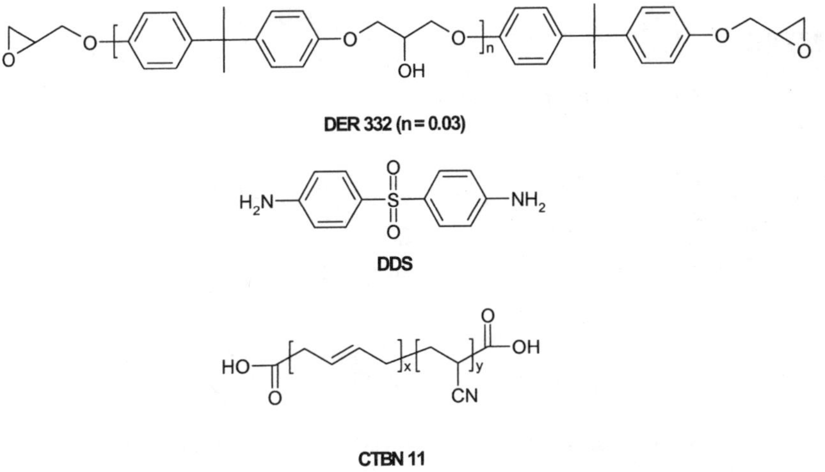

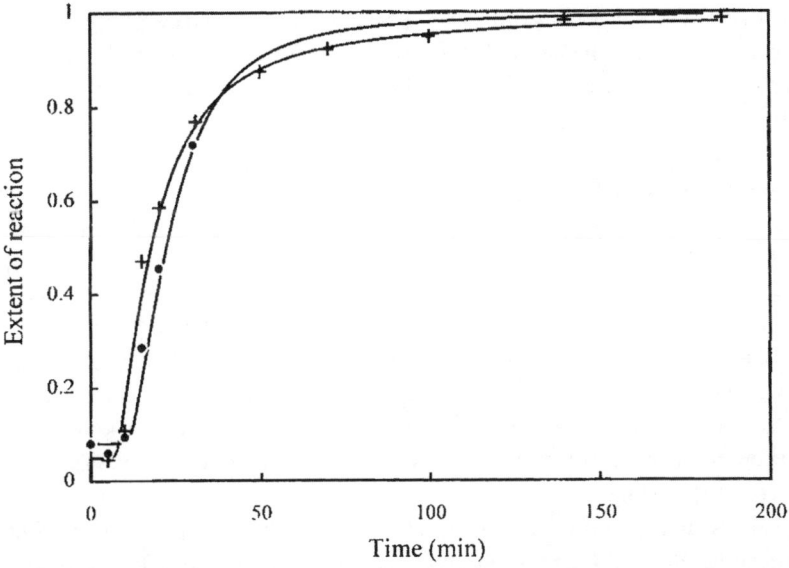

Fig. 26. Chemical structures of the components used in the formulation of modified epoxy resins. Reprinted from (1998) J Appl Polym Sci 68:543 [108] with permission

Fig. 27. Conversion vs time for DGEBA-DDS (*crosses*) and DGEBA-DDS-ETBN (*filled circles*) samples cured by microwave heating; lines represent the models prediction. Reprinted from (1998) J Appl Polym Sci 68:543 [108] with permission

The recent paper by Mijovic et al. [109] summarized their studies conducted on the mechanism and rate of chemical reactions in thermal and microwave fields. A number of nonpolymer-forming and polymer-forming mixtures of different func-

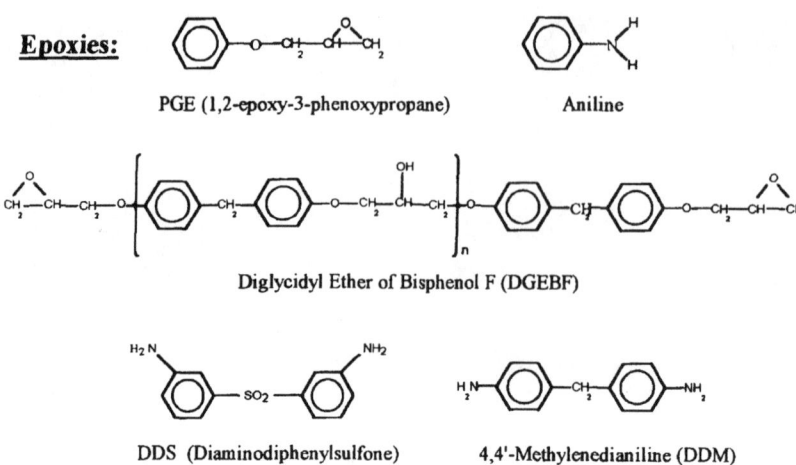

Fig. 28. Chemical structures of formulation components for cure of epoxy resins. Reprinted from (1998) Polym Adv Technol 9:231 [109] with permission

tionality and molecular architecture were investigated. In the case of epoxy systems, the reaction of phenyl glycidyl ether(PGE) and aniline as a control reference was investigated, whereas for epoxy crosslinking systems the reactions of diglycidyl ether of bisphenol F (DGEBF) with DDS and 4,4'-methylenedianiline (DDM) were chosen (Fig. 28).

The samples of reaction mixtures were held in cylindrical sample holders (10 mm diameter and 25 mm depth), which were placed in a waveguide operated under TE_{01} mode for microwave experiments. The sample temperature during cure was monitored by a fiber-optic thermometer while placing the fiber optic probe in the center of the sample. The temperature range investigated at this study was 70–120 °C and 140–190 °C for DGEBF-DDA and DGEBF-DDS, respectively. Similarly to the previous report on the rate and mechanisms of epoxy-amine reactions in thermal and microwave fields [110], it was found that the cure kinetics of the multifunctional epoxy formulation were identical under both thermal and microwave conditions (Figs. 29 and 30).

The reproducibility and reliability of data were verified by repeated runs, and the claims of accelerated kinetics due to the so-called microwave effect were unfounded; however, an important scientific research is still the issue of heat transfer in thermal and microwave fields [109].

More recently, Boey et al. reported the rate enhancement under microwave irradiation during investigation of the cure process of epoxy resins (DGEBA – Araldite GY6010) in the presence of three different types of crosslinking agents such as DDS, DDM, and mPDA [111]. Microwave curing of epoxy systems was carried out in a multi-mode cavity (300 mm × 298 mm × 202 mm) coupled

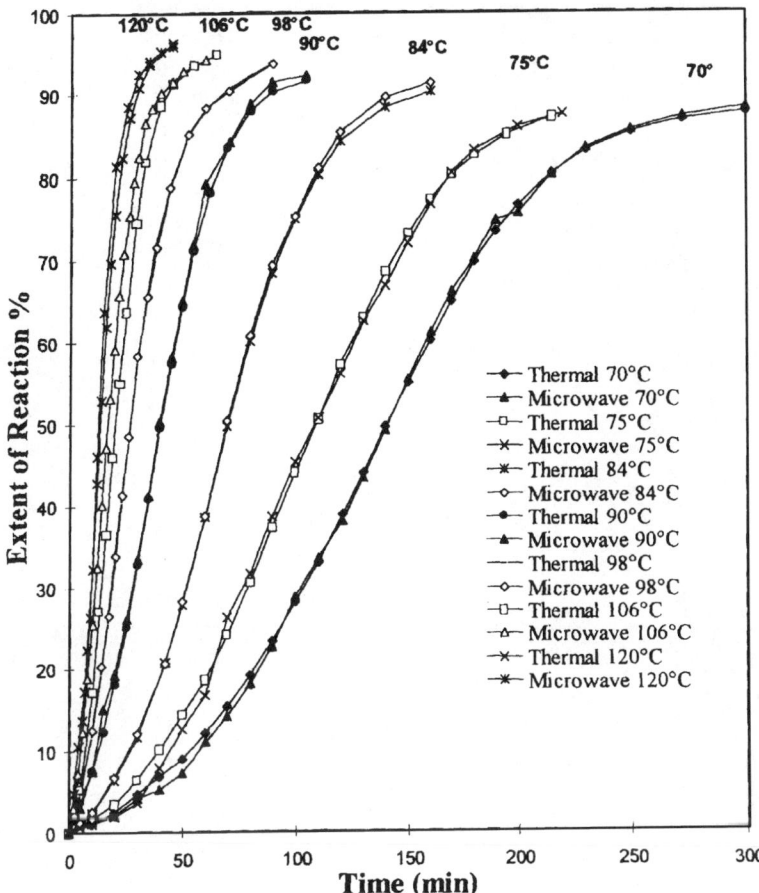

Fig. 29. Extent of reaction as a function of time in thermal and microwave fields for DGEBF-MDA reaction at different temperatures. Reprinted from (1998) Polym Adv Technol 9:231 [109] with permission

through a waveguide with a variable power generator. The results indicated that in all three cases the curing time was considerably shortened in contrast to thermal curing, where the DDS system took a longer time to completely cure. Under microwave conditions, at a low power level (i.e., 200 W), the DDM and mPDA systems appeared to give shorter curing times of 10–8 min, respectively, in comparison with the DDS system, which cured after 15 min. However, at higher power levels (i.e., 400 and 600 W), the curing time was almost the same: 2.5 min for mPDA and 3 min for both DDS and DDM systems (Tables 13 and 14) [111].

According to the next report, the observed rate enhancement was the result of the decrease in the lag time prior to incitation of crosslinking as well as the decrease

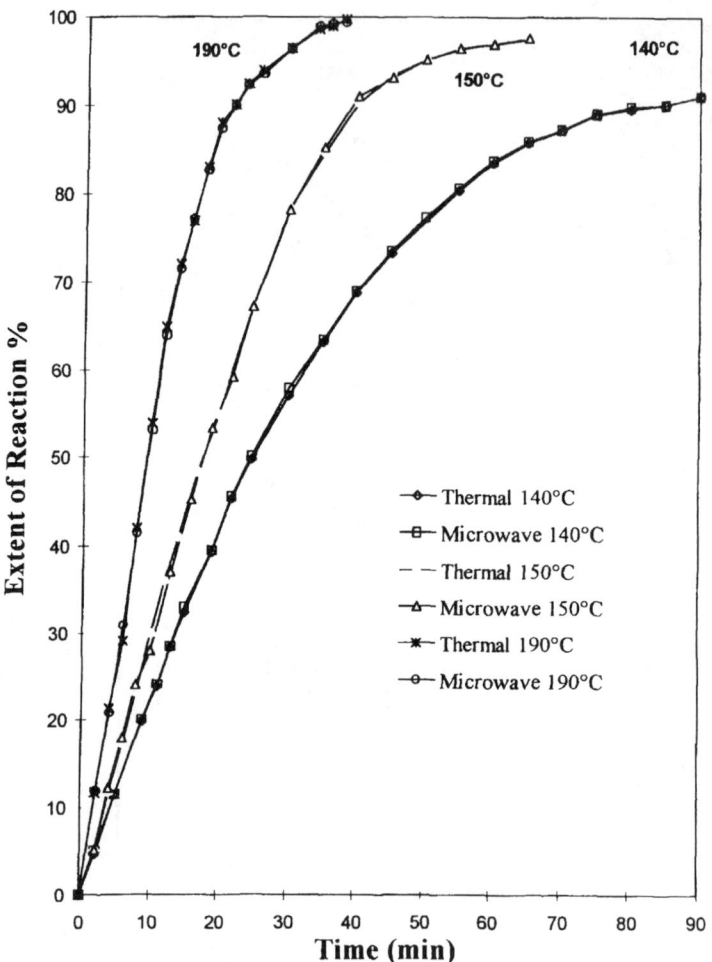

Fig. 30. Extent of reaction as a function of time in thermal and microwave fields for DGEBF-DDS reaction at different temperatures. Reprinted from (1998) Polym Adv Technol 9:231 [109] with permission

in the effective cure time [112]. Since a shortening of cure time is not different from a shortening achieved by a higher curing temperature, it is possible to obtain temperature equivalent values for the microwave cure, which are only virtual ones and are prepared for the purpose of analysis of the cure kinetics. The results are presented in Table 15. It is of interest to note that the equivalent temperatures obtained in all the cases were consistent and significantly higher than the maximum temperature measured in the samples, providing further support that the microwave curing is not merely thermally based [112].

Table 13. Maximum of specimen temperature (T_{spec}) of thermally treated epoxy/amine systems. Reprinted from (1999) Polym Test 18:93 [111] with permission

Thermal curing temperature (°C)	T_{spec} (°C)	Cure time (min)
DGEBA/DDS		
130	108	60
150	131	60
170	143	30
180	156	30
DGEBA/DDM		
110	102	20
130	135	15
150	137	5
DGEBA/mPDA		
100	110	20
120	138	12.5
150	139	5

Table 14. Maximum of specimen temperature (T_{spec}) of microwave treated epoxy/amine systems Reprinted from (1999) Polym Test 18:93 [111] with permission

Microwave curing temperature (°C)	T_{spec} (°C)	Cure time (min)
DGEBA/DDS		
200	141	15
300	165	10
400	184	5
500	225	5
600	235	3
DGEBA/DDM		
200	111	10
400	137	5
600	174	3
DGEBA/mPDA		
200	135	8
400	150	4
600	212	2.5

A comparison of thermal and microwave cure assumes a new dimension when the temperature distribution inside the sample is considered, and that is where the scientific challenge lies. The fundamental difference in the heat transfer during re-

Table 15. Equivalent cure temperature for microwave processing of epoxy/amine systems. Reprinted from (2000) Adv Polym Techn 19:194 [112] with permission

Microwave power (W)	Maximum temperature (°C)	Lag time (s)	Effective cure time (s)	Total cure time (s)
DGEBA/DDM system				
200	120	144	163	154
400	130	170	179	173
600	162	181	196	190
DGEBA/mPDA system				
200	135	150	153	152
400	151	171	172	170
600	178	193	190	191

active processing in thermal and microwave fields is that microwave energy, in contrast to thermal heating, is supplied directly to a large volume, thus avoiding the thermal lags associated with conduction and/or convection. Consequently, temperature gradients and the excessive heat build-up during thermal processing could be reduced by a microwave power control [109]. In large polymer composite structures high temperatures caused by exothermic resin cure can degrade the mechanical properties of the composites

For example, microwave heating was examined numerically and experimentally as an alternative to conventional thermal processing techniques and investigated by Thostenson and Chou for a glass/epoxy laminate [113]. A numerical simulation has been developed to predict the one-dimensional transient temperature profile of the composite during both microwave irradiation and conventional heating. Numerical and experimental results were presented for a glass/epoxy laminate with a thickness of 25 mm. The raw materials used for the experimental investigation were a bisphenol-F/epichlorohydrin epoxy resin (Shell Epon 862) and an aromatic diamine (Shell Epi-Cure W) as a curing agent. The microwave applicator was a cylindrical multi-mode cavity with an internal volume of 500 l. The large cavity was equipped with a mode stirrer and multiple microwave inputs to enhance the uniformity of the microwave field. The output power of the microwave source was varied continuously from 0 to 6 kW. The results showed that it was possible to cure thick glass/epoxy composites uniformly and eliminate temperature excursions due to chemical reactions during the cure cycle. In addition, through continuous wave feedback control of the microwave power, it was possible to monitor the cure behavior of the composite. In conclusion, microwave processing has the potential to increase the quality of thick-section composites and reduce the manufacturing costs through more efficient transfer of energy [113].

In the later stage, a calorimetric analysis (DSC) of the cure kinetics of the same glass/epoxy composite was conducted for both thermal and microwave curing

[114]. To develop a kinetic model for conventional thermal cure, isothermal experiments were conducted in the range of 135–175 °C, and the degree of cure as a function of time was calculated. A different approach was required to examine the cure kinetics under microwave conditions because the cure behavior was hard to be monitored in situ within a microwave cavity. Therefore, samples that were placed in a microwave cavity were heated rapidly to the desired cure temperature, and the cure was stopped at a specified time by removing the sample from the cavity and quenching it. Then the samples were analyzed by means of DSC to obtain the residual heat of reaction. It was found that the reaction rate was accelerated by the application of microwave irradiation. The results of the numerical simulation showed that it was possible to decrease the cycle time for processing of thick-section composites because microwave heating allowed for a better control of the spatial solidification of the laminate, which resulted in significantly reduced processing times and enhanced quality [114].

Finally, that work was summarized by Thostenson and Chou, who showed that both numerical and experimental results indicated that volumetric heating due to microwaves promoted an inside-out cure of the thick laminates and dramatically reduced the overall processing time [115]. Under conventional thermal conditions, to reduce thermal gradients, thick laminates were processed at lower cure temperature and heated with slow heating rates, resulting in excessive cure times. Outside-in curing of the autoclave processed composite resulted in visible matrix cracks, while cracks could not be seen in the microwave-processed composite. The formation of cure gradients within the two composites cured under both microwave and conventional conditions are presented in Fig. 31.

Although cure gradients exist in both composites treated under microwave and thermal conditions, the differences in the solidification behavior are seen. In the conventionally processed composite, the outside-in cure gradients are most significant during the early stages of the cure cycle, and the maximum cure rate for this epoxy resin system occurs at the beginning of cure. Therefore, it is critically important to initiate an inside-out cure at the beginning of the cure cycle. Reduced thermal gradients during the early stages of microwave curing allow for better control over solidification behavior of the resin. In conventional processing, very slow heating rates are required to reduce the thermal lag and heat the composite up to the temperature where additional heat is generated by the chemical reaction. Once additional heat is generated, it will help to promote the desired inside-out cure. To obtain an inside-out cure in conventional processing, the required cycle time was almost three times longer than in the case of microwave processing. Thus, the processing time can be drastically reduced to achieve the desired inside-out cure through the use of microwaves [115].

In another example, the fiber-glass/epoxy (Dow-Derakane 411–350) composite panels with 15 layers of glass fiber mats were cured under microwave irradiation [116]. The final panels (approximately 1.5 cm thick) were placed perpendicularly

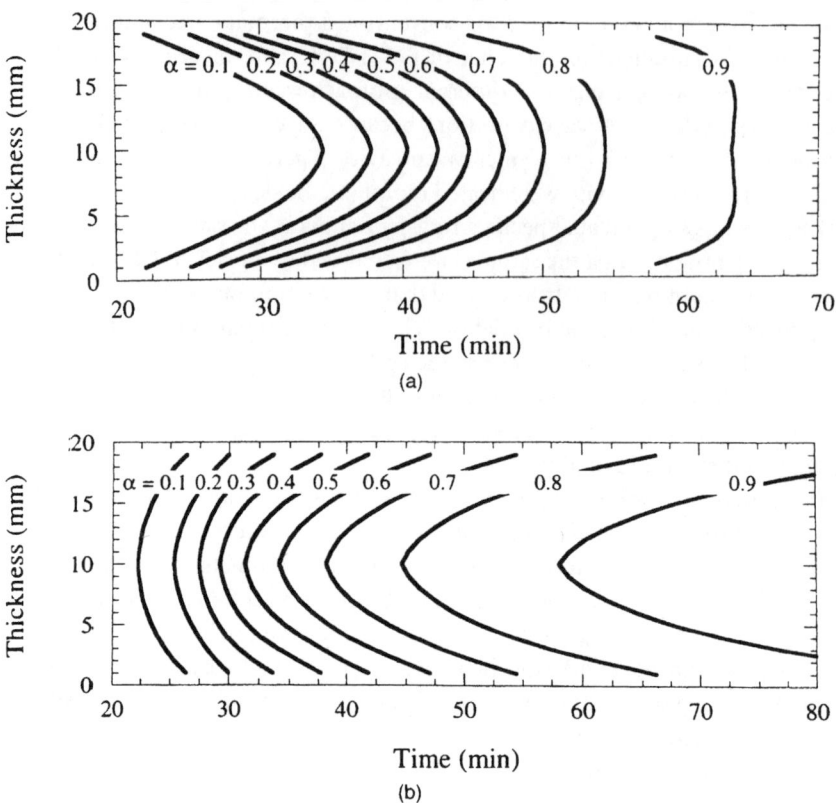

Fig. 31a,b. Formation of cure gradients with two laminates during: **a** conventional cures; **b** microwave cures. Reprinted from (2001) Polym Compos 22:197 [115] with permission

between a microwave source (1 W at 9 GHz) and receiver used to monitor the microwave energy absorption by the composite during the cure cycle. Temperature profiles at various locations across the surface were probed using thermocouples so that the resin cure temperature data could be collected during microwave processing. Using that temperature information, the potential for localized microwave-accelerated cure to reduce the occurrence of material degradation from resin over-temperature was evaluated. Consequently, it was demonstrated that the application of microwave-assisted cure techniques reduced material degradation and residual stress in the composite. In addition, to elucidate the influence of microwave irradiation on the temperature profiles, a theoretical model was presented.

The synthesis of elevated-molecular-weight (E-M) epoxy resins (solid epoxy resins) based on polyaddition of bisphenol-A and lower-molecular-weight (L-M) epoxy resin under microwave irradiation was recently described [117]. Usually, E-M epoxy resins are manufactured by the melt process in which liquid epoxy resins and organic compounds that contain two active hydrogen atoms (e.g., bisphenol

A) are combined in a reactor and heated for a couple of hours. The reaction is carried out at the temperature range of 140–180 °C under a nitrogen 'blanket' to minimize oxidative degradative reactions. In the reaction, known catalysts such as tertiary amines or phosphines, quaternary ammonium or phosphonium salts, and imidazoles are applied [118].

In the present chapter, all the syntheses were performed in a multi-mode microwave reactor (Plazmatronika) with the microwave frequency at 2.45 GHz and maximum power of 300 W. Bisphenol-A (11.649 g) and L-M epoxy resin (25.000 g) were mixed thoroughly with a catalyst, 2-methylimidazole (2-MI), and were irradiated for 16–90 min (Table 16).

The reaction temperature range of 140–180 °C was chosen because below 140 °C the viscosity of the reaction mixture, which increases during the process, was too high to stir the mixture. Above 180 °C, degradation of the resin resulted in strong darkening. Finally, it was found that the optimum conditions for the microwave process were 40 min at 160 °C with the catalyst content of 0.001 mol N in 2-MI/mol OH in bisphenol A, and microwave irradiation power of 90 W. Shortening

Table 16. Synthesis of E-M epoxy resins based on L-M epoxy resin (EV=0.57 mol/100 g), bisphenol-A and in the presence of 2-methylimidazole as a catalyst. Reprinted from [117] with permission

Reaction conditions	Temp. (°C)	Time (min)	Catalyst, content[a]	EV[b] (mol/100 g)	M_n; M_w; M_w/M_n
MW	180	65	0.5	0.109	2380; 4340; 1.85
Δ		80		0.101	2180; 4000; 1.84
MW	160	65	0.5	0.110	2140; 3780; 1.77
Δ		120		0.106	1790; 3130; 1.75
MW	140	150	0.5	0.114	1810; 3260; 1.80
Δ		280		0.114	1380; 2860; 2.08
MW	180	30	1	0.109	2180; 3990; 1.83
Δ		50		0.105	2250; 4250; 1.89
MW	160	40	1	0.113	2150; 3930; 1.83
Δ		80		0.111	2180; 4000; 1.84
MW	140	90	1	0.112	1470; 2580; 1.75
Δ		150		0.114	2020; 3760; 1.84
MW	180	16	5	0.105	2420; 4580; 1.89
Δ		35		0.100	2320; 4420; 1.91
MW	160	20	5	0.104	2470; 3390; 1.83
Δ		35		0.100	2380; 5010; 2.10
MW	140	25	5	0.110	1950; 3780; 1.94
Δ		55		0.113	2170; 3640; 1.68

[a] 10^{-3} mol N/mol OH bisphenol-A;
[b] EV – epoxy value

of the reaction time for all the processes performed in the microwave reactor in comparison to conventional heating was observed.

The characterization of the final product shows that the resins prepared by the microwave and conventional methods have comparable properties. In particular, GPC analysis showed that all the E-M epoxy resins have a similar molecular weight and molecular weight distribution. The main advantage of such a process performed in the microwave reactor was the reduction of the reaction time under microwave conditions (40 min, 160 °C) in comparison to thermal heating (80 min, 160 °C) [117].

In turn, Palumbo and Tempesti found that the morphology of a syntactic foam (hollow microspheres in a polymeric matrix) could be differently affected when cured under thermal and microwave conditions [119]. The prepolymer used in this study was DGEBA (DER 332) hardened with DDM. The glass microspheres (30 wt%) were given no special surface treatment and had wall thickness between 1 and 3 μm, with particle diameters ranging from 50 to 70 μm. The samples cured thermally were placed in an oven at 60 °C for 24 h. The microwave cure was conducted in a microwave circuit (TE_{01} mode) with 2.45 GHz frequency during a three-step cycle: preheating 2 h at 20 W, curing 1 h at 25 W and 1 h at 30 W. Under microwave irradiation, due to faster energy transfer, more efficient crosslinking effects were observed at the interface, leading to higher internodular crosslink density at the particle interface. As a consequence, slight but observable differences of mechanical behavior arise when curing the hollow glass microsphere filled epoxy resin composites in thermal and microwave conditions. The microwave-cured syntactic composites were found less ductile and a little more rigid due to a greater homogeneity [119].

3.2.4
Polyurethanes

In the middle of the 1980s the study of crosslinking of polyurethane resins by means of pulsed microwave irradiation was made by Jullien and Valot [120]. They investigated the behavior of a polyisocyanate-polyol mixture, i.e. 75% ethyl acetate solution of two prepolymers: triisocyanate (Desmodur L75) and polyester-polyalcohol (Desmophen 800) which were taken in stoichiometric amounts. The formation of polyurethane coatings from the same mixture and film hardness as a function of different pulse regimes were examined. In order to avoid the substrate effect, the prepolymer solution (1.4 g) was poured in hollowed PTFE plates that were same in size so the microwave cavity filling factor only varied as a function of the complex dielectric constant in the evolved material. The experimental set-up included a microwave generator (2.45 GHz, 3 kW), provided with an auxiliary pulse generator and a d.c. generator for the power level adjustment. During the polyurethane formation, the microwave average power was 30 W and the peak pulse power was varied from 200 to 2500 W. The pulse period was varied from 2 to 200 ms so that the pulse time was varied from 50 µs to 30 ms. It was found that the energy transfer by pulsed microwaves is more efficient than by an energy equivalent continuous wave value. Finally, it was concluded that thermal kinetics, film hardness and thus macromolecular networks were influenced by microwave pulse irradiation. Microwave cured polyurethane films were much harder than oven cured materials [120].

Later, the effect of microwave irradiation on the crosslinking of polyurethane resin, including diisocyanate and polyethertriol prepolymers, was investigated by Gourdenne et al. [121]. The resin mixtures consisted of two components, i.e., diisocyanate derived from 4,4'-diisocyanate diphenylmethane and a low viscosity polyethertriol, while no catalyst was used. All the samples of polyurethane resin were crosslinked inside a waveguide with TE_{01} propagation mode. The curing of the polyurethane resin samples irradiated at power values from 10 to 60 W led to final networks with mechanical properties of quality at least equivalent to those prepared under conventional thermal conditions. For example, the average elasticity modulus determined from uniaxial compression with samples (25 mm of height and 12.5 mm of diameter) was equal to 3120 MPa for curing under microwave conditions (1 h at 20 W level) and 2810 MPa for a thermal curing (8 h at 60 °C) [121].

The application of microwave irradiation to cure isocyanate/epoxy resins in the presence of N-(2-hydroxyalkyl)trialkylammonium halides was claimed to impart accelerations to both curing and post-curing kinetics with respect to conventional hot-air heating [122]. More recently, Parodi et al. presented further development of new class of catalysts that endow aromatic isocyanate/epoxy and aliphatic or cycloaliphatic epoxy/anhydride systems with a particular efficiency for microwave processability [123]. The catalysts belong to the family of N-(cyano-alkoxyalkyl)trialkylammonium halides, of the general formula:

$NC-R-O-R'-N^{(+)} R1R2R3 X^{(-)}$ and evaluation of the microwave enhancements was performed via isothermal comparative curing experiments under hot-air and microwave heating, respectively. All curing runs were performed on 3 g samples of various liquid resins that were placed in 5 ml PTFE (microwave transparent) beakers. Microwave irradiation was carried out in a cylindrical (diameter 18.5 cm) single-mode tunable microwave cavity. In both the thermal and microwave experiments, the sample temperature was monitored by a fiber-optic probe that was immersed at the center of samples (Table 17).

The data of Table 17 show the strong reaction enhancements of the specific catalysts impart under microwave heating to all of reactive systems examined. The gelation and vitrification times were lowered to one-eighth to one-tenth of those under hot-air heating with the same catalyst and its concentration. An ion-hopping conduction mechanism was recognized as the dominant source of the microwave absorption capacities of these catalysts [2].

Table 17. Isothermal curing times of aromatic isocyanate/epoxy and aliphatic epoxy/anhydride resin systems under conventional and microwave heating. Reprinted from (1999) Microwave heating and the acceleration of polymerization processes. In: Wlochowicz (ed) Polymers and liquid crystals Proceedings of SPIE – The International Society for Optical Engineering, vol 2 [2] with permission

Resin system, curing type (temperature)	Catalyst and concentration (mmol/100 g)	Curing time Gelation	Vitrification
System A: l-MDI/DGEBA 70:30 w/w	I-1[a]		
Conventional (66°C)	1.90	40 min	>80 min
Microwave (66°C)	1.90	?	15–20 min
System B: ERL-4299/(SA/Me-HHPA) 1:1 w/w	I-2[b]		
Conventional (90°C)	1.08	420 min	870 min
Microwave (90°C)	1.08	50 min	110 min
System C: ERL-4234/DDSA	I-3[c]		
Conventional (70°C)	1.09	25.5 h	63 h
Microwave (70°C)	1.09	3 h	6 h
System C: ERL-4234/DDSA	I-4[d]		
Conventional (85°C)	1.14	270 min	450 min
Microwave (85°C)	1.14	30 min	60 min
Microwave (85°C)	TBAI[e] 1.52	210 min	?

[a] N-[3-(2-Cyanoethoxy)propyl]-N,N-dimethyldecylammonium iodide
[b] 4-[3-(2-Cyano-ethoxy)ethyl]-4-butylmorpholinium iodide
[c] N-[2-(2-Cyanoethoxy)propyl]-N,N-dimethyl-decylammonium iodide
[d] 4-[3-(2-Cyanoethoxy)ethyl]-4-butylmorpholinium iodide
[e] Tetrabutylammonium iodide

3.3
Polymer Modification

The modification of oxetane based polymers with 4-(2-aminoethyl)morpholine (Fig. 32) under microwave irradiation was investigated by Hurduc et al. [124]. In a typical experiment, 0.1 g of the polymer together with 0.1–0.4 g of 4-(2-aminoethyl)morpholine in a DMF solution (10 ml) were exposed to microwave irradiation for 1–3 h in a stereo-mode applicator while temperature was monitored by a thermovision camera.

The highest degree of conversion under microwave irradiation reached 27% while in several syntheses under conventional conditions the conversions of oxetane rings were always lower and achieved maximum value of 18%.

The ring opening reaction of naphthyl acetic acid with epoxidized liquid natural rubber (ELNR) (Fig. 33) under microwave irradiation was studied by Huy et al. [125]. The reactions were carried out in a chlorobenzene solution (1.2 ml) in a

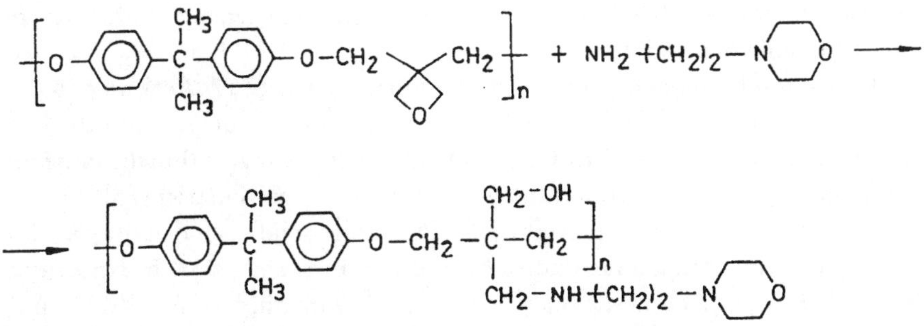

Fig. 32. Modification of oxetane based polymers with 4-(2-amino-ethyl)morpholine. Reprinted from (1996) Polym J 28:550 [124] with permission

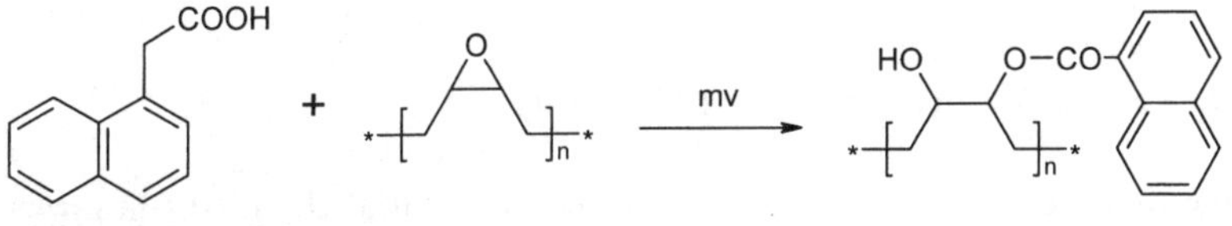

Fig. 33. Fixation of α-naphthyl acetic acid (NAA) onto epoxidized liquid natural rubber. Reprinted from (1996) J Mater Sci Pure Appl Chem A33:1957 [125] with permission

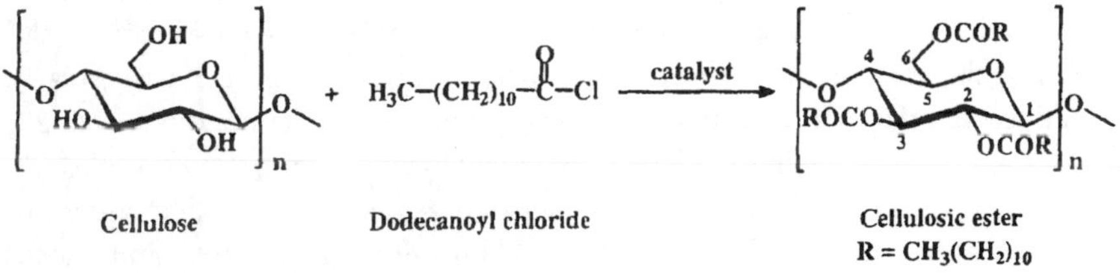

Fig. 34. Esterification of cellulose with dodecanoyl chloride under microwave conditions. Reprinted from (1999) C R Acad Sci Paris 2:75 [126] with permission

monomode Maxidigest Mx-3500 microwave reactor, in which the mixtures of naphthylacetic acid (2.5–10.2 mmol), ELNR (2.5–10.2 mmol), and TEBA (0.083–0.34 mmol) were irradiated for 5–15 min.

The conversion of epoxy groups under microwave irradiation at 110 °C was 44% (10 min), whereas under similar conventional conditions the conversion reached 33% after 24 h of heating.

Applying dodecanoyl chloride under microwave conditions, Krausz et al. demonstrated that the esterification of cellulose can be achieved within 6–9 min (Fig. 34) [126]. In a general procedure, a mixture of cellulose, acid chloride and catalyst (4-dimethylaminopyridine) was irradiated in an open vessel placed in a domestic microwave oven. Prior to the reaction, the homogeneity of the mixture was ensured by sonification.

The best results under microwave irradiation (i.e., yield of 17–30% and 2.1–2.5 degree of substitution) were obtained when the reaction mixture was adsorbed on inorganic supports such as K_2CO_3 or Al_2O_3. It was shown that within the same reaction time, under similar conventional conditions, no reaction occurred.

The surface oxidation of polyethylene (PE) under solid-state conditions with potassium permanganate was studied by Mallakpour et al. [127]. The oxidations were performed in a domestic microwave oven. The mixtures of 2.0 g (0.071 ml) of PE powder ground with 3.76 g (0.024 mol) of $KMnO_4$ were irradiated for 90 s. After the reaction, FT-IR analysis revealed the presence of hydroxyl as well as vinylic functional groups on the polymer surface whereas hydroperoxy groups were not detected.

3.4
Reactions on Polymer Matrices (Immobilized Reagents)

One of the most developed methods used in combinatorial chemistry libraries preparation is solid-phase organic synthesis (SPOS) based on the Merrifield method for peptide synthesis [128]. A great number of such libraries have been prepared on a solid support, generally a functionalized polystyrene resin cross-linked with a small amount of divinylbenzene. Recently, it was demonstrated that microwave irradiation can be applied to solid-phase immobilized reagents to reduce significantly the reaction time. Those readers who are interested in such processes we would like to refer to more extensive reviews published by Chamberlin et al. [129] and Kappe [130], while in this chapter we are giving most common examples.

Applying the original Merrifield method, Yu et al. demonstrated that under microwave irradiation a number of symmetrical anhydride or activated amino acids could be coupled onto functionalized resin in a DMF solution (6 min, 55 °C) [131]. It was demonstrated that no racemization occurred during the reaction, while all the reaction went to completion. The authors managed to synthesize heptapeptides and decapeptides with 99–100% completion, applying microwave irradiation for 4 min in each coupling. Under conventional conditions, the coupling yields were about 80% for every step (Fig. 35).

Suzuki coupling reaction of polymer-bound aryl halides with arylboronic acids and a palladium catalyst was described by Larhed et al. [132]. The reagents in a mixture of ethylene glycol dimethyl ether (DME) and water were irradiated (ca. 4 min) in a heavy-walled Pyrex tube (Fig. 36) placed in a single-mode microwave reactor.

Ugi four-component coupling reaction, in which an amine, aldehyde or ketone, carboxylic acid and isocyanide react to yield an α-acylamino amide, was performed by Nielsen et al. on a solid polymer support under microwave irradiation

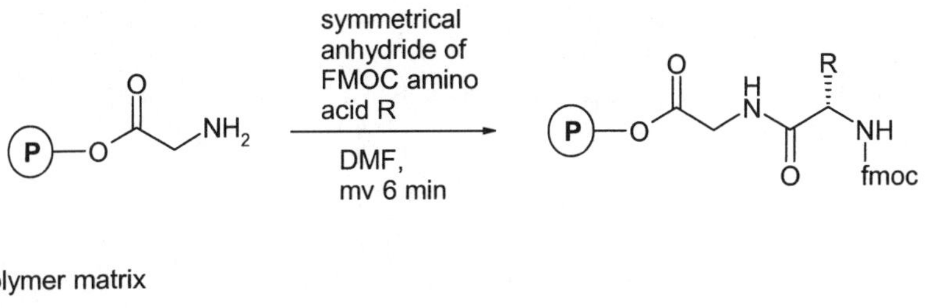

Fig. 35. Peptide synthesis on Merrifield resin under microwave conditions. Reprinted from (2002) J Comb Chem 4:95 [129] with permission

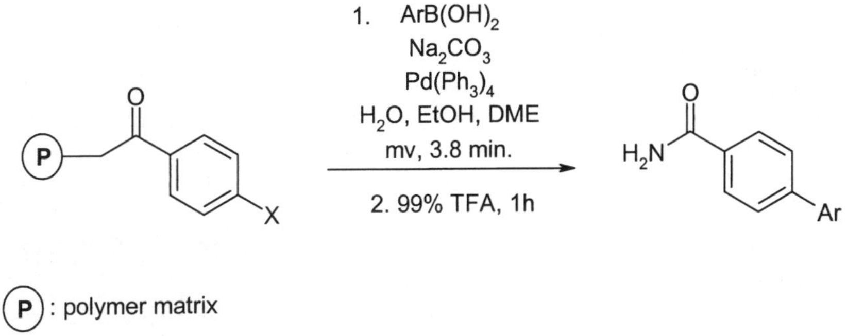

Fig. 36. Suzuki coupling reaction of polymer-bound aryl halides with arylboronic acids and a palladium catalyst. Reprinted from (2002) J Comb Chem 4:95 [129] with permission

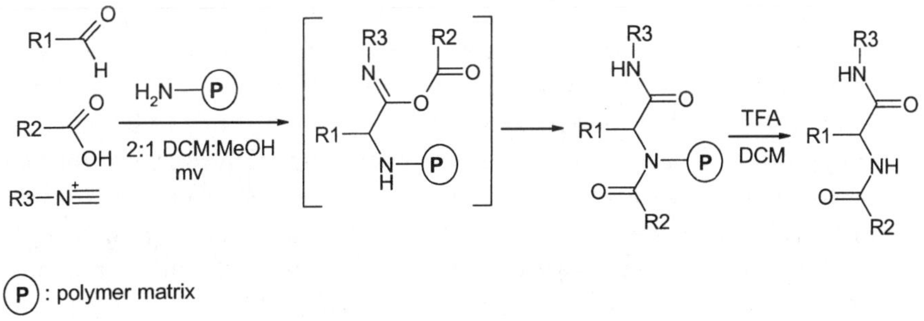

Fig. 37. General reaction for microwave-assisted Ugi Four Component Condensation. Reprinted from (1999) Tetrahedron Lett 40:3941 [133] with permission

[133]. The mixture of aldehydes, carboxylic acids, and isocyanides in a methylene dichloride (DCM) and methanol solution together with amine-bound resins were irradiated (5 min) in a single-mode reactor (Microwell 10) in 10-ml Teflon screw-capped vials to give 18 α-acylamino amides in 24–96% yield (Fig. 37). Usually, un-

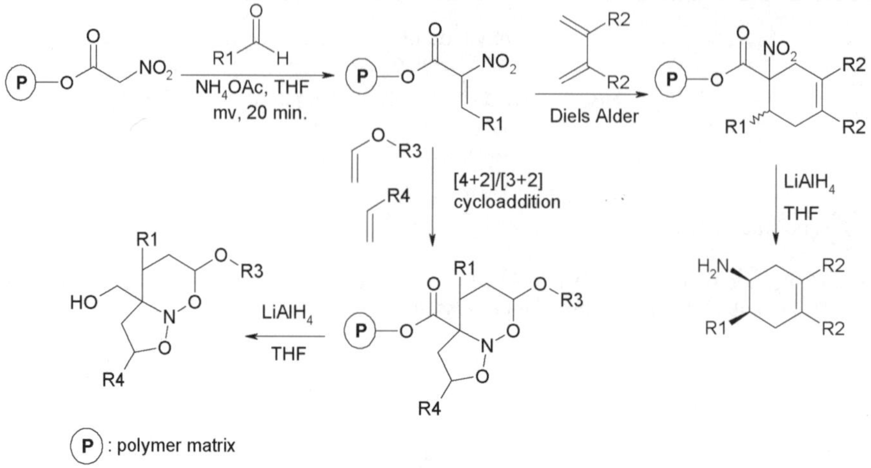

Fig. 38. Knoevenagel condensation between aldehydes and the Wang resin bound nitroacetic acid. Reprinted from (2000) Tetrahedron Lett 41:515 [134] with permission

der conventional thermal conditions reaction times for solid-phase Ugi reactions extend from 24 h to several days [133].

The preparation of resin-bound nitroalkenes via microwave-assisted Knoevenagel condensation of resin-bound nitroacetic acid with alkyl and aryl substituted aldehydes was described by Kustner and Scheeren [134]. The condensation reactions under microwave irradiation were achieved in 20 min at 350 W, whereas under conventional conditions they needed 17 h at room temperature. The potential of the resin-bound nitroalkenes for application in combinatorial chemistry was demonstrated by Diels-Alder reaction with 2,3-dimethylbutadiene and tandem reaction with ethyl vinyl ether and styrene (Fig. 38).

Claisen rearrangement of Merrifield resin derivatized with O-allyl phenolic ethers into ortho-allylic phenols under microwave irradiation was performed by Kumar et al. [135]. The reactions were carried out in a household microwave oven, in which samples were irradiated in an open Erlenmeyer flask for 4–6 min (600 W) to afford products in 84–92% yield. In comparison, under thermal conditions, similar yields were obtained after 10–16 h at 140 °C (Fig. 39).

Esterification reaction of polymer-supported alcohol with benzoic acid under microwave irradiation was investigated by Stadler and Kappe [136]. Activation of the carboxylic acid was carried out using diisopropylcarboimide either via the O-acylisourea or symmetrical protocols (Fig. 40). The reactions were carried out in a Milestone MLS ETHOS 1600 microwave reactor applying either open or sealed vessel systems. Significant rate enhancement was observed while the coupling of benzoic anhydride to resin was performed in 1-methyl-2-pyrrolidone as a solvent. Reaction times were reduced from 2–3 days using conventional conditions (at

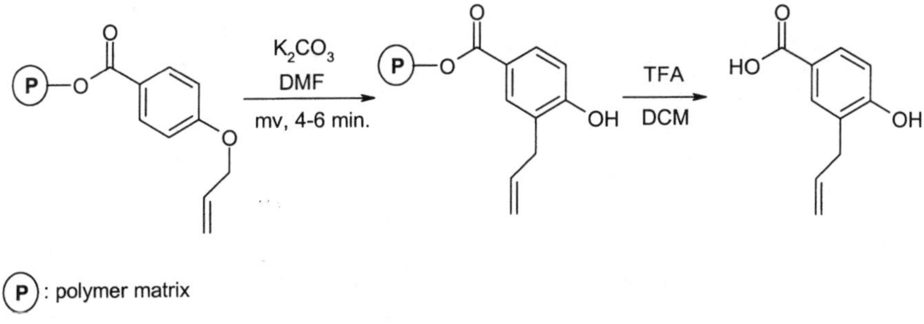

Fig. 39. Claisen rearrangement of Merrifield resin derivatized with O-allyl phenolic ethers into ortho-allylic phenols under microwave conditions. Reprinted from (2000) Synlett 1129 [135] with permission

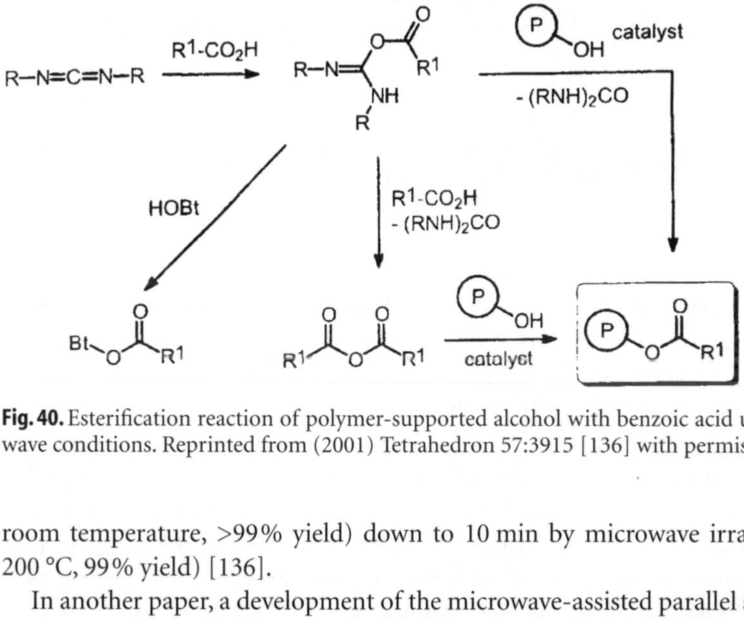

Fig. 40. Esterification reaction of polymer-supported alcohol with benzoic acid under microwave conditions. Reprinted from (2001) Tetrahedron 57:3915 [136] with permission

room temperature, >99% yield) down to 10 min by microwave irradiation (at 200 °C, 99% yield) [136].

In another paper, a development of the microwave-assisted parallel solid-phase synthesis of a collection of 21 polymer-bound enones was described. The two-step protocol involves initial high-speed acetoacetylation of polystyrene resins with a selection of seven common β-ketoesters. When microwave irradiation at 170 °C was employed, complete conversions were achieved within 1–10 min, a significant improvement over the conventional thermal method, which takes several hours for completion. Significant rate enhancements were also observed for the subsequent microwave-heated Knoevenagel condensations. Reaction times were reduced to 30–60 min at 125 °C in the microwave protocol compared to 1–2 days using conventional thermal conditions. Kinetic comparative studies indicate that the observed rate enhancements can be attributed to the rapid direct heating of the

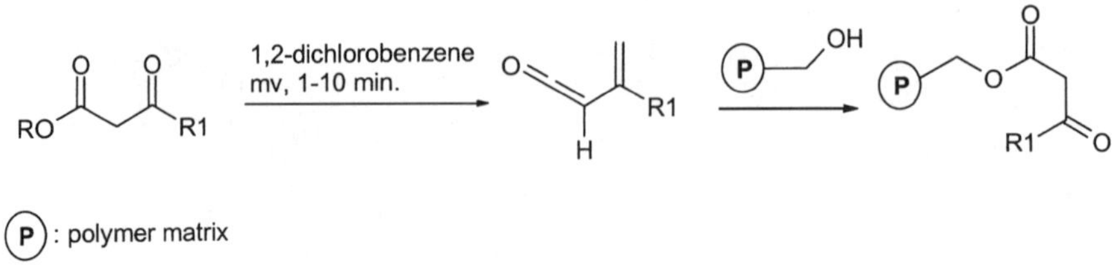

Fig. 41. Acetoacetylation of polystyrene resins with β-ketoesters under microwave conditions. Reprinted from (2002) J Comb Chem 4:154 [137] with permission

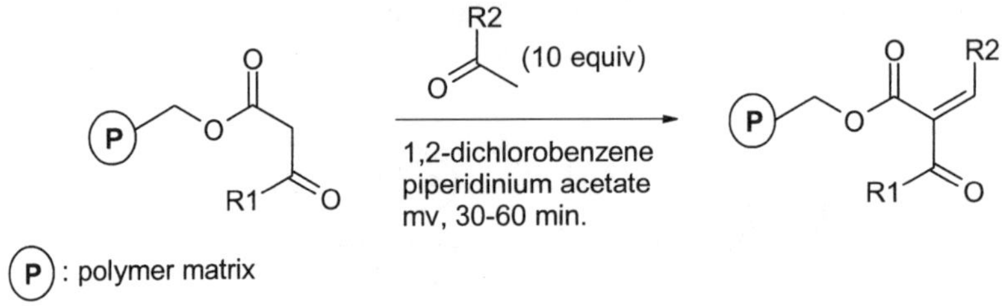

Fig. 42. Knoevenagel condensation with piperidinium acetate under microwave conditions. Reprinted from (2002) J Comb Chem 4:154 [137] with permission

solvent (1,2-dichlorobenzene) by microwaves rather than to any specific microwave effect. All the reactions were carried out in commercially available parallel reactors with on-line temperature measurement, designed specifically for use in multimode microwave cavities (Figs. 41 and 42) [137].

Poly(styrene-*co*-allyl alcohol) can be esterified under microwave solventless conditions to give the polymers that are effective precursors for the preparation of various nitrogen and oxygen heterocycles. The reaction were carried out in an open vessels that were placed in a domestic microwave oven and irradiated for 10–20 min. In some cases, it was possible to scale-up the microwave procedure from 1 to 10 g scale. Under conventional thermal conditions, most of these experiments were run for 4 h in refluxing toluene solutions (Fig. 43) [138].

3.5
Rubber Vulcanization

Vulcanization of rubber in the tire industry is the first industrial application of microwave irradiation for the processing of polymeric materials [139]. Improved product uniformity, reduced extrusion-line length, reduced scrap, improved cleanliness, enhanced process control and automation accomplished through the use of an appropriate combination of microwave, hot air, and/or infrared heating technologies are the main advantages of microwave carbon black-filled rubber

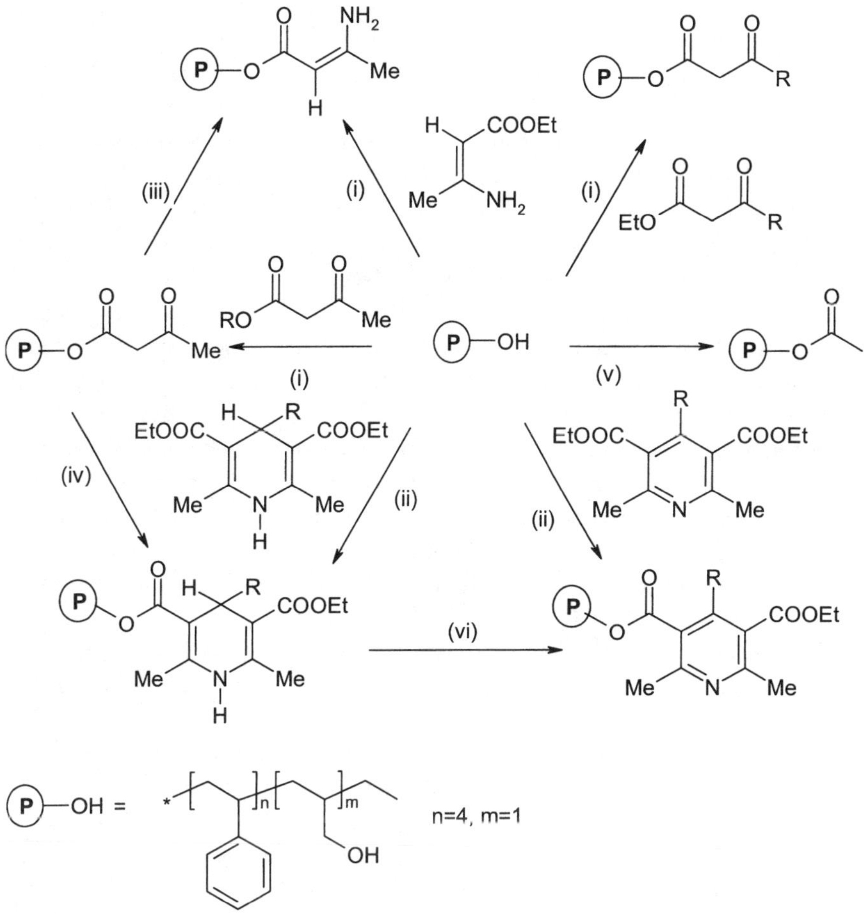

Fig. 43. Derivatization of poly(styrene-*co*-allyl alcohol) under microwave conditions. Reprinted from (1999) Tetrahedron 55:2687 [138] with permission

manufacturing [19]. As the result, microwave vulcanization extrusion lines are applied world-wide in the automotive and construction industries [140].

Unlike the black rubber materials, in the absence of conductive black fillers, the white and colored rubber compounds cannot be heated suitably and vulcanized under microwave irradiation. For this reason, only a small fraction of white and colored rubber compounds are processed by means of microwaves. The novel, specialty microwave heating sensitizers for microwave vulcanization of white and colored rubber compounds, their use, costs and benefits were recently discussed by Parodi [141].

3.6
Recycling

Solvolysis by glycols and alcohols is an established method for the chemical recycling of PET. Krzan investigated the use of microwave irradiation as the energy source in PET solvolysis reactions and the conditions that govern its effectiveness [142]. PET was obtained from washed used beverage bottles that were chopped into flakes varied from 1 to 15 mm. Solvolysis reagents used were reagent grade methanol, propylene glycol and polyethylene glycol. Zinc acetate was used as a catalyst. Degradation experiments were performed in a Milestone mLS 1200 microwave oven. A typical sample containing 0.75 g PET flakes, 0.1 g zinc acetate and 10.0 g of the solvolysis reagent was mixed in the PTFE vessel. Finally, the characterization of resulting solution was performed by gel permeation chromatography. It was found that complete PET degradation-solubilization was achieved in shortest time (4 min) in methanol, while more than 8 min are required in polyethylene glycol. The presented paper shows that the use of microwave radiation as the energy source in PET solvolysis resulted in the short reaction times (a few minutes) needed for complete PET degradation, compared with conventional heating methods (a few hours) [142].

In another paper, the glycolysis reactions of PET in diethylene glycol (DEG) and propylene glycol (PG) were monitored in terms of temperature and pressure [143]. A number of basic inorganic compounds ($NaHCO_3$, K_2CO_3, CaO, KH_2PO_4, $NaOCH_3$) were examined as potential catalysts. The reactions were carried out in a Milestone mLS Mega 1200 microwave oven equipped with temperature and pressure sensors inserted directly into the reaction vessel. For both DEG and PG, temperature and pressure in the reaction vessel were measured during the exposure of reaction mixtures to microwave irradiation (500 W) for 5 min. Maximum temperatures of about 240 and 200 °C and maximum pressures of about 1.0 and 1.6×10^5 Pa were detected for DEG and PG, respectively. The results appeared to be in agreement with average molecular weight of 556 g/mol reported for polyols obtained by depolymerization of PET with DEG under conventional heating. The method was compatible with the use of a wide range of glycol reagents as well as a variety of basic catalysts. The application of microwave irradiation was not limited to using only zinc acetate or other acetates as catalysts. With many other compounds, comparable effectiveness can be achieved. The main advantage of microwave application was a short reaction time for complete alcoholysis and glycolysis, in which PET degradation was achieved [143].

In a similar paper, the depolymerization of polyamide-6 (PA-6) was performed using microwave irradiation with phosphoric acid as a catalyst [144]. Besides its catalytic activity, phosphoric acid has a very high dipole moment, which makes it an excellent microwave absorbent. Reaction mixtures consisted of 10 g PA-6, 10 g water and 1–10 g concentrated H_3PO_4. Commercial PA-6 was ground to 0.5–

1 mm particle size and used without further preparation. Microwave irradiation of 200 W was applied for 23 min to the reaction mixture in a sealed reaction vessel in a Milestone Mega 1200 laboratory oven. After 15 min of irradiation, PA-6 was completely solubilized, and the solubilization efficiency increased linearly up to 6 g of added acid. However, the acidolysis was most effective with 7 g of the acid. The resulting product mixture consisted of more than 90% ε-aminocaproic acid (ACA) and its linear oligomers together with the minor part consisting of cyclic products. With longer irradiation time or more acid, there was a shift toward higher concentration of ACA and its dimer, as well as a decrease in the concentration of higher oligomers.

3.7
Other Applications

The application of inherently conducting polymers (ICPs) to welding thermoplastics was reported by Kathirgamanathan [145]; however, the use of ICPs as microwave absorbers for welding was suggested earlier by Epstein and MacDiarmid [146]. In a typical experiment, a small amount of ICP either as powder (20 mg) or tape (two strips with dimensions 20 mm × 2 mm × 40 µm) was placed onto the upper surface of the plastic (10 cm × 2 cm × 0.1 cm). The second piece was placed on the top and clamped. The clamp assembly was then placed in a household microwave oven with a total power of 500 W for 2–120 s. The results are summarized in Table 18. The method can be applied to any thermoplastic.

Microwave irradiation was examined as an alternative to conventional heating for joining of composite structures [147]. Through proper material selection, microwaves are able to penetrate the substrate materials and cure the adhesives in-situ. Selective heating with microwave is achieved by incorporating interlayer materials that have high dielectric loss properties relative to the substrate materials. A processing window for elevated temperature curing of an epoxy paste adhesive system was developed and composite joint systems were manufactured using conventional and microwave techniques and tested in shear. Joint systems were allowed to cure: C1 – at room temperature for ten days, C2 – at room temperature for five months, and C3 – at 90 °C for 1 h. Another joint system – M1 – was processed using microwaves (2.45 GHz, 6 kW) and was subjected to rapid heating 20 °C/min to a final temperature of 90 °C. Microwave curing resulted in enhanced shear strength samples M1 (10% higher than C1, 70% higher than C3) and less scattered experimental data [147].

Intrinsically conductive polyaniline (PANI) composite gaskets were used to microwave (2.45 GHz) weld high density polyethylene (HDPE) bars [148]. Two composite gaskets were made from a mixture of HDPE and PANI (50 and 60 wt%) powders in different proportions. The mixtures were compression molded in a hot (180 °C) press. Adiabatic heating experiments were used to estimate the internal

Table 18. Microwave welding of thermoplastics using inherently conducting polymers. Reprinted from (1993) Polymer 34:3105 [148] with permission

Substrate to welded	Conducting polymer[a]	Conductivity (S cm^{-1})	Time to weld (s)
Polycarbonate	PPPTS(p)	22.0±2.0	100±10
Polycarbonate	PAPTS(p)	9.0±1.0	120±5
Polycarbonate	A(t)	1.0±0.1	5±1
Polycarbonate	B(t)	12.0±3.0	2±1
Polycarbonate	C(t)	(5.7±0.3)×10^{-2}	40±2
Polypropylene	A(t)	1.0±0.1	5±1
Polypropylene	B(t)	12.0±3.0	2±1
Polypropylene	C(t)	(5.7±0.3)×10^{-2}	60±5
Polyethylene	PPPTS(p)	22.0±2.0	5±2
Polyethylene	PAPTS(p)	9.0±1.0	15±2
Polyethylene	A(t)	1.0±0.1	30±5
Polyethylene	B(t)	12.0±3.0	2±1

[a] PPPTS=polypyrrole *p*-toluenesulfonate, PAPTS=polyaniline *p*-toluenesulfonate, p=powder (thickness of powder on the weld 100 μm), t=tape, A(t)=non-woven polyester tape impregnated with PPPTS (thickness 110 μm), B(t)=microporous polyethylene impregnated with PPPTS (thickness 40 μm), C(t)=non-woven polyester impregnated with carbon black (110 μm)

heat generation. It was observed that the temperature rise rate (heat generation rate) decreased with increasing temperature. The decrease in heat generation rate was probably due to a reduction of the electrical conductivity of PANI at elevated temperatures. During microwave welding, the increase in the heating time resulted in the development of a thicker molten layer in the parts, which improved the joint strength. The maximum tensile joint strength was achieved using a 60 wt% PANI gasket with a heating time of 60 s and a welding pressure of 0.9 MPa. This resulted in a tensile weld strength of about 25 MPa, which equals the tensile strength of the bulk HDPE [148].

Ku et al. applied variable frequency microwave (VFM) facilities (2–18 GHz) for joining parts made of thermoplastic composites and found that a selective heating over a large volume at a high energy coupling efficiency could be obtained [149]. Five different thermoplastic polymer matrix composites were processed, including 33 wt% random carbon fiber (CF) or glass fiber (GF) reinforced polystyrene (PS-CF, PS-GF), low density polyethylene (LDPE-CF, LDPE-GF) and polyamide 66-GF. Bond strengths of lap joints of the composites were tested in shear, and the results were compared with those obtained using fixed frequency (2.45 GHz). The coupling agent used was a two part adhesive, i.e., 100% liquid epoxy resins and 8% amine hardener. Considering the quality of bond brought about by processing using different microwaves facilities, hot spots were found on the joint of LDPE-CF bound using a fixed frequency of 2.5 GHz. The joint of PS-CF processed by variable frequency was perfect. It could therefore be argued that VFM produces stronger bonds for the two materials, PS and LDPE, with excellent quality of joint properties [150].

4
Conclusions

In conclusion, polymer synthesis as well as processing can greatly benefit from the unique feature offered by modern microwave technology, which was demonstrated in many successful laboratory-scale applications. These can include such technical issues as shorter processing time, increased process yield, and temperature uniformity during polymerization and crosslinking. Although microwave energy is more expensive than electrical energy due to the low conversion efficiency from electrical energy (50% for 2.45 GHz and 85% for 915 MHz), efficiency of microwave heating is often much higher than conventional heating and more than compensates for the higher energy cost.

From the present paper it can be seen that a lot of factors have influence on polymer synthesis and processing under microwave conditions. Not all materials are suitable for microwave applications, and a special characteristic of every process has to be matched. Therefore, a real cross-disciplinary approach has to be considered to understand fully all the limitations and advantages of microwave processing. Improper application of microwave irradiation will usually lead to disappointments, while proper understanding and use of microwave power can bring greater benefits than would be expected. For example, three typical temperature-time stages can be observed during polymerization reactions. First, initial temperature rise by direct heating of monomer(s) where the temperature rises slowly. Second, a significant temperature peak with maximum temperature due to the exothermic reaction. Third, free convective cooling to an ambient temperature indicating the end of the exothermic reaction processes. Fast exothermic reaction heating usually accelerates the temperature rise and gradient inside the samples. Neither continuous microwave nor thermal processing can be effectively controlled in order to maintain constant temperature/time profile through the entire process. However, pulsed microwave heating can be used to control temperature and eliminate the exothermic temperature peak, maintaining the same temperature at the end of reaction [100]. Furthermore, fundamental difference in the heat transfer during material processing in thermal and microwave fields is that microwave energy, in contrast to thermal heating, is supplied directly to a large volume, thus avoiding the thermal lags associated with conduction and/or convection. Consequently, temperature gradients and the excessive heat build-up during thermal processing could be reduced by a microwave power control. Thus, a comparison of thermal and microwave processing assumes a new dimension when the temperature distribution inside the sample is considered, and that is where the scientific challenge lies [109].

The action of microwave irradiation on chemical reactions is still under debate, and some research groups have mentioned the existence of so-called non-thermal microwave effects, i.e. inadequate to the observed reaction temperatures' sudden acceleration of reaction rates [18]. Regardless of the type of activation (thermal) or kind of microwave effects (non-thermal), microwave energy has its own advantages which are still waiting to be understood fully and applied to chemical processes.

Acknowledgment The work was supported by The Polish State Committee for Scientific Research (KBN) through the grant 7T09B06521.

References

1. Wei JB, Shidaker T, Hawley MC (1996) Trends Polym Sci (TRIP) 4:18
2. Parodi F (1999) Microwave heating and the acceleration of polymerization processes. In: Wlochowicz A (ed) Polymers and liquid crystals. Proceedings of SPIE – The International Society for Optical Engineering, vol 4017, p 2
3. Thostenson ET, Chou T-W (1999) Composites A 30:1055
4. Mijovic J, Wijaya J (1990) Polym Compos 11:191
5. Jacob J, Chia LHL, Boey FYC (1995) J Mater Sci 30:5321
6. Metaxas AC, Meredith RJ (1983) Industrial microwave heating. Peter Perigrinus, London
7. Mingos DMP, Baghurst DR (1997) Applications of microwave dielectric heating effects to synthetic problems in chemistry. In: Kingston HM, Haswell SJ (eds) Microwave-enhanced chemistry. fundamentals, sample preparation, and applications (1997). American Chemical Society
8. Mingos DMP, Baghurst DR (1991) Chem Soc Rev 20:1
9. Gabriel C, Gabriel S, Grant EH, Halstread BS, Mingos DMP (1998) Chem Soc Rev 27:213
10. Parodi F (1996) Physics and chemistry of microwave processing. In: Agarwal SL, Russo S (eds) Comprehensive polymer science, 2nd supplement volume. Pergamon-Elsevier, Oxford
11. Abramovitch RA (1991) Org Prep Proc Int 23:683
12. Strauss CR, Trainor RW (1995) Aust J Chem 48:1665
13. Caddick S (1995) Tetrahedron 51:10,403
14. Loupy A, Petit A, Hamelin J, Texier-Boullet F, Jacquault P, Mathe D (1998) Synthesis 1213
15. Varma RS (1999) Green Chem 1:43
16. Bogdal D (1999) Wiad Chem 53:66
17. Lidstrom P, Tierney J, Wathey B, Westamn J (2001) Tetrahedron 57:9225
18. Perreux L, Loupy A (2001) Tetrahedron 57:2001
19. Commission on Engineering and Technical Systems (1994) Microwave processing of materials. Commission on Engineering and Technical Systems. National Academy Press, Washington DC
20. CEM Corporation homepage: http://www.cemsynthesis.com
21. Lambda Technologies, Inc. Homepage: http://www.microcure.com
22. Milestone, Inc. Homepage: http://www.milestonesci.com
23. PersonalChemistry AB. Homepage: http://www.personalchemistry.com
24. Plazmatronika SA. Homepage: http://www.plazmatronika.pl
25. Walter PJ, Chale S, Kingston HM (1997) Overview in microwave assisted sample preparation. In: Kingston HM, Haswell SJ (eds) Microwave-enhanced chemistry. Fundamentals, sample preparation, and applications. American Chemical Society

26. Bose AK, Manhas MS, Ganguly SN, Sharma AH, Banik BK, Bimal K (2002) Synthesis 11
27. Morcuende A, Valverde S, Harradon B (1994) Synlett 89
28. Gedye RN, Smith FE, Westaway KC (1988) Can J Chem 66:17
29. Giguere RJ, Bray TL, Duncan SM, Majetich G (1986) Tetrahedron Lett 27:4945
30. Majetich G, Hicks R (1995) Radiat Phys Chem 45:567
31. Bram G, Loupy A, Villemin D (1992) In: Smith K (ed) Solid supports and catalysts in organic chemistry. Ellis Harwood, London
32. Laszlo P (ed) (1987) Preparative chemistry using supported reagents. Academic Press, New York
33. Teffal M, Gourdenne A, (1983) Eur Polym J 19:543
34. Palacios J, Sierra M, Rodriguez P (1992) New Polym Mater 3:273
35. Chia LHL, Jacob J, Boey FYC (1995) J Mat Proc Techn 48:445
36. Jacob J, Chia LHL, Boey FYC (1997) J Appl Polym Sci 63:787
37. Albert P, Holderle M, Mulhaupt R, Janda R (1996) Acta Polym 47:74
38. Chia LHL, Jacob J, Boey FYC (1996) J Polym Sci A Polym Chem 34:2087
39. Palacios J, Velverde C (1996) New Polym Mat 5:93
40. Correa R, Gonzalez G, Dougar V (1998) Polymer 39:1471
41. Zhang G, Niu A, Peng S, Jiang M, Tu Y, Li M, Wu CH (2001) Accounts Chem Res 34:249
42. Zhang W, Gao J, Wu Ch (1997) Macromolecules 30:6388
43. Chen Y, Wang J, Zhang D, Cai R, Yu H, Su Ch, Huang Z-E (2000) Polymer 41:7877
44. Lu J, Zhu X, Ji S, Zhu J, Chen Z (1998) J Appl Polym Sci 68:1563
45. Lu J, Zhu X, Zhu J, Yu J (1997) J Appl Polym Sci 66:129
46. Lu J, Jiang Q, Zhu J, Wang F (2001) J Appl Polym Sci 79:312
47. Sundararajan G, Dhanalakshmi K (1997) Polym Bull 39:333
48. O'Brien WJ (1989) Dental materials properties and application. Quintessence Publishing, Chicago
49. Nishii M, (1968) J Osaka Dent Univ 2:23
50. Kimura H, Teraoka F (1984) Quintessence Dent Tech 9:547
51. Kimura H (1984) Osaka Dent Univ 24:21
52. Reitz PV, Sanders JL, Levin B (1985) Phys Prop Quint Int 8:547
53. Levin B, Sander JL, Reitz PV (1989) J. Prosthet Dent 61:381
54. Al Doori A, Hugget R, Bates JF, Brooks SC (1988) Dent Mater 4:25
55. Muhtarogullari IY, Dogan A, Muhtarogullari M, Usanmaz A (1999) J Appl Polym Sci 74:2971
56. Blagojevic A, Murphy VM (1999) J Oral Rehabil 26:804
57. Teraoka F, Takahashi J (2000) J Prosthet Dent 83:514
58. Urabe H, Nomura Y, Shirai K, Yoshioka M, Shintani H (1999) J Oral Rehabil 26:442
59. Ruyter IE, Oysaed H (1987) J Biomed Mater Res 21:11
60. Truong VT, Thomas FGV (1988) Aust Dent J 33:201
61. Thomas CJ, Webb BC (1995) Eur J Prosthodont Restor Dent 3:179
62. Polyzois GL, Handel RW, Stafford GD (1995) Eur J Prosthodont Restor Dent 3:183
63. Yunus N, Harrison A, Huggett R (1994) J Oral Rehabil 21:641
64. May KB, Shotwell JR, Koran A III, Wang R-F (1996) J Prosthet Dent 76:581
65. Bayasan A, Parker S, Wright PS (1998) J Prosthet Dent 79:182
66. Lewis DA, Summers JD, Ward TC, McGrath JE (1992) J Polym Sci A Polym Chem 30:1647
67. Imai Y (1996) A new facile and rapid synthesis of polyamides and polyimides by microwave assisted polycondensation. In: Hendrick JL, Labadie JW (eds) Step-growth polymers for high performance materials: new synthetic methods. ACS, p 421
68. Imai Y, Nemoto H, Watanabe S, Kakimoto M (1996) Polym J 28:256
69. Imai Y, Nemoto H, Kakimoto M (1996) J Polym Sci A Polym Chem 34:701
70. Imai Y (1996) React Funct Polym 30:3
71. Imai Y (1999) Adv Polym Sci 140:1
72. Park KH, Watanabe S, Kakimoto M, Imai Y (1993) Polym J 25:209
73. Chen J, Chen Q, Yu X (1996) J Appl Polym Sci 62:2135

74. Pourjavadi A, Zamanalu MR, Zohurian-Mehr MJ (1999) Angew Makromol Chem 269:54
75. Mallakpour SE, Hajipour AR, Faghihi K (2001) Eur Polym J 37:119
76. Mallakpour SE, Hajipour AR, Zamanlou MR (2001) J Polym Sci A Polym Chem 39:177
77. Mallakpour SE, Hajipour AR, Faghihi K (2000) Polym Int 49:1388
78. Mallakpour SE, Hajipour AR, Khoee S (2000) J Appl Polym Sci 77:3003
79. Mallakpour SE, Hajipour AR, Khoee S (2000) J Polym Sci A Polym Chem 38:1154
80. Liu Y, Sun XD, Xie XQ, Scola DA (1998) J Polym Sci A Polym Chem 36:2653
81. Liu Y, Xiao Y, Sun X, Scola DA (1999) J Appl Polym Sci 73:2391
82. Fang X, Hutcheon R, Scola DA (2000) J Polym Sci A Polym Chem 38:2526
83. Fang X, Scola DA (1999) J Polym Sci A Polym Chem 37:4616
84. Fang X, Hutcheon R, Scola DA (2000) J Polym Sci A Polym Chem 38:1379
85. Hurduc N, Abdelylah D, Buisine JM, Decock P, Surpateanu G (1997) Eur Polym J 33:187
86. Chatti S, Bortolussi M, Loupy A, Blais JC, Bogdal D, Majdoub M (2002) Eur Polym J 38:1851
87. Chatti S, Bortolussi M, Loupy A, Blais JC, Bogdal D, Roger P (2003) J Appl Polym Sci (accepted)
88. Fleche G, Huchette M (1986) Starch 38:26
89. Mallon FK, Ray WH (1998) J Appl Polym Sci 69:1203
90. Keki S, Bodnar I, Borda J, Deak G, Zsuga M (2001) Macromol Rapid Commun 22:1063
91. Hottong U, Wei J, Dhulipala R, Hawley MC (1991) Proc 93rd Annu Meeting AmerCeramic Soc, p 587
92. Johnson MS, Rudd CD, Hill DJ (1998) Composites A 29A:71
93. Hill DJ, Rudd CD, Johnson MS (1998) J Microwave Power Electromagn Energy 33:216
94. Williams NH (1967) J Microwave Power 2:123
95. Beldjoudi N, Gourdenne A (1988) Eur Polym J 24:53
96. Baziard Y, Gourdenne A (1988) Eur Polym J 24:873
97. Lewis DA, Hedrick JC, McGrath JE, Ward TC (1987) Polym Prepr 28:330
98. Beldjoudi N, Gourdenne A (1988) Eur Polym J 24:265
99. Thuillier FM, Jullien H (1989) Makromol Chem Macromol Symp 25:63
100. Jow J, DeLong JD, Hawley MC (1989) SAMPE Quart 20:46
101. Levita G, Livi A, Rolla PA, Culicchi C (1996) J Polym Sci B Polym Phys 34:2731
102. Casalini R, Corezzi S, Livi A, Levita G, Rolla PA (1997) J Appl Polym Sci 65:17
103. Marand E, Baker HR, Graybeal JD (1992) Macromolecules 25:2243
104. Delmotte M, Jullien H, Ollivon M (1991) Eur Polym J 27:371
105. Forysth A, Whittaker AK (1999) J Appl Polym Sci 74:2917
106. Wei J, Hawley MC, DeLong JD (1993) Polym Eng Sci 33:1132
107. Jordan C, Galy J, Pascault J-P (1995) Polym Eng Sci 35:233
108. Hedreul C, Galy J, Dupuy J, Delmotte M, More C (1998) J Appl Polym Sci 68:543
109. Mijovic J, Corso WV, Nicolais L, d'Ambrosio G (1998) Polym Adv Technol 9:231
110. Mijovic J, Fishbain A, Wijaya J (1992) Macromolecules 25:986
111. Boey FYC, Yap BH, Chia L (1999) Polym Test 18:93
112. Boey FYC, Rath SK (2000) Adv Polym Techn 19:194
113. Thostenson ET, Chou T-W (1997) Proceedings of the 12th Annual Meeting of the American Society of Composites, Dearborn, p 931
114. Thostenson ET, Chou T-W (1998) Proceeedings of the 13th Annual Meeting of the American Society of Composites, Baltimore
115. Thostenson ET, Chou T-W (2001) Polym Compos 22:197
116. Shull PJ, Hurley DH, Spicer JWM, Spicer JB (2000) Polym Eng Sci 40:1857
117. Bogdal D, Pielichowski J, Penczek P, Gorczyk J, Kowalski G (2002) Polimery 48:842
118. Bryan E (1994) Chemistry and technology of epoxy resins. Chapman and Hall, London
119. Palumbo M, Tempesti E (1998) Acta Polym 49:482
120. Jullien H, Valot H (1985) Polymer 26:506
121. Silinski B, Kuzmycz C, Gourdenne A (1987) Eur Polym J 23:273

122. Parodi F, Belgiovine C, Zannoni C (1994) US Patent 5,288,833
123. Parodi F, Gerbelli R, De Meuse M (1996) US Patent 5,489,664
124. Hurduc N, Buisine J-M, Decock P, Talewee J, Surpateneau G (1996) Polym J 28:550
125. Huy HT, Buu TN, Dung TTK, Han TN, Qaung PV (1996) J Mater Sci Pure Appl Chem A33:1957
126. Satge C, Verneuil B, Branland P, Granet R, Krausz P, Rozier J, Petit C (2002) Carbohydr Polymers 49:373
127. Mallakpour SE, Hajipour AR, Mahdavian AR, Zadhoush A, Ali-Hosseini F (2001) Eur Polym J 37:1199
128. Merrifield RB (1985) Angew Chem Int Ed 24:799
129. Lew A, Krutzik PO, Hart ME, Chamberlin AR (2002) J Comb Chem 4:95
130. Kappe CO (2002) Curr Opin Chem Biol 6:314
131. Yu HM, Chen ST, Wang KT (1992) J Org Chem 57:4781
132. Larhed M, Lindeberg G, Hallberg A (1996) Tetrahedron Lett 37:8219
133. Hoel AM, Nielsen J (1999) Tetrahedron Lett 40:3941
134. Kustner GJ, Scheeren HW (2000) Tetrahedron Lett 41:515
135. Kumar HMS, Anjaneyulu S, Reddy BVS, Yadav JS (2000) Synlett 1129
136. Stadler A, Kappe CO (2001) Tetrahedron 57:3915
137. Strohmeier GA, Kappe CO (2002) J Comb Chem 4:154
138. Vanden Eynde JJ, Rutot D (1999) Tetrahedron 55:2687
139. Chabinsky IJ (1983) Elastomerics 115:17
140. Krieger B (1992) Polym. Mat. Sci. Eng. 66:339
141. Parodi F. Technical Brochure. http://www.fpchem.com/fap_6a2-en.html
142. Krzan A (1998) J Appl Polym Sci 69:1115
143. Krzan A (1999) Polym Adv Technol 10:603
144. Klun U, Krzan A (2000) Polymer 41:4361
145. Kathirgamanathan P (1993) Polymer 34:3105
146. Epstein AJ, MacDiarmid AG (1991) In: Salaneck WR, Clark DT (eds) Science and application of conducting polymers. Adam Hilger, London
147. Thostenson ET, Chou T-W (1999) Advances in aerospace materials and structures, Newaz G (ed), ASME (AD-Vol. 58) New York, p 89
148. Wu C-Y, Benatar A (1997) Polym Eng Sci 37:738
149. Ku HS, MacRobert M, Siores E, Ball JAR (2000) Plast Rubber Compos 29:278
150. Ku HS, Siu F, Siores E, Ball JAR, Blickblau AS (2001) J Mater Process Tech 113:184

Editor: Karel Dušek
Received: October 2002

Author Index Volumes 101–163

Author Index Volumes 1–100 see Volume 100

de, Abajo, J. and *de la Campa, J.G.*: Processable Aromatic Polyimides. Vol. 140, pp. 23-60.
Adolf, D. B. see Ediger, M. D.: Vol. 116, pp. 73-110.
Aharoni, S. M. and *Edwards, S. F.*: Rigid Polymer Networks. Vol. 118, pp. 1-231.
Albertsson, A.-C., Varma, I. K.: Aliphatic Polyesters: Synthesis, Properties and Applications. Vol. 157, pp. 99–138.
Albertsson, A.-C. see Edlund, U.: Vol. 157, pp. 53-98.
Albertsson, A.-C. see Söderqvist Lindblad, M.: Vol. 157, pp. 139–161.
Albertsson, A.-C. see Stridsberg, K. M.: Vol. 157, pp. 27–51.
Améduri, B., Boutevin, B. and *Gramain, P.*: Synthesis of Block Copolymers by Radical Polymerization and Telomerization. Vol. 127, pp. 87-142.
Améduri, B. and *Boutevin, B.*: Synthesis and Properties of Fluorinated Telechelic Monodispersed Compounds. Vol. 102, pp. 133-170.
Amselem, S. see Domb, A. J.: Vol. 107, pp. 93-142.
Andrady, A. L.: Wavelenght Sensitivity in Polymer Photodegradation. Vol. 128, pp. 47-94.
Andreis, M. and *Koenig, J. L.*: Application of Nitrogen-15 NMR to Polymers. Vol. 124, pp. 191-238.
Angiolini, L. see Carlini, C.: Vol. 123, pp. 127-214.
Anjum, N. see Gupta, B.: Vol. 162, pp. 37-63.
Anseth, K. S., Newman, S. M. and *Bowman, C. N.*: Polymeric Dental Composites: Properties and Reaction Behavior of Multimethacrylate Dental Restorations. Vol. 122, pp. 177-218.
Antonietti, M. see Cölfen, H.: Vol. 150, pp. 67-187.
Armitage, B. A. see O'Brien, D. F.: Vol. 126, pp. 53-58.
Arndt, M. see Kaminski, W.: Vol. 127, pp. 143-187.
Arnold Jr., F. E. and *Arnold, F. E.*: Rigid-Rod Polymers and Molecular Composites. Vol. 117, pp. 257-296.
Arora, M. see Kumar, M.N.V.R.: Vol. 160, pp. 45-118.
Arshady, R.: Polymer Synthesis via Activated Esters: A New Dimension of Creativity in Macromolecular Chemistry. Vol. 111, pp. 1-42.

Bahar, I., Erman, B. and *Monnerie, L.*: Effect of Molecular Structure on Local Chain Dynamics: Analytical Approaches and Computational Methods. Vol. 116, pp. 145-206.
Ballauff, M. see Dingenouts, N.: Vol. 144, pp. 1-48.
Baltá-Calleja, F. J., González Arche, A., Ezquerra, T. A., Santa Cruz, C., Batallón, F., Frick, B. and *López Cabarcos, E.*: Structure and Properties of Ferroelectric Copolymers of Poly(vinylidene) Fluoride. Vol. 108, pp. 1-48.
Barnes, M. D. see Otaigbe, J.U.: Vol. 154, pp. 1-86.
Barshtein, G. R. and *Sabsai, O. Y.*: Compositions with Mineralorganic Fillers. Vol. 101, pp.1-28.
Baschnagel, J., Binder, K., Doruker, P., Gusev, A. A., Hahn, O., Kremer, K., Mattice, W. L., Müller-Plathe, F., Murat, M., Paul, W., Santos, S., Sutter, U. W., Tries, V.: Bridging the Gap Between Atomistic and Coarse-Grained Models of Polymers: Status and Perspectives. Vol. 152, pp. 41-156.
Batallán, F. see Baltá-Calleja, F. J.: Vol. 108, pp. 1-48.
Batog, A. E., Pet'ko, I. P., Penczek, P.: Aliphatic-Cycloaliphatic Epoxy Compounds and Polymers. Vol. 144, pp. 49-114.

Barton, J. see Hunkeler, D.: Vol. 112, pp. 115-134.
Bell, C. L. and *Peppas, N. A.*: Biomedical Membranes from Hydrogels and Interpolymer Complexes. Vol. 122, pp. 125-176.
Bellon-Maurel, A. see Calmon-Decriaud, A.: Vol. 135, pp. 207-226.
Bennett, D. E. see O'Brien, D. F.: Vol. 126, pp. 53-84.
Berry, G.C.: Static and Dynamic Light Scattering on Moderately Concentraded Solutions: Isotropic Solutions of Flexible and Rodlike Chains and Nematic Solutions of Rodlike Chains. Vol. 114, pp. 233-290.
Bershtein, V. A. and *Ryzhov, V. A.*: Far Infrared Spectroscopy of Polymers. Vol. 114, pp. 43-122.
Bhargava R., Wang S.-Q., Koenig J.L: FTIR Microspectroscopy of Polymeric Systems. Vol. 163, pp. 137-191.
Bigg, D. M.: Thermal Conductivity of Heterophase Polymer Compositions. Vol. 119, pp. 1-30.
Binder, K.: Phase Transitions in Polymer Blends and Block Copolymer Melts: Some Recent Developments. Vol. 112, pp. 115-134.
Binder, K.: Phase Transitions of Polymer Blends and Block Copolymer Melts in Thin Films. Vol. 138, pp. 1-90.
Binder, K. see Baschnagel, J.: Vol. 152, pp. 41-156.
Bird, R. B. see Curtiss, C. F.: Vol. 125, pp. 1-102.
Biswas, M. and *Mukherjee, A.*: Synthesis and Evaluation of Metal-Containing Polymers. Vol. 115, pp. 89-124.
Biswas, M. and *Sinha Ray, S.*: Recent Progress in Synthesis and Evaluation of Polymer-Montmorillonite Nanocomposites. Vol. 155, pp. 167-221.
Bogdal, D., Penczek, P., Pielichowski, J., Prociak, A.: Microwave Assisted Synthesis, Crosslinking, and Processing of Polymeric Materials. Vol. 163, pp. 193-263.
Bolze, J. see Dingenouts, N.: Vol. 144, pp. 1-48.
Bosshard, C.: see Gubler, U.: Vol. 158, pp. 123-190.
Boutevin, B. and *Robin, J. J.*: Synthesis and Properties of Fluorinated Diols. Vol. 102. pp. 105-132.
Boutevin, B. see Amédouri, B.: Vol. 102, pp. 133-170.
Boutevin, B. see Améduri, B.: Vol. 127, pp. 87-142.
Bowman, C. N. see Anseth, K. S.: Vol. 122, pp. 177-218.
Boyd, R. H.: Prediction of Polymer Crystal Structures and Properties. Vol. 116, pp. 1-26.
Briber, R. M. see Hedrick, J. L.: Vol. 141, pp. 1-44.
Bronnikov, S. V., Vettegren, V. I. and *Frenkel, S. Y.*: Kinetics of Deformation and Relaxation in Highly Oriented Polymers. Vol. 125, pp. 103-146.
Brown, H. R. see Creton, C.: Vol. 156, pp. 53-135.
Bruza, K. J. see Kirchhoff, R. A.: Vol. 117, pp. 1-66.
Budkowski, A.: Interfacial Phenomena in Thin Polymer Films: Phase Coexistence and Segregation. Vol. 148, pp. 1-112.
Burban, J. H. see Cussler, E. L.: Vol. 110, pp. 67-80.
Burchard, W.: Solution Properties of Branched Macromolecules. Vol. 143, pp. 113-194.

Calmon-Decriaud, A., Bellon-Maurel, V., Silvestre, F.: Standard Methods for Testing the Aerobic Biodegradation of Polymeric Materials. Vol 135, pp. 207-226.
Cameron, N. R. and *Sherrington, D. C.*: High Internal Phase Emulsions (HIPEs)-Structure, Properties and Use in Polymer Preparation. Vol. 126, pp. 163-214.
de la Campa, J. G. see de Abajo, , J.: Vol. 140, pp. 23-60.
Candau, F. see Hunkeler, D.: Vol. 112, pp. 115-134.
Canelas, D. A. and *DeSimone, J. M.*: Polymerizations in Liquid and Supercritical Carbon Dioxide. Vol. 133, pp. 103-140.
Canva, M., Stegeman, G. I.: Quadratic Parametric Interactions in Organic Waveguides. Vol. 158, pp. 87-121.
Capek, I.: Kinetics of the Free-Radical Emulsion Polymerization of Vinyl Chloride. Vol. 120, pp. 135-206.

Capek, I.: Radical Polymerization of Polyoxyethylene Macromonomers in Disperse Systems. Vol. 145, pp. 1-56.
Capek, I.: Radical Polymerization of Polyoxyethylene Macromonomers in Disperse Systems. Vol. 146, pp. 1-56.
Capek, I. and *Chern, C.-S.*: Radical Polymerization in Direct Mini-Emulsion Systems. Vol. 155, pp. 101-166.
Carlesso, G. see Prokop, A.: Vol. 160, pp. 119-174.
Carlini, C. and *Angiolini, L.*: Polymers as Free Radical Photoinitiators. Vol. 123, pp. 127-214.
Carter, K. R. see Hedrick, J. L.: Vol. 141, pp. 1-44.
Casas-Vazquez, J. see Jou, D.: Vol. 120, pp. 207-266.
Chandrasekhar, V.: Polymer Solid Electrolytes: Synthesis and Structure. Vol 135, pp. 139-206.
Chang, J.Y. see Han, M. J.: Vol. 153, pp. 1-36.
Chang, T.: Recent Advances in Liquid Chromatography Analysis of Synthetic Polymers. Vol. 163, pp. 1-60.
Charleux, B., Faust R.: Synthesis of Branched Polymers by Cationic Polymerization. Vol. 142, pp. 1-70.
Chen, P. see Jaffe, M.: Vol. 117, pp. 297-328.
Chern, C.-S. see Capek, I.: Vol. 155, pp. 101-166.
Chevolot, Y. see Mathieu, H. J.: Vol. 162, pp. 1-35.
Choe, E.-W. see Jaffe, M.: Vol. 117, pp. 297-328.
Chow, T. S.: Glassy State Relaxation and Deformation in Polymers. Vol. 103, pp. 149-190.
Chung, S.-J. see Lin, T.-C.: Vol. 161, pp. 157-193
Chung, T.-S. see Jaffe, M.: Vol. 117, pp. 297-328.
Cölfen, H. and *Antonietti, M.*: Field-Flow Fractionation Techniques for Polymer and Colloid Analysis. Vol. 150, pp. 67-187.
Comanita, B. see Roovers, J.: Vol. 142, pp. 179-228.
Connell, J. W. see Hergenrother, P. M.: Vol. 117, pp. 67-110.
Creton, C., Kramer, E. J., Brown, H. R., Hui, C.-Y.: Adhesion and Fracture of Interfaces Between Immiscible Polymers: From the Molecular to the Continuum Scale. Vol. 156, pp. 53-135.
Criado-Sancho, M. see Jou, D.: Vol. 120, pp. 207-266.
Curro, J.G. see Schweizer, K.S.: Vol. 116, pp. 319-378.
Curtiss, C. F. and *Bird, R. B.*: Statistical Mechanics of Transport Phenomena: Polymeric Liquid Mixtures. Vol. 125, pp. 1-102.
Cussler, E. L., Wang, K. L. and *Burban, J. H.*: Hydrogels as Separation Agents. Vol. 110, pp. 67-80.

Dalton, L. Nonlinear Optical Polymeric Materials: From Chromophore Design to Commercial Applications. Vol. 158, pp. 1-86.
Davidson, J.M. see Prokop, A.: Vol. 160, pp.119-174.
DeSimone, J. M. see Canelas D. A.: Vol. 133, pp. 103-140.
DiMari, S. see Prokop, A.: Vol. 136, pp. 1-52.
Dimonie, M. V. see Hunkeler, D.: Vol. 112, pp. 115-134.
Dingenouts, N., Bolze, J., Pötschke, D., Ballauf, M.: Analysis of Polymer Latexes by Small-Angle X-Ray Scattering. Vol. 144, pp. 1-48.
Dodd, L. R. and *Theodorou, D. N.*: Atomistic Monte Carlo Simulation and Continuum Mean Field Theory of the Structure and Equation of State Properties of Alkane and Polymer Melts. Vol. 116, pp. 249-282.
Doelker, E.: Cellulose Derivatives. Vol. 107, pp. 199-266.
Dolden, J. G.: Calculation of a Mesogenic Index with Emphasis Upon LC-Polyimides. Vol. 141, pp. 189-245.
Domb, A. J., Amselem, S., Shah, J. and *Maniar, M.*: Polyanhydrides: Synthesis and Characterization. Vol.107, pp. 93-142.
Domb, A.J. see Kumar, M.N.V.R.: Vol. 160, pp. 45-118.
Doruker, P. see Baschnagel, J.: Vol. 152, pp. 41-156.
Dubois, P. see Mecerreyes, D.: Vol. 147, pp. 1-60.

Dubrovskii, S. A. see Kazanskii, K. S.: Vol. 104, pp. 97-134.
Dunkin, I. R. see Steinke, J.: Vol. 123, pp. 81-126.
Dunson, D. L. see McGrath, J. E.: Vol. 140, pp. 61-106.

Eastmond, G. C.: Poly(ε-caprolactone) Blends. Vol. 149, pp. 59-223.
Economy, J. and *Goranov, K.:* Thermotropic Liquid Crystalline Polymers for High Performance Applications. Vol. 117, pp. 221-256.
Ediger, M. D. and *Adolf, D. B.:* Brownian Dynamics Simulations of Local Polymer Dynamics. Vol. 116, pp. 73-110.
Edlund, U. Albertsson, A.-C.: Degradable Polymer Microspheres for Controlled Drug Delivery. Vol. 157, pp. 53-98.
Edwards, S. F. see Aharoni, S. M.: Vol. 118, pp. 1-231.
Endo, T. see Yagci, Y.: Vol. 127, pp. 59-86.
Engelhardt, H. and *Grosche, O.:* Capillary Electrophoresis in Polymer Analysis. Vol. 150, pp. 189-217.
Erman, B. see Bahar, I.: Vol. 116, pp. 145-206.
Ewen, B, Richter, D.: Neutron Spin Echo Investigations on the Segmental Dynamics of Polymers in Melts, Networks and Solutions. Vol. 134, pp. 1-130.
Ezquerra, T. A. see Baltá-Calleja, F. J.: Vol. 108, pp. 1-48.

Faust, R. see Charleux, B: Vol. 142, pp. 1-70.
Fekete, E see Pukánszky, B: Vol. 139, pp. 109-154.
Fendler, J.H.: Membrane-Mimetic Approach to Advanced Materials. Vol. 113, pp. 1-209.
Fetters, L. J. see Xu, Z.: Vol. 120, pp. 1-50.
Förster, S. and *Schmidt, M.:* Polyelectrolytes in Solution. Vol. 120, pp. 51-134.
Freire, J. J.: Conformational Properties of Branched Polymers: Theory and Simulations. Vol. 143, pp. 35-112.
Frenkel, S. Y. see Bronnikov, S. V.: Vol. 125, pp. 103-146.
Frick, B. see Baltá-Calleja, F. J.: Vol. 108, pp. 1-48.
Fridman, M. L.: see Terent'eva, J. P.: Vol. 101, pp. 29-64.
Fukui, K. see Otaigbe, J. U.: Vol. 154, pp. 1-86.
Funke, W.: Microgels-Intramolecularly Crosslinked Macromolecules with a Globular Structure. Vol. 136, pp. 137-232.

Galina, H.: Mean-Field Kinetic Modeling of Polymerization: The Smoluchowski Coagulation Equation. Vol. 137, pp. 135-172.
Ganesh, K. see Kishore, K.: Vol. 121, pp. 81-122.
Gaw, K. O. and *Kakimoto, M.:* Polyimide-Epoxy Composites. Vol. 140, pp. 107-136.
Geckeler, K. E. see Rivas, B.: Vol. 102, pp. 171-188.
Geckeler, K. E.: Soluble Polymer Supports for Liquid-Phase Synthesis. Vol. 121, pp. 31-80.
Gehrke, S. H.: Synthesis, Equilibrium Swelling, Kinetics Permeability and Applications of Environmentally Responsive Gels. Vol. 110, pp. 81-144.
de Gennes, P.-G.: Flexible Polymers in Nanopores. Vol. 138, pp. 91-106.
Giannelis, E.P., Krishnamoorti, R., Manias, E.: Polymer-Silicate Nanocomposites: Model Systems for Confined Polymers and Polymer Brushes. Vol. 138, pp. 107-148.
Godovsky, D. Y.: Device Applications of Polymer-Nanocomposites. Vol. 153, pp. 163-205.
Godovsky, D. Y.: Electron Behavior and Magnetic Properties Polymer-Nanocomposites. Vol. 119, pp. 79-122.
González Arche, A. see Baltá-Calleja, F. J.: Vol. 108, pp. 1-48.
Goranov, K. see Economy, J.: Vol. 117, pp. 221-256.
Gramain, P. see Améduri, B.: Vol. 127, pp. 87-142.
Grest, G.S.: Normal and Shear Forces Between Polymer Brushes. Vol. 138, pp. 149-184.
Grigorescu, G, Kulicke, W.-M.: Prediction of Viscoelastic Properties and Shear Stability of Polymers in Solution. Vol. 152, p. 1-40.
Grosberg, A. and *Nechaev, S.:* Polymer Topology. Vol. 106, pp. 1-30.
Grosche, O. see Engelhardt, H.: Vol. 150, pp. 189-217.

Grubbs, R., Risse, W. and *Novac, B.:* The Development of Well-defined Catalysts for Ring-Opening Olefin Metathesis. Vol. 102, pp. 47-72.
Gubler, U., Bosshard, C.: Molecular Design for Third-Order Nonlinear Optics. Vol. 158, pp. 123-190.
van Gunsteren, W. F. see Gusev, A. A.: Vol. 116, pp. 207-248.
Gupta, B., Anjum, N.: Plasma and Radiation-Induced Graft Modification of Polymers for Biomedical Applications. Vol. 162, pp. 37-63.
Gusev, A. A., Müller-Plathe, F., van Gunsteren, W. F. and *Suter, U. W.:* Dynamics of Small Molecules in Bulk Polymers. Vol. 116, pp. 207-248.
Gusev, A. A. see Baschnagel, J.: Vol. 152, pp. 41-156.
Guillot, J. see Hunkeler, D.: Vol. 112, pp. 115-134.
Guyot, A. and *Tauer, K.:* Reactive Surfactants in Emulsion Polymerization. Vol. 111, pp. 43-66.

Hadjichristidis, N., Pispas, S., Pitsikalis, M., Iatrou, H., Vlahos, C.: Asymmetric Star Polymers Synthesis and Properties. Vol. 142, pp. 71-128.
Hadjichristidis, N. see Xu, Z.: Vol. 120, pp. 1-50.
Hadjichristidis, N. see Pitsikalis, M.: Vol. 135, pp. 1-138.
Hahn, O. see Baschnagel, J.: Vol. 152, pp. 41-156.
Hakkarainen, M.: Aliphatic Polyesters: Abiotic and Biotic Degradation and Degradation Products. Vol. 157, pp. 1-26.
Hall, H. K. see Penelle, J.: Vol. 102, pp. 73-104.
Hamley, I. W.: Crystallization in Block Copolymers. Vol. 148, pp. 113-138.
Hammouda, B.: SANS from Homogeneous Polymer Mixtures: A Unified Overview. Vol. 106, pp. 87-134.
Han, M.J. and *Chang, J.Y.:* Polynucleotide Analogues. Vol. 153, pp. 1-36.
Harada, A.: Design and Construction of Supramolecular Architectures Consisting of Cyclodextrins and Polymers. Vol. 133, pp. 141-192.
Haralson, M. A. see Prokop, A.: Vol. 136, pp. 1-52.
Hassan, C.M. and *Peppas, N.A.:* Structure and Applications of Poly(vinyl alcohol) Hydrogels Produced by Conventional Crosslinking or by Freezing/Thawing Methods. Vol. 153, pp. 37-65.
Hawker, C. J. Dentritic and Hyperbranched Macromolecules – Precisely Controlled Macromolecular Architectures. Vol. 147, pp. 113-160.
Hawker, C. J. see Hedrick, J. L.: Vol. 141, pp. 1-44.
He, G. S. see Lin, T.-C.: Vol. 161, pp. 157-193.
Hedrick, J. L., Carter, K. R., Labadie, J. W., Miller, R. D., Volksen, W., Hawker, C. J., Yoon, D. Y., Russell, T. P., McGrath, J. E., Briber, R. M.: Nanoporous Polyimides. Vol. 141, pp. 1-44.
Hedrick, J. L., Labadie, J. W., Volksen, W. and *Hilborn, J. G.:* Nanoscopically Engineered Polyimides. Vol. 147, pp. 61-112.
Hedrick, J. L. see Hergenrother, P. M.: Vol. 117, pp. 67-110.
Hedrick, J. L. see Kiefer, J.: Vol. 147, pp. 161-247.
Hedrick, J.L. see McGrath, J. E.: Vol. 140, pp. 61-106.
Heinrich, G. and *Klüppel, M.:* Recent Advances in the Theory of Filler Networking in Elastomers. Vol. 160, pp. 1-44.
Heller, J.: Poly (Ortho Esters). Vol. 107, pp. 41-92.
Hemielec, A. A. see Hunkeler, D.: Vol. 112, pp. 115-134.
Hergenrother, P. M., Connell, J. W., Labadie, J. W. and *Hedrick, J. L.:* Poly(arylene ether)s Containing Heterocyclic Units. Vol. 117, pp. 67-110.
Hernández-Barajas, J. see Wandrey, C.: Vol. 145, pp. 123-182.
Hervet, H. see Léger, L.: Vol. 138, pp. 185-226.
Hilborn, J. G. see Hedrick, J. L.: Vol. 147, pp. 61-112.
Hilborn, J. G. see Kiefer, J.: Vol. 147, pp. 161-247.
Hiramatsu, N. see Matsushige, M.: Vol. 125, pp. 147-186.
Hirasa, O. see Suzuki, M.: Vol. 110, pp. 241-262.
Hirotsu, S.: Coexistence of Phases and the Nature of First-Order Transition in Poly-N-isopropylacrylamide Gels. Vol. 110, pp. 1-26.

Höcker, H. see *Klee, D.*: Vol. 149, pp. 1-57.
Hornsby, P.: Rheology, Compoundind and Processing of Filled Thermoplastics. Vol. 139, pp. 155-216.
Hui, C.-Y. see *Creton, C.*: Vol. 156, pp. 53-135
Hult, A., Johansson, M., Malmström, E.: Hyperbranched Polymers. Vol. 143, pp. 1-34.
Hunkeler, D., Candau, F., Pichot, C., Hemielec, A. E., Xie, T. Y., Barton, J., Vaskova, V., Guillot, J., Dimonie, M. V., Reichert, K. H.: Heterophase Polymerization: A Physical and Kinetic Comparision and Categorization. Vol. 112, pp. 115-134.
Hunkeler, D. see *Macko, T.*: Vol. 163, pp. 61-136.
Hunkeler, D. see *Prokop, A.*: Vol. 136, pp. 1-52; 53-74.
Hunkeler, D see *Wandrey, C.*: Vol. 145, pp. 123-182.

Iatrou, H. see *Hadjichristidis, N.*: Vol. 142, pp. 71-128.
Ichikawa, T. see *Yoshida, H.*: Vol. 105, pp. 3-36.
Ihara, E. see *Yasuda, H.*: Vol. 133, pp. 53-102.
Ikada, Y. see *Uyama, Y.*: Vol. 137, pp. 1-40.
Ilavsky, M.: Effect on Phase Transition on Swelling and Mechanical Behavior of Synthetic Hydrogels. Vol. 109, pp. 173-206.
Imai, Y.: Rapid Synthesis of Polyimides from Nylon-Salt Monomers. Vol. 140, pp. 1-23.
Inomata, H. see *Saito, S.*: Vol. 106, pp. 207-232.
Inoue, S. see *Sugimoto, H.*: Vol. 146, pp. 39-120.
Irie, M.: Stimuli-Responsive Poly(N-isopropylacrylamide), Photo- and Chemical-Induced Phase Transitions. Vol. 110, pp. 49-66.
Ise, N. see *Matsuoka, H.*: Vol. 114, pp. 187-232.
Ito, K., Kawaguchi, S,:Poly(macronomers), Homo- and Copolymerization. Vol. 142, pp. 129-178.
Ivanov, A. E. see *Zubov, V. P.*: Vol. 104, pp. 135-176.

Jacob, S. and *Kennedy, J.*: Synthesis, Characterization and Properties of OCTA-ARM Polyisobutylene-Based Star Polymers. Vol. 146, pp. 1-38.
Jaffe, M., Chen, P., Choe, E.-W., Chung, T.-S. and *Makhija, S.*: High Performance Polymer Blends. Vol. 117, pp. 297-328.
Jancar, J.: Structure-Property Relationships in Thermoplastic Matrices. Vol. 139, pp. 1-66.
Jen, A. K-Y. see *Kajzar, F.*: Vol. 161, pp. 1-85.
Jerôme, R.: see *Mecerreyes, D.*: Vol. 147, pp. 1-60.
Jiang, M., Li, M., Xiang, M. and *Zhou, H.*: Interpolymer Complexation and Miscibility and Enhancement by Hydrogen Bonding. Vol. 146, pp. 121-194.
Jin, J.: see *Shim, H.-K.*: Vol. 158, pp. 191-241.
Jo, W. H. and *Yang, J. S.*: Molecular Simulation Approaches for Multiphase Polymer Systems. Vol. 156, pp. 1-52.
Johansson, M. see *Hult, A.*: Vol. 143, pp. 1-34.
Joos-Müller, B. see *Funke, W.*: Vol. 136, pp. 137-232.
Jou, D., Casas-Vazquez, J. and *Criado-Sancho, M.*: Thermodynamics of Polymer Solutions under Flow: Phase Separation and Polymer Degradation. Vol. 120, pp. 207-266.

Kaetsu, I.: Radiation Synthesis of Polymeric Materials for Biomedical and Biochemical Applications. Vol. 105, pp. 81-98.
Kaji, K. see *Kanaya, T.*: Vol. 154, pp. 87-141.
Kajzar, F., Lee, K.-S., Jen, A.K.-Y.: Polymeric Materials and their Orientation Techniques for Second-Order Nonlinear Optics.Vol. 161, pp. 1-85.
Kakimoto, M. see *Gaw, K. O.*: Vol. 140, pp. 107-136.
Kaminski, W. and *Arndt, M.*: Metallocenes for Polymer Catalysis. Vol. 127, pp. 143-187.
Kammer, H. W., Kressler, H. and *Kummerloewe, C.*: Phase Behavior of Polymer Blends - Effects of Thermodynamics and Rheology. Vol. 106, pp. 31-86.
Kanaya, T. and *Kaji, K.*: Dynamcis in the Glassy State and Near the Glass Transition of Amorphous Polymers as Studied by Neutron Scattering. Vol. 154, pp. 87-141.

Kandyrin, L. B. and *Kuleznev, V. N.*: The Dependence of Viscosity on the Composition of Concentrated Dispersions and the Free Volume Concept of Disperse Systems. Vol. 103, pp. 103-148.
Kaneko, M. see Ramaraj, R.: Vol. 123, pp. 215-242.
Kang, E. T., Neoh, K. G. and *Tan, K. L.*: X-Ray Photoelectron Spectroscopic Studies of Electroactive Polymers. Vol. 106, pp. 135-190.
Karlsson, S. see Söderqvist Lindblad, M.: Vol. 157, pp. 139-161.
Kato, K. see Uyama, Y.: Vol. 137, pp. 1-40.
Kawaguchi, S. see Ito, K.: Vol. 142, p 129-178.
Kazanskii, K. S. and *Dubrovskii, S. A.*: Chemistry and Physics of „Agricultural" Hydrogels. Vol. 104, pp. 97-134.
Kennedy, J. P. see Jacob, S.: Vol. 146, pp. 1-38.
Kennedy, J. P. see Majoros, I.: Vol. 112, pp. 1-113.
Khokhlov, A., Starodybtzev, S. and *Vasilevskaya, V.*: Conformational Transitions of Polymer Gels: Theory and Experiment. Vol. 109, pp. 121-172.
Kiefer, J., Hedrick J. L. and *Hiborn, J. G.*: Macroporous Thermosets by Chemically Induced Phase Separation. Vol. 147, pp. 161-247.
Kilian, H. G. and *Pieper, T.*: Packing of Chain Segments. A Method for Describing X-Ray Patterns of Crystalline, Liquid Crystalline and Non-Crystalline Polymers. Vol. 108, pp. 49-90.
Kim, J. see Quirk, R.P.: Vol. 153, pp. 67-162.
Kim, K.-S. see Lin, T.-C.: Vol. 161, pp. 157-193.
Kippelen, B. and *Peyghambarian, N.*: Photorefractive Polymers and their Applications. Vol. 161, pp. 87-156.
Kishore, K. and *Ganesh, K.*: Polymers Containing Disulfide, Tetrasulfide, Diselenide and Ditelluride Linkages in the Main Chain. Vol. 121, pp. 81-122.
Kitamaru, R.: Phase Structure of Polyethylene and Other Crystalline Polymers by Solid-State ^{13}C/MNR. Vol. 137, pp 41-102.
Klee, D. and *Höcker, H.*: Polymers for Biomedical Applications: Improvement of the Interface Compatibility. Vol. 149, pp. 1-57.
Klier, J. see Scranton, A. B.: Vol. 122, pp. 1-54.
Klüppel, M. see Heinrich, G.: Vol. 160, pp 1-44.
Kobayashi, S., Shoda, S. and *Uyama, H.*: Enzymatic Polymerization and Oligomerization. Vol. 121, pp. 1-30.
Köhler, W. and *Schäfer, R.*: Polymer Analysis by Thermal-Diffusion Forced Rayleigh Scattering. Vol. 151, pp. 1-59.
Koenig, J. L. see Bhargava, R.: Vol. 163, pp. 137-191.
Koenig, J. L. see Andreis, M.: Vol. 124, pp. 191-238.
Koike, T.: Viscoelastic Behavior of Epoxy Resins Before Crosslinking. Vol. 148, pp. 139-188.
Kokufuta, E.: Novel Applications for Stimulus-Sensitive Polymer Gels in the Preparation of Functional Immobilized Biocatalysts. Vol. 110, pp. 157-178.
Konno, M. see Saito, S.: Vol. 109, pp. 207-232.
Kopecek, J. see Putnam, D.: Vol. 122, pp. 55-124.
Koßmehl, G. see Schopf, G.: Vol. 129, pp. 1-145.
Kozlov, E. see Prokop, A.: Vol. 160, pp. 119-174.
Kramer, E. J. see Creton, C.: Vol. 156, pp. 53-135.
Kremer, K. see Baschnagel, J.: Vol. 152, pp. 41-156.
Kressler, J. see Kammer, H. W.: Vol. 106, pp. 31-86.
Kricheldorf, H. R.: Liquid-Cristalline Polyimides. Vol. 141, pp. 83-188.
Krishnamoorti, R. see Giannelis, E.P.: Vol. 138, pp. 107-148.
Kirchhoff, R. A. and *Bruza, K. J.*: Polymers from Benzocyclobutenes. Vol. 117, pp. 1-66.
Kuchanov, S. I.: Modern Aspects of Quantitative Theory of Free-Radical Copolymerization. Vol. 103, pp. 1-102.
Kuchanov, S. I.: Principles of Quantitive Description of Chemical Structure of Synthetic Polymers. Vol. 152, p. 157-202.

Kudaibergennow, S.E.: Recent Advances in Studying of Synthetic Polyampholytes in Solutions. Vol. 144, pp. 115-198.
Kuleznev, V. N. see Kandyrin, L. B.: Vol. 103, pp. 103-148.
Kulichkhin, S. G. see Malkin, A. Y.: Vol. 101, pp. 217-258.
Kulicke, W.-M. see Grigorescu, G.: Vol. 152, p. 1-40.
Kumar, M.N.V.R., Kumar, N., Domb, A.J. and Arora, M.: Pharmaceutical Polyme-ric Controlled Drug Delivery Systems. Vol. 160, pp. 45-118.
Kumar, N. see Kumar M.N.V.R.: Vol. 160, pp. 45-118.
Kummerloewe, C. see Kammer, H. W.: Vol. 106, pp. 31-86.
Kuznetsova, N. P. see Samsonov, G. V.: Vol. 104, pp. 1-50.*Labadie, J. W.* see Hergenrother, P. M.: Vol. 117, pp. 67-110.

Labadie, J. W. see Hedrick, J. L.: Vol. 141, pp. 1-44.
Labadie, J. W. see Hedrick, J. L.: Vol. 147, pp. 61-112.
Lamparski, H. G. see O´Brien, D. F.: Vol. 126, pp. 53-84.
Laschewsky, A.: Molecular Concepts, Self-Organisation and Properties of Polysoaps. Vol. 124, pp. 1-86.
Laso, M. see Leontidis, E.: Vol. 116, pp. 283-318.
Lazár, M. and RychlΩ, R.: Oxidation of Hydrocarbon Polymers. Vol. 102, pp. 189-222.
Lechowicz, J. see Galina, H.: Vol. 137, pp. 135-172.
Léger, L., Raphaël, E., Hervet, H.: Surface-Anchored Polymer Chains: Their Role in Adhesion and Friction. Vol. 138, pp. 185-226.
Lenz, R. W.: Biodegradable Polymers. Vol. 107, pp. 1-40.
Leontidis, E., de Pablo, J. J., Laso, M. and *Suter, U. W.:* A Critical Evaluation of Novel Algorithms for the Off-Lattice Monte Carlo Simulation of Condensed Polymer Phases. Vol. 116, pp. 283-318.
Lee, B. see Quirk, R.P: Vol. 153, pp. 67-162.
Lee, K.-S. see Kajzar, F.: Vol. 161, pp. 1-85.
Lee, Y. see Quirk, R.P: Vol. 153, pp. 67-162.
Leónard,, D. see Mathieu, H. J.: Vol. 162, pp. 1-35.
Lesec, J. see Viovy, J.-L.: Vol. 114, pp. 1-42.
Li, M. see Jiang, M.: Vol. 146, pp. 121-194.
Liang, G. L. see Sumpter, B. G.: Vol. 116, pp. 27-72.
Lienert, K.-W.: Poly(ester-imide)s for Industrial Use. Vol. 141, pp. 45-82.
Lin, J. and *Sherrington, D. C.:* Recent Developments in the Synthesis, Thermostability and Liquid Crystal Properties of Aromatic Polyamides. Vol. 111, pp. 177-220.
Lin, T.-C., Chung, S.-J., Kim, K.-S., Wang, X., He, G. S., Swiatkiewicz, J., Pudavar, H. E. and Prasad, P. N.: Organics and Polymers with High Two-Photon Activities and their Applications. Vol. 161, pp. 157-193.
Liu, Y. see Söderqvist Lindblad, M.: Vol. 157, pp. 139–161
López Cabarcos, E. see Baltá-Calleja, F. J.: Vol. 108, pp. 1-48.

Macko, T. and Hunkeler, D.: Liquid Chromatography under Critical and Limiting Conditions: A Survey of Experimental Systems for Synthetic Polymers. Vol. 163, pp. 61-136.
Majoros, I., Nagy, A. and *Kennedy, J. P:* Conventional and Living Carbocationic Polymerizations United. I. A Comprehensive Model and New Diagnostic Method to Probe the Mechanism of Homopolymerizations. Vol. 112, pp. 1-113.
Makhija, S. see Jaffe, M.: Vol. 117, pp. 297-328.
Malmström, E. see Hult, A.: Vol. 143, pp. 1-34.
Malkin, A. Y. and *Kulichkhin, S. G.:* Rheokinetics of Curing. Vol. 101, pp. 217-258.
Maniar, M. see Domb, A. J.: Vol. 107, pp. 93-142.
Manias, E., see Giannelis, E.P.: Vol. 138, pp. 107-148.
Mashima, K., Nakayama, Y. and *Nakamura, A.:* Recent Trends in Polymerization of a-Olefins Catalyzed by Organometallic Complexes of Early Transition Metals. Vol. 133, pp. 1-52.
Mathew, D. see Reghunadhan Nair, C.P.: Vol. 155, pp. 1-99.

Mathieu, H. J., Chevolot, Y, Ruiz-Taylor, L. and Leónard, D.: Engineering and Characterization of Polymer Surfaces for Biomedical Applications. Vol. 162, pp. 1-35.
Matsumoto, A.: Free-Radical Crosslinking Polymerization and Copolymerization of Multivinyl Compounds. Vol. 123, pp. 41-80.
Matsumoto, A. see Otsu, T.: Vol. 136, pp. 75-138.
Matsuoka, H. and Ise, N.: Small-Angle and Ultra-Small Angle Scattering Study of the Ordered Structure in Polyelectrolyte Solutions and Colloidal Dispersions. Vol. 114, pp. 187-232.
Matsushige, K., Hiramatsu, N. and Okabe, H.: Ultrasonic Spectroscopy for Polymeric Materials. Vol. 125, pp. 147-186.
Mattice, W. L. see Rehahn, M.: Vol. 131/132, pp. 1-475.
Mattice, W. L. see Baschnagel, J.: Vol. 152, p. 41-156.
Mays, W. see Xu, Z.: Vol. 120, pp. 1-50.
Mays, J.W. see Pitsikalis, M.: Vol.135, pp. 1-138.
McGrath, J. E. see Hedrick, J. L.: Vol. 141, pp. 1-44.
McGrath, J. E., Dunson, D. L., Hedrick, J. L.: Synthesis and Characterization of Segmented Polyimide-Polyorganosiloxane Copolymers. Vol. 140, pp. 61-106.
McLeish, T.C. B., Milner, S. T.: Entangled Dynamics and Melt Flow of Branched Polymers. Vol. 143, pp. 195-256.
Mecerreyes, D., Dubois, P. and Jerôme, R.: Novel Macromolecular Architectures Based on Aliphatic Polyesters: Relevance of the „Coordination-Insertion" Ring-Opening Polymerization. Vol. 147, pp. 1 -60.
Mecham, S. J. see McGrath, J. E.: Vol. 140, pp. 61-106.
Mikos, A. G. see Thomson, R. C.: Vol. 122, pp. 245-274.
Milner, S. T. see McLeish, T. C. B.: Vol. 143, pp. 195-256.
Mison, P. and Sillion, B.: Thermosetting Oligomers Containing Maleimides and Nadiimides End-Groups. Vol. 140, pp. 137-180.
Miyasaka, K.: PVA-Iodine Complexes: Formation, Structure and Properties. Vol. 108. pp. 91-130.
Miller, R. D. see Hedrick, J. L.: Vol. 141, pp. 1-44.
Monnerie, L. see Bahar, I.: Vol. 116, pp. 145-206.
Morishima, Y.: Photoinduced Electron Transfer in Amphiphilic Polyelectrolyte Systems. Vol. 104, pp. 51 96.
Morton M. see Quirk, R.P: Vol. 153, pp. 67-162
Mours, M. see Winter, H. H.: Vol. 134, pp. 165-234.
Müllen, K. see Scherf, U.: Vol. 123, pp. 1-40.
Müller-Plathe, F. see Gusev, A. A.: Vol. 116, pp. 207-248.
Müller-Plathe, F. see Baschnagel, J.: Vol. 152, p. 41-156.
Mukerherjee, A. see Biswas, M.: Vol. 115, pp. 89-124.
Murat, M. see Baschnagel, J.: Vol. 152, p. 41-156.
Mylnikov, V.: Photoconducting Polymers. Vol. 115, pp. 1-88.

Nagy, A. see Majoros, I.: Vol. 112, pp. 1-11.
Nakamura, A. see Mashima, K.: Vol. 133, pp. 1-52.
Nakayama, Y. see Mashima, K.: Vol. 133, pp. 1-52.
Narasinham, B., Peppas, N. A.: The Physics of Polymer Dissolution: Modeling Approaches and Experimental Behavior. Vol. 128, pp. 157-208.
Nechaev, S. see Grosberg, A.: Vol. 106, pp. 1-30.
Neoh, K. G. see Kang, E. T.: Vol. 106, pp. 135-190.
Newman, S. M. see Anseth, K. S.: Vol. 122, pp. 177-218.
Nijenhuis, K. te: Thermoreversible Networks. Vol. 130, pp. 1-252.
Ninan, K.N. see Reghunadhan Nair, C. P.: Vol. 155, pp. 1-99.
Noid, D. W. see Otaigbe, J.U.: Vol. 154, pp. 1-86.
Noid, D. W. see Sumpter, B. G.: Vol. 116, pp. 27-72.
Novac, B. see Grubbs, R.: Vol. 102, pp. 47-72.
Novikov, V. V. see Privalko, V. P.: Vol. 119, pp. 31-78.

O'Brien, D. F., Armitage, B. A., Bennett, D. E. and *Lamparski, H. G.*: Polymerization and Domain Formation in Lipid Assemblies. Vol. 126, pp. 53-84.
Ogasawara, M.: Application of Pulse Radiolysis to the Study of Polymers and Polymerizations. Vol.105, pp. 37-80.
Okabe, H. see Matsushige, K.: Vol. 125, pp. 147-186.
Okada, M.: Ring-Opening Polymerization of Bicyclic and Spiro Compounds. Reactivities and Polymerization Mechanisms. Vol. 102, pp. 1-46.
Okano, T.: Molecular Design of Temperature-Responsive Polymers as Intelligent Materials. Vol. 110, pp. 179-198.
Okay, O. see Funke, W.: Vol. 136, pp. 137-232.
Onuki, A.: Theory of Phase Transition in Polymer Gels. Vol. 109, pp. 63-120.
Osad'ko, I.S.: Selective Spectroscopy of Chromophore Doped Polymers and Glasses. Vol. 114, pp. 123-186.
Otaigbe, J. U., Barnes, M. D., Fukui, K., Sumpter, B. G., Noid, D. W.: Generation, Characterization, and Modeling of Polymer Micro- and Nano-Particles. Vol. 154, pp. 1-86.
Otsu, T., Matsumoto, A.: Controlled Synthesis of Polymers Using the Iniferter Technique: Developments in Living Radical Polymerization. Vol. 136, pp. 75-138.

de Pablo, J. J. see Leontidis, E.: Vol. 116, pp. 283-318.
Padias, A. B. see Penelle, J.: Vol. 102, pp. 73-104.
Pascault, J.-P. see Williams, R. J. J.: Vol. 128, pp. 95-156.
Pasch, H.: Analysis of Complex Polymers by Interaction Chromatography. Vol. 128, pp. 1-46.
Pasch, H.: Hyphenated Techniques in Liquid Chromatography of Polymers. Vol. 150, pp. 1-66.
Paul, W. see Baschnagel, J.: Vol. 152, p. 41-156.
Penczek, P. see Batog, A. E.: Vol. 144, pp. 49-114.
Penczek, P. see Bogdal, D.: Vol. 163, pp. 193-263.
Penelle, J., Hall, H. K., Padias, A. B. and *Tanaka, H.*: Captodative Olefins in Polymer Chemistry. Vol. 102, pp. 73-104.
Peppas, N. A. see Bell, C. L.: Vol. 122, pp. 125-176.
Peppas, N.A. see Hassan, C.M.: Vol. 153, pp. 37-65
Peppas, N. A. see Narasimhan, B.: Vol. 128, pp. 157-208.
Pet'ko, I. P. see Batog, A. E.: Vol. 144, pp. 49-114.
Pheyghambarian, N. see Kippelen, B.: Vol. 161, pp. 87-156.
Pichot, C. see Hunkeler, D.: Vol. 112, pp. 115-134.
Pielichowski, J. see Bogdal, D.: Vol. 163, pp. 193-263.
Pieper, T. see Kilian, H. G.: Vol. 108, pp. 49-90.
Pispas, S. see Pitsikalis, M.: Vol. 135, pp. 1-138.
Pispas, S. see Hadjichristidis: Vol. 142, pp. 71-128.
Pitsikalis, M., Pispas, S., Mays, J. W., Hadjichristidis, N.: Nonlinear Block Copolymer Architectures. Vol. 135, pp. 1-138.
Pitsikalis, M. see Hadjichristidis: Vol. 142, pp. 71-128.
Pötschke, D. see Dingenouts, N.: Vol 144, pp. 1-48.
Pokrovskii, V. N.: The Mesoscopic Theory of the Slow Relaxation of Linear Macromolecules. Vol. 154, pp. 143-219.
Pospíšil, J.: Functionalized Oligomers and Polymers as Stabilizers for Conventional Polymers. Vol. 101, pp. 65-168.
Pospíšil, J.: Aromatic and Heterocyclic Amines in Polymer Stabilization. Vol. 124, pp. 87-190.
Powers, A. C. see Prokop, A.: Vol. 136, pp. 53-74.
Prasad, P. N. see Lin, T.-C.: Vol. 161, pp. 157-193.
Priddy, D. B.: Recent Advances in Styrene Polymerization. Vol. 111, pp. 67-114.
Priddy, D. B.: Thermal Discoloration Chemistry of Styrene-co-Acrylonitrile. Vol. 121, pp. 123-154.
Privalko, V. P. and *Novikov, V. V.*: Model Treatments of the Heat Conductivity of Heterogeneous Polymers. Vol. 119, pp 31-78.

Prociak, A see Bogdal, D.: Vol. 163, pp. 193-263
Prokop, A., Hunkeler, D., Powers, A. C., Whitesell, R. R., Wang, T. G.: Water Soluble Polymers for Immunoisolation II: Evaluation of Multicomponent Microencapsulation Systems. Vol. 136, pp. 53-74.
Prokop, A., Hunkeler, D., DiMari, S., Haralson, M. A., Wang, T. G.: Water Soluble Polymers for Immunoisolation I: Complex Coacervation and Cytotoxicity. Vol. 136, pp. 1-52.
Prokop, A., Kozlov, E., Carlesso, G. and Davidsen, J.M.: Hydrogel-Based Colloidal Polymeric System for Protein and Drug Delivery: Physical and Chemical Characterization, Permeability Control and Applications. Vol. 160, pp. 119-174.
Pruitt, L. A.: The Effects of Radiation on the Structural and Mechanical Properties of Medical Polymers. Vol. 162, pp. 65-95.
Pudavar, H. E. see Lin, T.-C.: Vol. 161, pp. 157-193.
Pukánszky, B. and Fekete, E.: Adhesion and Surface Modification. Vol. 139, pp. 109-154.
Putnam, D. and Kopecek, J.: Polymer Conjugates with Anticancer Acitivity. Vol. 122, pp. 55-124.

Quirk, R.P. and Yoo, T., Lee, Y., M., Kim, J. and Lee, B.: Applications of 1,1-Diphenylethylene Chemistry in Anionic Synthesis of Polymers with Controlled Structures. Vol. 153, pp. 67-162.

Ramaraj, R. and Kaneko, M.: Metal Complex in Polymer Membrane as a Model for Photosynthetic Oxygen Evolving Center. Vol. 123, pp. 215-242.
Rangarajan, B. see Scranton, A. B.: Vol. 122, pp. 1-54.
Ranucci, E. see Söderqvist Lindblad, M.: Vol. 157, pp. 139-161.
Raphaël, E. see Léger, L.: Vol. 138, pp. 185-226.
Reddinger, J. L. and Reynolds, J. R.: Molecular Engineering of π-Conjugated Polymers. Vol. 145, pp. 57-122.
Reghunadhan Nair, C.P., Mathew, D. and Ninan, K.N., : Cyanate Ester Resins, Recent Developments. Vol. 155, pp. 1-99.
Reichert, K. H. see Hunkeler, D.: Vol. 112, pp. 115-134.
Rehahn, M., Mattice, W. L., Suter, U. W.: Rotational Isomeric State Models in Macromolecular Systems. Vol. 131/132, pp. 1-475.
Reynolds, J.R. see Reddinger, J. L.: Vol. 145, pp. 57-122.
Richter, D. see Ewen, B.: Vol. 134, pp.1-130.
Risse, W. see Grubbs, R.: Vol. 102, pp. 47-72.
Rivas, B. L. and Geckeler, K. E.: Synthesis and Metal Complexation of Poly(ethyleneimine) and Derivatives. Vol. 102, pp. 171-188.
Robin, J. J. see Boutevin, B.: Vol. 102, pp. 105-132.
Roe, R.-J.: MD Simulation Study of Glass Transition and Short Time Dynamics in Polymer Liquids. Vol. 116, pp. 111-114.
Roovers, J., Comanita, B.: Dendrimers and Dendrimer-Polymer Hybrids. Vol. 142, pp 179-228.
Rothon, R. N.: Mineral Fillers in Thermoplastics: Filler Manufacture and Characterisation. Vol. 139, pp. 67-108.
Rozenberg, B. A. see Williams, R. J. J.: Vol. 128, pp. 95-156.
Ruckenstein, E.: Concentrated Emulsion Polymerization. Vol. 127, pp. 1-58.
Ruiz-Taylor, L. see Mathieu, H. J.: Vol. 162, pp. 1-35.
Rusanov, A. L.: Novel Bis (Naphtalic Anhydrides) and Their Polyheteroarylenes with Improved Processability. Vol. 111, pp. 115-176.
Russel, T. P. see Hedrick, J. L.: Vol. 141, pp. 1-44.
Rychlý, J. see Lazár, M.: Vol. 102, pp. 189-222.
Ryner, M. see Stridsberg, K. M.: Vol. 157, pp. 27–51.
Ryzhov, V. A. see Bershtein, V. A.: Vol. 114, pp. 43-122.

Sabsai, O. Y. see Barshtein, G. R.: Vol. 101, pp. 1-28.
Saburov, V. V. see Zubov, V. P.: Vol. 104, pp. 135-176.
Saito, S., Konno, M. and Inomata, H.: Volume Phase Transition of N-Alkylacrylamide Gels. Vol. 109, pp. 207-232.

Samsonov, G. V. and *Kuznetsova, N. P.*: Crosslinked Polyelectrolytes in Biology. Vol. 104, pp. 1-50.
Santa Cruz, C. see Baltá-Calleja, F. J.: Vol. 108, pp. 1-48.
Santos, S. see Baschnagel, J.: Vol. 152, p. 41-156.
Sato, T. and *Teramoto, A.*: Concentrated Solutions of Liquid-Christalline Polymers. Vol. 126, pp. 85-162.
Schäfer R. see Köhler, W.: Vol. 151, pp. 1-59.
Scherf, U. and *Müllen, K.*: The Synthesis of Ladder Polymers. Vol. 123, pp. 1-40.
Schmidt, M. see Förster, S.: Vol. 120, pp. 51-134.
Scholz, M.: Effects of Ion Radiation on Cells and Tissues. Vol. 162, pp. 97-158.
Schopf, G. and *Koßmehl, G.*: Polythiophenes - Electrically Conductive Polymers. Vol. 129, pp. 1-145.
Schweizer, K. S.: Prism Theory of the Structure, Thermodynamics, and Phase Transitions of Polymer Liquids and Alloys. Vol. 116, pp. 319-378.
Scranton, A. B., Rangarajan, B. and *Klier, J.*: Biomedical Applications of Polyelectrolytes. Vol. 122, pp. 1-54.
Sefton, M. V. and *Stevenson, W. T. K.*: Microencapsulation of Live Animal Cells Using Polycrylates. Vol. 107, pp. 143-198.
Shamanin, V. V.: Bases of the Axiomatic Theory of Addition Polymerization. Vol. 112, pp. 135-180.
Sheiko, S. S.: Imaging of Polymers Using Scanning Force Microscopy: From Superstructures to Individual Molecules. Vol. 151, pp. 61-174.
Sherrington, D. C. see Cameron, N. R., Vol. 126, pp. 163-214.
Sherrington, D. C. see Lin, J.: Vol. 111, pp. 177-220.
Sherrington, D. C. see Steinke, J.: Vol. 123, pp. 81-126.
Shibayama, M. see Tanaka, T.: Vol. 109, pp. 1-62.
Shiga, T.: Deformation and Viscoelastic Behavior of Polymer Gels in Electric Fields. Vol. 134, pp. 131-164.
Shim, H.-K., Jin, J.: Light-Emitting Characteristics of Conjugated Polymers. Vol. 158, pp. 191-241.
Shoda, S. see Kobayashi, S.: Vol. 121, pp. 1-30.
Siegel, R. A.: Hydrophobic Weak Polyelectrolyte Gels: Studies of Swelling Equilibria and Kinetics. Vol. 109, pp. 233-268.
Silvestre, F. see Calmon-Decriaud, A.: Vol. 207, pp. 207-226.
Sillion, B. see Mison, P.: Vol. 140, pp. 137-180.
Singh, R. P. see Sivaram, S.: Vol. 101, pp. 169-216.
Sinha Ray, S. see Biswas, M: Vol. 155, pp. 167-221.
Sivaram, S. and *Singh, R. P.*: Degradation and Stabilization of Ethylene-Propylene Copolymers and Their Blends: A Critical Review. Vol. 101, pp. 169-216.
Söderqvist Lindblad, M., Liu, Y., Albertsson, A.-C., Ranucci, E., Karlsson, S.: Polymer from Renewable Resources. Vol. 157, pp. 139–161
Starodybtzev, S. see Khokhlov, A.: Vol. 109, pp. 121-172.
Stegeman, G. I.: see Canva, M.: Vol. 158, pp. 87-121.
Steinke, J., Sherrington, D. C. and *Dunkin, I. R.*: Imprinting of Synthetic Polymers Using Molecular Templates. Vol. 123, pp. 81-126.
Stenzenberger, H. D.: Addition Polyimides. Vol. 117, pp. 165-220.
Stevenson, W. T. K. see Sefton, M. V.: Vol. 107, pp. 143-198.
Stridsberg, K. M., Ryner, M., Albertsson, A.-C.: Controlled Ring-Opening Polymerization: Polymers with Designed Macromoleculars Architecture. Vol. 157, pp. 27–51.
Suematsu, K.: Recent Progress of Gel Theory: Ring, Excluded Volume, and Dimension. Vol. 156, pp. 136-214.
Sumpter, B. G., Noid, D. W., Liang, G. L. and *Wunderlich, B.*: Atomistic Dynamics of Macromolecular Crystals. Vol. 116, pp. 27-72.
Sumpter, B. G. see Otaigbe, J.U.: Vol. 154, pp. 1-86.
Sugimoto, H. and *Inoue, S.*: Polymerization by Metalloporphyrin and Related Complexes. Vol. 146, pp. 39-120.

Suter, U. W. see Gusev, A. A.: Vol. 116, pp. 207-248.
Suter, U. W. see Leontidis, E.: Vol. 116, pp. 283-318.
Suter, U. W. see Rehahn, M.: Vol. 131/132, pp. 1-475.
Suter, U. W. see Baschnagel, J.: Vol. 152, p. 41-156.
Suzuki, A.: Phase Transition in Gels of Sub-Millimeter Size Induced by Interaction with Stimuli. Vol. 110, pp. 199-240.
Suzuki, A. and *Hirasa, O.*: An Approach to Artifical Muscle by Polymer Gels due to Micro-Phase Separation. Vol. 110, pp. 241-262.
Swiatkiewicz, J. see Lin, T.-C.: Vol. 161, pp. 157-193.

Tagawa, S.: Radiation Effects on Ion Beams on Polymers. Vol. 105, pp. 99-116.
Tan, K. L. see Kang, E. T.: Vol. 106, pp. 135-190.
Tanaka, H. and *Shibayama, M.*: Phase Transition and Related Phenomena of Polymer Gels. Vol. 109, pp. 1-62.
Tanaka, T. see Penelle, J.: Vol. 102, pp. 73-104.
Tauer, K. see Guyot, A.: Vol. 111, pp. 43-66.
Teramoto, A. see Sato, T.: Vol. 126, pp. 85-162.
Terent´eva, J. P. and *Fridman, M. L.*: Compositions Based on Aminoresins. Vol. 101, pp. 29-64.
Theodorou, D. N. see Dodd, L. R.: Vol. 116, pp. 249-282.
Thomson, R. C., Wake, M. C., Yaszemski, M. J. and *Mikos, A. G.*: Biodegradable Polymer Scaffolds to Regenerate Organs. Vol. 122, pp. 245-274.
Tokita, M.: Friction Between Polymer Networks of Gels and Solvent. Vol. 110, pp. 27-48.
Tries, V. see Baschnagel, J:. Vol. 152, p. 41-156.
Tsuruta, T.: Contemporary Topics in Polymeric Materials for Biomedical Applications. Vol. 126, pp. 1-52.

Uyama, H. see Kobayashi, S.: Vol. 121, pp. 1-30.
Uyama, Y: Surface Modification of Polymers by Grafting. Vol. 137, pp. 1-40.

Varma, I. K. see Albertsson, A.-C.: Vol. 157, pp. 99-138.
Vasilevskaya, V. see Khokhlov, A.: Vol. 109, pp. 121-172.
Vaskova, V. see Hunkeler, D.: Vol.:112, pp. 115-134.
Verdugo, P.: Polymer Gel Phase Transition in Condensation-Decondensation of Secretory Products. Vol. 110, pp. 145-156.
Vettegren, V. I.: see Bronnikov, S. V.: Vol. 125, pp. 103-146.
Viovy, J.-L. and *Lesec, J.*: Separation of Macromolecules in Gels: Permeation Chromatography and Electrophoresis. Vol. 114, pp. 1-42.
Vlahos, C. see Hadjichristidis, N.: Vol. 142, pp. 71-128.
Volksen, W.: Condensation Polyimides: Synthesis, Solution Behavior, and Imidization Characteristics. Vol. 117, pp. 111-164.
Volksen, W. see Hedrick, J. L.: Vol. 141, pp. 1-44.
Volksen, W. see Hedrick, J. L.: Vol. 147, pp. 61-112.

Wake, M. C. see Thomson, R. C.: Vol. 122, pp. 245-274.
Wandrey C., Hernández-Barajas, J. and *Hunkeler, D.*: Diallyldimethylammonium Chloride and its Polymers. Vol. 145, pp. 123-182.
Wang, K. L. see Cussler, E. L.: Vol. 110, pp. 67-80.
Wang, S.-Q.: Molecular Transitions and Dynamics at Polymer/Wall Interfaces: Origins of Flow Instabilities and Wall Slip. Vol. 138, pp. 227-276.
Wang, S.-Q. see Bhargava, R.: Vol. 163, pp. 137-191.
Wang, T. G. see Prokop, A.: Vol. 136, pp.1-52; 53-74.
Wang, X. see Lin, T.-C.: Vol. 161, pp. 157-193.
Whitesell, R. R. see Prokop, A.: Vol. 136, pp. 53-74.
Williams, R. J. J., Rozenberg, B. A., Pascault, J.-P.: Reaction Induced Phase Separation in Modified Thermosetting Polymers. Vol. 128, pp. 95-156.

Winter, H. H., Mours, M.: Rheology of Polymers Near Liquid-Solid Transitions. Vol. 134, pp. 165-234.
Wu, C.: Laser Light Scattering Characterization of Special Intractable Macromolecules in Solution. Vol 137, pp. 103-134.
Wunderlich, B. see Sumpter, B. G.: Vol. 116, pp. 27-72.

Xiang, M. see Jiang, M.: Vol. 146, pp. 121-194.
Xie, T. Y. see Hunkeler, D.: Vol. 112, pp. 115-134.
Xu, Z., Hadjichristidis, N., Fetters, L. J. and *Mays, J. W.*: Structure/Chain-Flexibility Relationships of Polymers. Vol. 120, pp. 1-50.

Yagci, Y. and *Endo, T.*: N-Benzyl and N-Alkoxy Pyridium Salts as Thermal and Photochemical Initiators for Cationic Polymerization. Vol. 127, pp. 59-86.
Yannas, I. V.: Tissue Regeneration Templates Based on Collagen-Glycosaminoglycan Copolymers. Vol. 122, pp. 219-244.
Yang, J. S. see Jo, W. H.: Vol. 156, pp. 1-52.
Yamaoka, H.: Polymer Materials for Fusion Reactors. Vol. 105, pp. 117-144.
Yasuda, H. and *Ihara, E.*: Rare Earth Metal-Initiated Living Polymerizations of Polar and Nonpolar Monomers. Vol. 133, pp. 53-102.
Yaszemski, M. J. see Thomson, R. C.: Vol. 122, pp. 245-274.
Yoo, T. see Quirk, R.P.: Vol. 153, pp. 67-162.
Yoon, D. Y. see Hedrick, J. L.: Vol. 141, pp. 1-44.
Yoshida, H. and *Ichikawa, T.*: Electron Spin Studies of Free Radicals in Irradiated Polymers. Vol. 105, pp. 3-36.

Zhou, H. see Jiang, M.: Vol. 146, pp. 121-194.
Zubov, V. P., Ivanov, A. E. and *Saburov, V. V.*: Polymer-Coated Adsorbents for the Separation of Biopolymers and Particles. Vol. 104, pp. 135-176.

Subject Index

Acousto-optic tunable filters 180
Acrylic resins 214, 215
Adsorli 66
AIBN 207
Alcohol ethoxylate, 2D-LC 35
Allylthiourea 213
4-(2-Aminoethyl)morpholine 248
Apertures, microspectroscopy 145
Aperturing, redundant 145
Applicator, multi-mode 202
–, traveling wave 203
Array detectors 140
ATR microspectroscopy 144
Attenuated total reflection 144

Backward wave tubes 199
Binarized images 170
Bis(4-aminophenyl)-3,4-diphenylthiophene 219
3,3-Bis(chloromethyl)oxetane 226
Bisnadimide 222
Bisphenol A 244, 245
Bisphenol F 242
Bisphenols 226
Block copolymers, LCCC 29
Boron nitride 152
BSA 155
BTDA 217
Bump bonds, failing 166
Butadiene-acrylonitrile random copolymer 236
Butyl acrylate 43

Calixarene 103
ε-Caprolactone 225
Carbon black 159
Carbon fiber 225
Carbon-clad zirconia column 51

CCD 5, 25
Cellulose, esterification 249
Chain architecture, chromatographic separation 18
Chemical composition distribution (CCD) 5, 25
Chromatography 1
Claisen rearrangement 252
Complex permittivity 198
Composition distribution 24
Conditions of entropy-enthalpy compensation (CEEC) 65
Conducting polymers 257
N-(Cyano-alkoxyalkyl)trialkylammonium halides 247

DBTM 213
DDM 238, 239, 246
DDS 217, 236, 238, 239
Defects, microscopic 148
Dental materials 214
Denture base resins 216
Desorli 66
Detection limits 147
Detectors, FTIR 137
DGEBA 236, 246
– / 3DCM 235
Diacid chlorides 220
4,4'-Diamino-3,3'-dimethyldicyclohexyl methane 234
4,4'-Diaminodiphenyl methane 156
Dibenzyl maleate 213
Dibutyltin maleate (DBTM) 213
Dielectric constant/heating 198
Diffusion 155, 173
Diglycidyl ether of bisphenol A 233
4,4'-Diisocyanate diphenylmethane 247
4-Dimethylaminopyridine 249

Diphenols 221
4,4'-Diphenyl methane diisocyanate 156
Disk-seal triodes 199
Dodecanoyl chloride 249
Dynamic integral desorption isotherm 28

E7 157
Element detector 140
Elution strength 114–116
Encapsulation 174
EPDM 71
Epoxy resins 102, 233, 244, 258
Ethylene glycol dimethyl ether 250
Ethylene oxide 151
Ethylene-acrylic acid 154
Ethylenediamine 233
Ethylene-propylene-diene rubber 43
Ethylene-propylene-dimer (EPDM) 71
Exclusion 6

Fibers 150, 160
Fillers 160
Fluorescence optical microscopy 153
Focal plane array 137, 161
Fourier transform, SEC 47
FTIR, errors, artifacts 145, 165
–, imaging 161
–, spectrometer 162
–, contaminants/defects 171
FTIR mapping 140, 141
FTIR microspectroscopy,
  additives/contaminants 147
–, instrumentation 141, 163, 179–183
–, rapid-scan interferometer 164
–, sampling techniques 142, 165
–, single polymer fibers 150
–, step-scan interferometer 163
Fullerene-C60 213

Glass fiber 231
Glass-graphite fibers 223

Hadamard transform infrared
  microscopy 179
HEMA 205
HPLC 11
2-Hydroxy-ethyl methacrylate
  (HEMA) 205

IC 5
Infrared diode-based imaging 184
–, imaging, near field 183
–, microspectroscopy 139, 140
Interaction chromatography 5

Interferometer 141
IR camera 227
–, microscope, schematic 142
–, microspectroscopy 139, 140
Isoidide 228
Isosorbide 228

Kaprolaktone-diol 104
Klystrons 199
Knoevenagel condensation 252, 253
KPS 208, 209, 211

Lactic acid 231
Laminate films 154, 172
LC 1,4ff
–, at the critical adsorption point (LC-CAP) 9, 10
–, at the critical condition (LCCC) 9, 10, 64, 67, 89
–, at the point of exclusion-adsorption transition (LC-PEAT) 9, 10
LC-CAP 9, 10
LCCC 9, 10, 64, 67, 89
–, multiblock copolymers 29
LC-PEAT 9, 10
Limiting conditions 106, 121–123
Linear polymers 12
Liquid chromatography (see also LC) 1, 4
Liquid crystals 157, 158, 175
Local orientation 156
Loss factor 198
Loss tangent 198

MA 207
Magnetrons 198
Maleic anhydride 213
MALS 15
Mark-Houwink constants 61, 115, 116
Methyl acrylate (MA) 207
Methyl methacrylate (MMA) 206, 214, 215
N-Methyl-2-pyrrolidone 219
4,4'-Methylenedianiline (DDM) 238, 239, 246
2-Methylimidazole 245
Microspectroscopy apertures 145
–, FTIR 137
–, IR 139
Microwave applicators 201
–, generators 198
–, irradiation 193
–, welding 258
MMA 206, 214, 215
Molecular weight distribution 4, 12
MORE 204

Subject Index

Multi-angle light scattering   15
Multiblock copolymers, LCCC   29
Multilayered polymer systems   154
MWD   4, 12

Nadic end-capped imide   221
Nafion-H   157
Nanoparticles   211
Naphthyl acetic acid   248
Near field mapping   183
Nitroacetic acid   252
5-Norbornene-2,3-dicarboxylic acid monomethyl ester   221
NVC   213

Optical pyrometer   204
3,4'-Oxydianiline   221
4,4'-Oxydianiline   152

PAA   155
PAM   110
PB   21, 22
PBA   70
PBGA   91
PDEGA   90, 92
PDLCs   158, 175
PDMA   26
PDMS   39, 103
PE   18, 89, 110, 117, 155, 250
PEG   30, 97–99, 111, 155
PEMA   52, 76, 119
–, SEC/LCCC   53
–, TGIC   52
PEO   11, 93–96, 111, 172
PEO-b-PLLA, RPLC   30
PEO-b-PPO   35
Permeation   6
PET   204, 229, 230
–, recycling   256
Phase separation, FTIR   177
PHB   152
Phenol   214
Phenyl glycidyl ether (PGE)   238
Phenylacetylenes   213
m-Phenylenediamine   234
Photothermal imaging   182
Phthalimide   152
PI   32, 34, 69, 110
PMMA   16, 51, 70–73, 107–110, 117–119, 124, 151, 153, 172, 214
–, separation   27
PMMA-b-PtBMA   32
PMMA-g-PS   45
Polarity index   114, 115

Pollulan   111
Poly(acrylic acid)   155
Poly(n-butyl methacrylate)   75, 116, 157
Poly(t-butyl methacrylate)   74, 75, 116, 118
Poly(butylene glycol adipate)   91
Poly(butylene terephthalate)   101
Poly(ε-caprolactam)   225
Poly(decyl methacrylate)   26, 73, 74, 116
Poly(diethylene glycol adipate)   90, 92
Poly(dimethylsiloxane)   39
Poly(ethyl methacrylate)   52, 76, 119
Poly(ethylene glycol)   30, 97–99, 111, 155
Poly(ethylene oxide)   93–96, 111
Poly(ethylene terephthalate)   204, 229, 230
Poly(hydroxybutyrate-co-valerate)   152
Poly(lactic-co-glycolic acid)   174
Poly(methyl methacrylate)   16, 51, 70–73, 107–110, 117–119, 124, 151, 153, 172, 214
Poly(α-methyl styrene)   174
Poly(MMA-co-BA)   41
Poly(oxypropylene adipate)   92
Poly(oxypropylene glycol)   100
Poly(phenylacetylene)   214
Poly(propylene glycol)   99, 100
Poly(propylene glycol adipate)   92
Poly(S-co-EA)   43
Poly(S-co-MMA)   46
Poly(styrene)-b-poly(ethene-alt-propene)   207
Poly(styrene-co-acrylonitrile)   153
Poly(styrene-co-allyl alcohol)   254
Poly(tetrahydrofuran)   93, 118, 155
Poly(tetramethylene oxide)glycol di-p-aminobenzoate   219
Poly(urea)   156
Poly(urethane)   154
Poly(vinyl acetate)   111, 119
Poly(vinyl alcohol)   152
Poly(n-vinyl carbazole)   213
Poly(vinyl chloride)   88, 119
Poly(2-vinyl pyridine)   103, 116, 119
Poly(vinyl pyrrolidone)   103
Poly(vinylidene fluoride)   151
Polyacetals   89
Polyacrylamide (PAM)   110
Polyamic acid   217
Polyamide 6   256
Polyamide 66   229
Polyamide 66-GF   258
Polyamides   89, 217
Polyamine-650   219
Polyaniline   257
Polybutadiene (PB)   67, 68
Polybutylacrylate (PBA)   70

Polycarbonate 101
Polydimethylsiloxane 103
Polydispersity 13
Polyesters 90–92, 111, 227
Polyethers 226, 229
Polyethertriol 247
Polyethylene 18, 89, 110, 117, 250
–, high density 257
–, low density 258
–, prosthetic components 151
Polyimides 217
–, RP-46 222
Polyisoprene 32, 34, 69, 110
Polylactide 105
Polymer degradation 149, 151
Polymer dispersed liquid crystals 175
Polymer-liquid systems 174
Polyoxyphenylene 100
Polyphenylenesulfone 104, 105
Polypropylene 110
Polystyrene 7, 76–88, 106, 107, 116, 118, 209, 211, 253
Polysulfone 103
Polyurethanes 246
–, films 247
Potassium persulfate 208
Power grid tubes 199
PP 153, 155
PPOA 92
PPOG 100
PS 7, 76–88, 106, 107, 116, 118
–, LCCC 49
–, separation 27
–, TGIC 48
PSAN 153
PS-*b*-PBMA 31
PS-*b*-PI, 2D-LC 38
–, LCCC 33, 36
PS-*b*-PMMA 31
PtBMA 74, 75, 116, 118
PtBMA-*b*-PMMA 32
PTFE 204
–, grafts 150
PTHF 93, 118, 155
PVA 152
PVC 88, 119, 172
PVDF 151
PVK 213
PVP 103
Pyromellitic dianhydride 152

Random copolymers, characterization 37
Real permittivity 198

Resin transfer molding 232
Ring polymers, separation/chromatograms 22
RP-46 223
Rubber vulcanization 254

Saccharin 152
Scanning probe microspectroscopy 182
SEC 4
Seeding agents 152
SEP 207
SFC 86, 87
Silica column, C18 7
Size exclusion chromatography (SEC) 4
Sodium dodecyl sulfate 213
Sodium lauryl sulfate 209
Solid state focal plane array imaging 180
Solid-phase organic synthesis 250
Solubility parameter 112-115
Spatial resolution visualization 140
Stereoregularity 1
Styrene 208, 211
–, and maleic anhydride 155
Supercritical fluid chromatography (SFC) 86, 87
Suzuki coupling 250
Synchrotron infrared microscopy 180
Synolac 6345 233

TBAB 227
TEBA 249
TEGDMA 216
Temperature gradient interaction chromatography (TGIC) 17, 69, 76, 81, 88, 97, 119
Tetrabutylammonium bromide 228
TGIC 17, 69, 76, 81, 88, 97, 119
Thermocouples 204
Thermometers, fiber-optic 204, 208
Thermovision 227
TLC 75, 85, 86, 92, 100-102
Transverse electric/magnetic mode 200
Traveling wave tubes 199
Triethylene glycol dimethacrylate 216

Vermiculite 208
N-Vinylcarbazole 213
Vinyltoluene 231

Welding thermoplastics 257

Zinc acetate 257

Be the first to know
with the new online notification service

Springer Alert

You decide how we keep you up to date on new publications:
- Select a specialist field within a subject area
- Take your pick from various information formats
- Choose how often you'd like to be informed

And receive customised information to suit your needs

http://www.springer.de/alert

Register now

and then you are one click away from a world of chemistry information!

Come and visit Springer's Chemistry Online Library

http://www.springer.de/chem

Springer

Printing: Saladruck Berlin
Binding: Stürtz AG, Würzburg